VOLUME FIVE HUNDRED AND EIGHTEEN

METHODS IN ENZYMOLOGY

Fluorescence Fluctuation Spectroscopy (FFS), Part A

METHODS IN ENZYMOLOGY

VOLUME FIVE HUNDRED AND EIGHTEEN

METHODS IN ENZYMOLOGY

Fluorescence Fluctuation Spectroscopy (FFS), Part A

Edited by

SERGEY Y. TETIN

Abbott Diagnostics Division
Abbott Laboratories
Abbott Park, IL, USA

AMSTERDAM • BOSTON • HEIDELBERG • LONDON
NEW YORK • OXFORD • PARIS • SAN DIEGO
SAN FRANCISCO • SINGAPORE • SYDNEY • TOKYO
Academic Press is an imprint of Elsevier

Academic Press is an imprint of Elsevier
525 B Street, Suite 1900, San Diego, CA 92101-4495, USA
225 Wyman Street, Waltham, MA 02451, USA
The Boulevard, Langford Lane, Kidlington, Oxford, OX51GB, UK
32, Jamestown Road, London NW1 7BY, UK
Radarweg 29, PO Box 211, 1000 AE Amsterdam, The Netherlands

First edition 2013

ISBN: 978-0-12-388422-0
ISSN: 0076-6879

Printed and bound in United States of America
13 14 15 16 11 10 9 8 7 6 5 4 3 2 1

CONTENTS

CONTRIBUTORS

Neil Anthony
Department of Physics, Emory University, Atlanta, Georgia, USA

Keith Berland
Department of Physics, Emory University, Atlanta, Georgia, USA

Yan Chen
School of Physics and Astronomy, University of Minnesota, Minneapolis, Minnesota, USA

Michelle A. Digman
Laboratory for Fluorescence Dynamics, Department of Biomedical Engineering, and Department of Development and Cell Biology, University of California, Irvine, California, USA

Elliot L. Elson
Department of Biochemistry and Molecular Biophysics, Washington University School of Medicine, Campus Box 8231, St. Louis, Missouri, USA

Jörg Enderlein
Third Institute of Physics–Biophysics, Georg-August-University Göttingen, Friedrich-Hund-Platz, Göttingen, Germany

Enrico Gratton
Laboratory for Fluorescence Dynamics, Department of Biomedical Engineering, and Department of Development and Cell Biology, University of California, Irvine, California, USA

Ingo Gregor
Third Institute of Physics–Biophysics, Georg-August-University Göttingen, Friedrich-Hund-Platz, Göttingen, Germany

Jelle Hendrix
Physical Chemistry, Department of Chemistry, Munich Center for Integrated Protein Science (CiPSM) and Center for Nanoscience (CeNS), Ludwig-Maximilians-Universität München, Munich, Germany, and Laboratory of Photochemistry and Spectroscopy, Department of Chemistry, University of Leuven, Celestijnenlaan, Leuven, Belgium

Jolene Johnson
School of Physics and Astronomy, University of Minnesota, Minneapolis, Minnesota, USA

Don C. Lamb
Physical Chemistry, Department of Chemistry, Munich Center for Integrated Protein Science (CiPSM) and Center for Nanoscience (CeNS), Ludwig-Maximilians-Universität München, Munich, Germany, and Department of Physics, University of Illinois at Urbana-Champaign, Urbana, Illinois, USA

Patrick Macdonald
Department of Biomedical Engineering, University of Minnesota, Minneapolis, Minnesota, USA

Joachim D. Mueller
Department of Biomedical Engineering, and School of Physics and Astronomy, University of Minnesota, Minneapolis, Minnesota, USA

Christoph Pieper
Third Institute of Physics–Biophysics, Georg-August-University Göttingen, Friedrich-Hund-Platz, Göttingen, Germany

Petra Schwille[1]
Biophysics/BIOTEC, Technische Universität Dresden, Tatzberg 47-51, Dresden, Germany

Robert H. Singer
Department of Anatomy and Structural Biology, and Gruss-Lipper Biophotonic Center, Albert Einstein College of Medicine, Bronx, New York, USA

Elizabeth Smith
School of Physics and Astronomy, University of Minnesota, Minneapolis, Minnesota, USA

Milka Stakic
Laboratory for Fluorescence Dynamics, Department of Biomedical Engineering, University of California, Irvine, California, USA

Kerstin Weiß
Third Institute of Physics–Biophysics, Georg-August-University Göttingen, Friedrich-Hund-Platz, Göttingen, Germany

Thomas Weidemann[1]
Biophysics/BIOTEC, Technische Universität Dresden, Tatzberg 47-51, Dresden, Germany

Paul W. Wiseman
Department of Physics, and Department of Chemistry, McGill University, Montreal, Quebec, Canada

Bin Wu
Department of Anatomy and Structural Biology, and Gruss-Lipper Biophotonic Center, Albert Einstein College of Medicine, Bronx, New York, USA

[1] Current address: Max Planck Institute of Biochemistry, Am Klopferspitz 18, 82152 Martinsried, Germany

PREFACE

Fluorescence Fluctuation Spectroscopy (FFS) is a growing family of methods for studying molecular motions and interactions. Introduced as Fluorescence Correlation Spectroscopy (FCS) 40 years ago, this elegant technique has evolved into a large group of powerful tools of modern fluorescence spectroscopy. FFS has found numerous applications in biophysical and biochemical studies of proteins, nucleic acids, and membranes. In combination with fluorescence imaging microscopy, FFS provides unique opportunities for characterizing dynamic processes in living cells and has been increasingly used by cell and systems biologists. Significant progress in the technology was achieved in the past decade, and the number of new FFS methods is still growing. The attraction of FFS relates to fact that it is based on statistical analysis of the fluctuations occurring at the molecular level. Therefore, these methods can resolve molecular populations and provide additional information about biological systems.

The chapters in this two-volume publication are written by the authors and first-hand users of the methods under discussion. Part A is dedicated predominately to FFS methods. It contains well-established and also recently introduced experimental techniques and data analysis algorithms. Part B continues with FFS methods and also includes various FFS applications *in vivo* and *in vitro*. Here the reader will find specific examples and detailed experimental protocols developed for several biological systems. The purpose of these chapters is to show the power and potentials of the technology in biological research.

I am very thankful to all the authors who participated in this project and wrote excellent chapters that made my editorial work easy. I am expressing my special gratitude to Elliot Elson, whose seminal work on FCS began in the early 1970s, for writing a historical note and current systemization of the FFS methods in addition to his chapter on FCS conceptual basis and theory. I am also very thankful to Shaun Gamble from Elsevier for his great assistance in organizing these volumes. On a personal note, I want to express my immense gratefulness to my wife Luda for her devotion, serenity, and constant support to all my work.

FFS offers a variety of powerful techniques for studying biological molecules *in vitro* and *in vivo*. I hope that these MIE volumes will be used by many new and experienced researchers alike.

Sergey Y. Tetin

METHODS IN ENZYMOLOGY

VOLUME I. Preparation and Assay of Enzymes
Edited by SIDNEY P. COLOWICK AND NATHAN O. KAPLAN

VOLUME II. Preparation and Assay of Enzymes
Edited by SIDNEY P. COLOWICK AND NATHAN O. KAPLAN

VOLUME III. Preparation and Assay of Substrates
Edited by SIDNEY P. COLOWICK AND NATHAN O. KAPLAN

VOLUME IV. Special Techniques for the Enzymologist
Edited by SIDNEY P. COLOWICK AND NATHAN O. KAPLAN

VOLUME V. Preparation and Assay of Enzymes
Edited by SIDNEY P. COLOWICK AND NATHAN O. KAPLAN

VOLUME VI. Preparation and Assay of Enzymes *(Continued)*
Preparation and Assay of Substrates
Special Techniques
Edited by SIDNEY P. COLOWICK AND NATHAN O. KAPLAN

VOLUME VII. Cumulative Subject Index
Edited by SIDNEY P. COLOWICK AND NATHAN O. KAPLAN

VOLUME VIII. Complex Carbohydrates
Edited by ELIZABETH F. NEUFELD AND VICTOR GINSBURG

VOLUME IX. Carbohydrate Metabolism
Edited by WILLIS A. WOOD

VOLUME X. Oxidation and Phosphorylation
Edited by RONALD W. ESTABROOK AND MAYNARD E. PULLMAN

VOLUME XI. Enzyme Structure
Edited by C. H. W. HIRS

VOLUME XII. Nucleic Acids (Parts A and B)
Edited by LAWRENCE GROSSMAN AND KIVIE MOLDAVE

VOLUME XIII. Citric Acid Cycle
Edited by J. M. LOWENSTEIN

VOLUME XIV. Lipids
Edited by J. M. LOWENSTEIN

Volume XLVIII. Enzyme Structure (Part F)
Edited by C. H. W. Hirs and Serge N. Timasheff

Volume XLIX. Enzyme Structure (Part G)
Edited by C. H. W. Hirs and Serge N. Timasheff

Volume L. Complex Carbohydrates (Part C)
Edited by Victor Ginsburg

Volume LI. Purine and Pyrimidine Nucleotide Metabolism
Edited by Patricia A. Hoffee and Mary Ellen Jones

Volume LII. Biomembranes (Part C: Biological Oxidations)
Edited by Sidney Fleischer and Lester Packer

Volume LIII. Biomembranes (Part D: Biological Oxidations)
Edited by Sidney Fleischer and Lester Packer

Volume LIV. Biomembranes (Part E: Biological Oxidations)
Edited by Sidney Fleischer and Lester Packer

Volume LV. Biomembranes (Part F: Bioenergetics)
Edited by Sidney Fleischer and Lester Packer

Volume LVI. Biomembranes (Part G: Bioenergetics)
Edited by Sidney Fleischer and Lester Packer

Volume LVII. Bioluminescence and Chemiluminescence
Edited by Marlene A. DeLuca

Volume LVIII. Cell Culture
Edited by William B. Jakoby and Ira Pastan

Volume LIX. Nucleic Acids and Protein Synthesis (Part G)
Edited by Kivie Moldave and Lawrence Grossman

Volume LX. Nucleic Acids and Protein Synthesis (Part H)
Edited by Kivie Moldave and Lawrence Grossman

Volume 61. Enzyme Structure (Part H)
Edited by C. H. W. Hirs and Serge N. Timasheff

Volume 62. Vitamins and Coenzymes (Part D)
Edited by Donald B. McCormick and Lemuel D. Wright

Volume 63. Enzyme Kinetics and Mechanism (Part A: Initial Rate and Inhibitor Methods)
Edited by Daniel L. Purich

Volume 64. Enzyme Kinetics and Mechanism
(Part B: Isotopic Probes and Complex Enzyme Systems)
Edited by Daniel L. Purich

Volume 65. Nucleic Acids (Part I)
Edited by Lawrence Grossman and Kivie Moldave

Volume 66. Vitamins and Coenzymes (Part E)
Edited by Donald B. McCormick and Lemuel D. Wright

Volume 67. Vitamins and Coenzymes (Part F)
Edited by Donald B. McCormick and Lemuel D. Wright

Volume 68. Recombinant DNA
Edited by Ray Wu

Volume 69. Photosynthesis and Nitrogen Fixation (Part C)
Edited by Anthony San Pietro

Volume 70. Immunochemical Techniques (Part A)
Edited by Helen Van Vunakis and John J. Langone

Volume 71. Lipids (Part C)
Edited by John M. Lowenstein

Volume 72. Lipids (Part D)
Edited by John M. Lowenstein

Volume 73. Immunochemical Techniques (Part B)
Edited by John J. Langone and Helen Van Vunakis

Volume 74. Immunochemical Techniques (Part C)
Edited by John J. Langone and Helen Van Vunakis

Volume 75. Cumulative Subject Index Volumes XXXI, XXXII, XXXIV–LX
Edited by Edward A. Dennis and Martha G. Dennis

Volume 76. Hemoglobins
Edited by Eraldo Antonini, Luigi Rossi-Bernardi, and Emilia Chiancone

Volume 77. Detoxication and Drug Metabolism
Edited by William B. Jakoby

Volume 78. Interferons (Part A)
Edited by Sidney Pestka

Volume 79. Interferons (Part B)
Edited by Sidney Pestka

Volume 80. Proteolytic Enzymes (Part C)
Edited by Laszlo Lorand

Volume 81. Biomembranes (Part H: Visual Pigments and Purple Membranes, I)
Edited by Lester Packer

Volume 82. Structural and Contractile Proteins (Part A: Extracellular Matrix)
Edited by Leon W. Cunningham and Dixie W. Frederiksen

Volume 83. Complex Carbohydrates (Part D)
Edited by Victor Ginsburg

Volume 84. Immunochemical Techniques (Part D: Selected Immunoassays)
Edited by John J. Langone and Helen Van Vunakis

Volume 85. Structural and Contractile Proteins (Part B: The Contractile Apparatus and the Cytoskeleton)
Edited by Dixie W. Frederiksen and Leon W. Cunningham

Volume 86. Prostaglandins and Arachidonate Metabolites
Edited by William E. M. Lands and William L. Smith

Volume 87. Enzyme Kinetics and Mechanism (Part C: Intermediates, Stereo-chemistry, and Rate Studies)
Edited by Daniel L. Purich

Volume 88. Biomembranes (Part I: Visual Pigments and Purple Membranes, II)
Edited by Lester Packer

Volume 89. Carbohydrate Metabolism (Part D)
Edited by Willis A. Wood

Volume 90. Carbohydrate Metabolism (Part E)
Edited by Willis A. Wood

Volume 91. Enzyme Structure (Part I)
Edited by C. H. W. Hirs and Serge N. Timasheff

Volume 92. Immunochemical Techniques (Part E: Monoclonal Antibodies and General Immunoassay Methods)
Edited by John J. Langone and Helen Van Vunakis

VOLUME 107. Posttranslational Modifications (Part B)
Edited by FINN WOLD AND KIVIE MOLDAVE

VOLUME 108. Immunochemical Techniques (Part G: Separation and Characterization of Lymphoid Cells)
Edited by GIOVANNI DI SABATO, JOHN J. LANGONE, AND HELEN VAN VUNAKIS

VOLUME 109. Hormone Action (Part I: Peptide Hormones)
Edited by LUTZ BIRNBAUMER AND BERT W. O'MALLEY

VOLUME 110. Steroids and Isoprenoids (Part A)
Edited by JOHN H. LAW AND HANS C. RILLING

VOLUME 111. Steroids and Isoprenoids (Part B)
Edited by JOHN H. LAW AND HANS C. RILLING

VOLUME 112. Drug and Enzyme Targeting (Part A)
Edited by KENNETH J. WIDDER AND RALPH GREEN

VOLUME 113. Glutamate, Glutamine, Glutathione, and Related Compounds
Edited by ALTON MEISTER

VOLUME 114. Diffraction Methods for Biological Macromolecules (Part A)
Edited by HAROLD W. WYCKOFF, C. H. W. HIRS, AND SERGE N. TIMASHEFF

VOLUME 115. Diffraction Methods for Biological Macromolecules (Part B)
Edited by HAROLD W. WYCKOFF, C. H. W. HIRS, AND SERGE N. TIMASHEFF

VOLUME 116. Immunochemical Techniques
(Part H: Effectors and Mediators of Lymphoid Cell Functions)
Edited by GIOVANNI DI SABATO, JOHN J. LANGONE, AND HELEN VAN VUNAKIS

VOLUME 117. Enzyme Structure (Part J)
Edited by C. H. W. HIRS AND SERGE N. TIMASHEFF

VOLUME 118. Plant Molecular Biology
Edited by ARTHUR WEISSBACH AND HERBERT WEISSBACH

VOLUME 119. Interferons (Part C)
Edited by SIDNEY PESTKA

VOLUME 120. Cumulative Subject Index Volumes 81–94, 96–101

VOLUME 121. Immunochemical Techniques (Part I: Hybridoma Technology and Monoclonal Antibodies)
Edited by JOHN J. LANGONE AND HELEN VAN VUNAKIS

Volume 166. Branched-Chain Amino Acids
Edited by Robert Harris and John R. Sokatch

Volume 167. Cyanobacteria
Edited by Lester Packer and Alexander N. Glazer

Volume 168. Hormone Action (Part K: Neuroendocrine Peptides)
Edited by P. Michael Conn

Volume 169. Platelets: Receptors, Adhesion, Secretion (Part A)
Edited by Jacek Hawiger

Volume 170. Nucleosomes
Edited by Paul M. Wassarman and Roger D. Kornberg

Volume 171. Biomembranes (Part R: Transport Theory: Cells and Model Membranes)
Edited by Sidney Fleischer and Becca Fleischer

Volume 172. Biomembranes (Part S: Transport: Membrane Isolation and Characterization)
Edited by Sidney Fleischer and Becca Fleischer

Volume 173. Biomembranes [Part T: Cellular and Subcellular Transport: Eukaryotic (Nonepithelial) Cells]
Edited by Sidney Fleischer and Becca Fleischer

Volume 174. Biomembranes [Part U: Cellular and Subcellular Transport: Eukaryotic (Nonepithelial) Cells]
Edited by Sidney Fleischer and Becca Fleischer

Volume 175. Cumulative Subject Index Volumes 135–139, 141–167

Volume 176. Nuclear Magnetic Resonance (Part A: Spectral Techniques and Dynamics)
Edited by Norman J. Oppenheimer and Thomas L. James

Volume 177. Nuclear Magnetic Resonance (Part B: Structure and Mechanism)
Edited by Norman J. Oppenheimer and Thomas L. James

Volume 178. Antibodies, Antigens, and Molecular Mimicry
Edited by John J. Langone

Volume 179. Complex Carbohydrates (Part F)
Edited by Victor Ginsburg

VOLUME 180. RNA Processing (Part A: General Methods)
Edited by JAMES E. DAHLBERG AND JOHN N. ABELSON

VOLUME 181. RNA Processing (Part B: Specific Methods)
Edited by JAMES E. DAHLBERG AND JOHN N. ABELSON

VOLUME 182. Guide to Protein Purification
Edited by MURRAY P. DEUTSCHER

VOLUME 183. Molecular Evolution: Computer Analysis of Protein and Nucleic Acid Sequences
Edited by RUSSELL F. DOOLITTLE

VOLUME 184. Avidin-Biotin Technology
Edited by MEIR WILCHEK AND EDWARD A. BAYER

VOLUME 185. Gene Expression Technology
Edited by DAVID V. GOEDDEL

VOLUME 186. Oxygen Radicals in Biological Systems (Part B: Oxygen Radicals and Antioxidants)
Edited by LESTER PACKER AND ALEXANDER N. GLAZER

VOLUME 187. Arachidonate Related Lipid Mediators
Edited by ROBERT C. MURPHY AND FRANK A. FITZPATRICK

VOLUME 188. Hydrocarbons and Methylotrophy
Edited by MARY E. LIDSTROM

VOLUME 189. Retinoids (Part A: Molecular and Metabolic Aspects)
Edited by LESTER PACKER

VOLUME 190. Retinoids (Part B: Cell Differentiation and Clinical Applications)
Edited by LESTER PACKER

VOLUME 191. Biomembranes (Part V: Cellular and Subcellular Transport: Epithelial Cells)
Edited by SIDNEY FLEISCHER AND BECCA FLEISCHER

VOLUME 192. Biomembranes (Part W: Cellular and Subcellular Transport: Epithelial Cells)
Edited by SIDNEY FLEISCHER AND BECCA FLEISCHER

VOLUME 193. Mass Spectrometry
Edited by JAMES A. MCCLOSKEY

VOLUME 194. Guide to Yeast Genetics and Molecular Biology
Edited by CHRISTINE GUTHRIE AND GERALD R. FINK

VOLUME 195. Adenylyl Cyclase, G Proteins, and Guanylyl Cyclase
Edited by ROGER A. JOHNSON AND JACKIE D. CORBIN

VOLUME 196. Molecular Motors and the Cytoskeleton
Edited by RICHARD B. VALLEE

VOLUME 197. Phospholipases
Edited by EDWARD A. DENNIS

VOLUME 198. Peptide Growth Factors (Part C)
Edited by DAVID BARNES, J. P. MATHER, AND GORDON H. SATO

VOLUME 199. Cumulative Subject Index Volumes 168–174, 176–194

VOLUME 200. Protein Phosphorylation (Part A: Protein Kinases: Assays, Purification, Antibodies, Functional Analysis, Cloning, and Expression)
Edited by TONY HUNTER AND BARTHOLOMEW M. SEFTON

VOLUME 201. Protein Phosphorylation (Part B: Analysis of Protein Phosphorylation, Protein Kinase Inhibitors, and Protein Phosphatases)
Edited by TONY HUNTER AND BARTHOLOMEW M. SEFTON

VOLUME 202. Molecular Design and Modeling: Concepts and Applications (Part A: Proteins, Peptides, and Enzymes)
Edited by JOHN J. LANGONE

VOLUME 203. Molecular Design and Modeling: Concepts and Applications (Part B: Antibodies and Antigens, Nucleic Acids, Polysaccharides, and Drugs)
Edited by JOHN J. LANGONE

VOLUME 204. Bacterial Genetic Systems
Edited by JEFFREY H. MILLER

VOLUME 205. Metallobiochemistry (Part B: Metallothionein and Related Molecules)
Edited by JAMES F. RIORDAN AND BERT L. VALLEE

VOLUME 206. Cytochrome P450
Edited by MICHAEL R. WATERMAN AND ERIC F. JOHNSON

VOLUME 207. Ion Channels
Edited by BERNARDO RUDY AND LINDA E. IVERSON

Factors and Inhibitors)
Edited by Laszlo Lorand and Kenneth G. Mann

Volume 223. Proteolytic Enzymes in Coagulation, Fibrinolysis, and Complement Activation (Part B: Complement Activation, Fibrinolysis, and Nonmammalian Blood Coagulation Factors)
Edited by Laszlo Lorand and Kenneth G. Mann

Volume 224. Molecular Evolution: Producing the Biochemical Data
Edited by Elizabeth Anne Zimmer, Thomas J. White, Rebecca L. Cann, and Allan C. Wilson

Volume 225. Guide to Techniques in Mouse Development
Edited by Paul M. Wassarman and Melvin L. DePamphilis

Volume 226. Metallobiochemistry (Part C: Spectroscopic and Physical Methods for Probing Metal Ion Environments in Metalloenzymes and Metalloproteins)
Edited by James F. Riordan and Bert L. Vallee

Volume 227. Metallobiochemistry (Part D: Physical and Spectroscopic Methods for Probing Metal Ion Environments in Metalloproteins)
Edited by James F. Riordan and Bert L. Vallee

Volume 228. Aqueous Two-Phase Systems
Edited by Harry Walter and Göte Johansson

Volume 229. Cumulative Subject Index Volumes 195–198, 200–227

Volume 230. Guide to Techniques in Glycobiology
Edited by William J. Lennarz and Gerald W. Hart

Volume 231. Hemoglobins (Part B: Biochemical and Analytical Methods)
Edited by Johannes Everse, Kim D. Vandegriff, and Robert M. Winslow

Volume 232. Hemoglobins (Part C: Biophysical Methods)
Edited by Johannes Everse, Kim D. Vandegriff, and Robert M. Winslow

Volume 233. Oxygen Radicals in Biological Systems (Part C)
Edited by Lester Packer

Volume 234. Oxygen Radicals in Biological Systems (Part D)
Edited by Lester Packer

Volume 235. Bacterial Pathogenesis (Part A: Identification and Regulation of Virulence Factors)
Edited by Virginia L. Clark and Patrik M. Bavoil

Volume 236. Bacterial Pathogenesis (Part B: Integration of Pathogenic Bacteria with Host Cells)
Edited by Virginia L. Clark and Patrik M. Bavoil

Volume 237. Heterotrimeric G Proteins
Edited by Ravi Iyengar

Volume 238. Heterotrimeric G-Protein Effectors
Edited by Ravi Iyengar

Volume 239. Nuclear Magnetic Resonance (Part C)
Edited by Thomas L. James and Norman J. Oppenheimer

Volume 240. Numerical Computer Methods (Part B)
Edited by Michael L. Johnson and Ludwig Brand

Volume 241. Retroviral Proteases
Edited by Lawrence C. Kuo and Jules A. Shafer

Volume 242. Neoglycoconjugates (Part A)
Edited by Y. C. Lee and Reiko T. Lee

Volume 243. Inorganic Microbial Sulfur Metabolism
Edited by Harry D. Peck, Jr., and Jean LeGall

Volume 244. Proteolytic Enzymes: Serine and Cysteine Peptidases
Edited by Alan J. Barrett

Volume 245. Extracellular Matrix Components
Edited by E. Ruoslahti and E. Engvall

Volume 246. Biochemical Spectroscopy
Edited by Kenneth Sauer

Volume 247. Neoglycoconjugates (Part B: Biomedical Applications)
Edited by Y. C. Lee and Reiko T. Lee

Volume 248. Proteolytic Enzymes: Aspartic and Metallo Peptidases
Edited by Alan J. Barrett

Volume 249. Enzyme Kinetics and Mechanism (Part D: Developments in Enzyme Dynamics)
Edited by Daniel L. Purich

Volume 250. Lipid Modifications of Proteins
Edited by Patrick J. Casey and Janice E. Buss

Volume 251. Biothiols (Part A: Monothiols and Dithiols, Protein Thiols, and Thiyl Radicals)
Edited by Lester Packer

Volume 252. Biothiols (Part B: Glutathione and Thioredoxin; Thiols in Signal Transduction and Gene Regulation)
Edited by Lester Packer

Volume 253. Adhesion of Microbial Pathogens
Edited by Ron J. Doyle and Itzhak Ofek

Volume 254. Oncogene Techniques
Edited by Peter K. Vogt and Inder M. Verma

Volume 255. Small GTPases and Their Regulators (Part A: Ras Family)
Edited by W. E. Balch, Channing J. Der, and Alan Hall

Volume 256. Small GTPases and Their Regulators (Part B: Rho Family)
Edited by W. E. Balch, Channing J. Der, and Alan Hall

Volume 257. Small GTPases and Their Regulators (Part C: Proteins Involved in Transport)
Edited by W. E. Balch, Channing J. Der, and Alan Hall

Volume 258. Redox-Active Amino Acids in Biology
Edited by Judith P. Klinman

Volume 259. Energetics of Biological Macromolecules
Edited by Michael L. Johnson and Gary K. Ackers

Volume 260. Mitochondrial Biogenesis and Genetics (Part A)
Edited by Giuseppe M. Attardi and Anne Chomyn

Volume 261. Nuclear Magnetic Resonance and Nucleic Acids
Edited by Thomas L. James

Volume 262. DNA Replication
Edited by Judith L. Campbell

Volume 263. Plasma Lipoproteins (Part C: Quantitation)
Edited by William A. Bradley, Sandra H. Gianturco, and Jere P. Segrest

Volume 264. Mitochondrial Biogenesis and Genetics (Part B)
Edited by Giuseppe M. Attardi and Anne Chomyn

Volume 265. Cumulative Subject Index Volumes 228, 230–262

Volume 313. Antisense Technology
(Part A: General Methods, Methods of Delivery, and RNA Studies)
Edited by M. Ian Phillips

Volume 314. Antisense Technology (Part B: Applications)
Edited by M. Ian Phillips

Volume 315. Vertebrate Phototransduction and the Visual Cycle
(Part A)
Edited by Krzysztof Palczewski

Volume 316. Vertebrate Phototransduction and the Visual Cycle (Part B)
Edited by Krzysztof Palczewski

Volume 317. RNA–Ligand Interactions (Part A: Structural Biology
Methods)
Edited by Daniel W. Celander and John N. Abelson

Volume 318. RNA–Ligand Interactions (Part B: Molecular Biology
Methods)
Edited by Daniel W. Celander and John N. Abelson

Volume 319. Singlet Oxygen, UV-A, and Ozone
Edited by Lester Packer and Helmut Sies

Volume 320. Cumulative Subject Index Volumes 290–319

Volume 321. Numerical Computer Methods (Part C)
Edited by Michael L. Johnson and Ludwig Brand

Volume 322. Apoptosis
Edited by John C. Reed

Volume 323. Energetics of Biological Macromolecules (Part C)
Edited by Michael L. Johnson and Gary K. Ackers

Volume 324. Branched-Chain Amino Acids (Part B)
Edited by Robert A. Harris and John R. Sokatch

Volume 325. Regulators and Effectors of Small GTPases
(Part D: Rho Family)
Edited by W. E. Balch, Channing J. Der, and Alan Hall

Volume 326. Applications of Chimeric Genes and Hybrid Proteins
(Part A: Gene Expression and Protein Purification)
Edited by Jeremy Thorner, Scott D. Emr, and John N. Abelson

VOLUME 327. Applications of Chimeric Genes and Hybrid Proteins (Part B: Cell Biology and Physiology)
Edited by JEREMY THORNER, SCOTT D. EMR, AND JOHN N. ABELSON

VOLUME 328. Applications of Chimeric Genes and Hybrid Proteins (Part C: Protein–Protein Interactions and Genomics)
Edited by JEREMY THORNER, SCOTT D. EMR, AND JOHN N. ABELSON

VOLUME 329. Regulators and Effectors of Small GTPases (Part E: GTPases Involved in Vesicular Traffic)
Edited by W. E. BALCH, CHANNING J. DER, AND ALAN HALL

VOLUME 330. Hyperthermophilic Enzymes (Part A)
Edited by MICHAEL W. W. ADAMS AND ROBERT M. KELLY

VOLUME 331. Hyperthermophilic Enzymes (Part B)
Edited by MICHAEL W. W. ADAMS AND ROBERT M. KELLY

VOLUME 332. Regulators and Effectors of Small GTPases (Part F: Ras Family I)
Edited by W. E. BALCH, CHANNING J. DER, AND ALAN HALL

VOLUME 333. Regulators and Effectors of Small GTPases (Part G: Ras Family II)
Edited by W. E. BALCH, CHANNING J. DER, AND ALAN HALL

VOLUME 334. Hyperthermophilic Enzymes (Part C)
Edited by MICHAEL W. W. ADAMS AND ROBERT M. KELLY

VOLUME 335. Flavonoids and Other Polyphenols
Edited by LESTER PACKER

VOLUME 336. Microbial Growth in Biofilms (Part A: Developmental and Molecular Biological Aspects)
Edited by RON J. DOYLE

VOLUME 337. Microbial Growth in Biofilms (Part B: Special Environments and Physicochemical Aspects)
Edited by RON J. DOYLE

VOLUME 338. Nuclear Magnetic Resonance of Biological Macromolecules (Part A)
Edited by THOMAS L. JAMES, VOLKER DÖTSCH, AND ULI SCHMITZ

VOLUME 339. Nuclear Magnetic Resonance of Biological Macromolecules (Part B)
Edited by THOMAS L. JAMES, VOLKER DÖTSCH, AND ULI SCHMITZ

Volume 354. Enzyme Kinetics and Mechanisms (Part F: Detection and Characterization of Enzyme Reaction Intermediates)
Edited by Daniel L. Purich

Volume 355. Cumulative Subject Index Volumes 321–354

Volume 356. Laser Capture Microscopy and Microdissection
Edited by P. Michael Conn

Volume 357. Cytochrome P450, Part C
Edited by Eric F. Johnson and Michael R. Waterman

Volume 358. Bacterial Pathogenesis (Part C: Identification, Regulation, and Function of Virulence Factors)
Edited by Virginia L. Clark and Patrik M. Bavoil

Volume 359. Nitric Oxide (Part D)
Edited by Enrique Cadenas and Lester Packer

Volume 360. Biophotonics (Part A)
Edited by Gerard Marriott and Ian Parker

Volume 361. Biophotonics (Part B)
Edited by Gerard Marriott and Ian Parker

Volume 362. Recognition of Carbohydrates in Biological Systems (Part A)
Edited by Yuan C. Lee and Reiko T. Lee

Volume 363. Recognition of Carbohydrates in Biological Systems (Part B)
Edited by Yuan C. Lee and Reiko T. Lee

Volume 364. Nuclear Receptors
Edited by David W. Russell and David J. Mangelsdorf

Volume 365. Differentiation of Embryonic Stem Cells
Edited by Paul M. Wassauman and Gordon M. Keller

Volume 366. Protein Phosphatases
Edited by Susanne Klumpp and Josef Krieglstein

Volume 367. Liposomes (Part A)
Edited by Nejat Düzgüneş

Volume 368. Macromolecular Crystallography (Part C)
Edited by Charles W. Carter, Jr., and Robert M. Sweet

Volume 369. Combinational Chemistry (Part B)
Edited by Guillermo A. Morales and Barry A. Bunin

VOLUME 370. RNA Polymerases and Associated Factors (Part C)
Edited by SANKAR L. ADHYA AND SUSAN GARGES

VOLUME 371. RNA Polymerases and Associated Factors (Part D)
Edited by SANKAR L. ADHYA AND SUSAN GARGES

VOLUME 372. Liposomes (Part B)
Edited by NEJAT DÜZGÜNEŞ

VOLUME 373. Liposomes (Part C)
Edited by NEJAT DÜZGÜNEŞ

VOLUME 374. Macromolecular Crystallography (Part D)
Edited by CHARLES W. CARTER, JR., AND ROBERT W. SWEET

VOLUME 375. Chromatin and Chromatin Remodeling Enzymes (Part A)
Edited by C. DAVID ALLIS AND CARL WU

VOLUME 376. Chromatin and Chromatin Remodeling Enzymes (Part B)
Edited by C. DAVID ALLIS AND CARL WU

VOLUME 377. Chromatin and Chromatin Remodeling Enzymes (Part C)
Edited by C. DAVID ALLIS AND CARL WU

VOLUME 378. Quinones and Quinone Enzymes (Part A)
Edited by HELMUT SIES AND LESTER PACKER

VOLUME 379. Energetics of Biological Macromolecules (Part D)
Edited by JO M. HOLT, MICHAEL L. JOHNSON, AND GARY K. ACKERS

VOLUME 380. Energetics of Biological Macromolecules (Part E)
Edited by JO M. HOLT, MICHAEL L. JOHNSON, AND GARY K. ACKERS

VOLUME 381. Oxygen Sensing
Edited by CHANDAN K. SEN AND GREGG L. SEMENZA

VOLUME 382. Quinones and Quinone Enzymes (Part B)
Edited by HELMUT SIES AND LESTER PACKER

VOLUME 383. Numerical Computer Methods (Part D)
Edited by LUDWIG BRAND AND MICHAEL L. JOHNSON

VOLUME 384. Numerical Computer Methods (Part E)
Edited by LUDWIG BRAND AND MICHAEL L. JOHNSON

VOLUME 385. Imaging in Biological Research (Part A)
Edited by P. MICHAEL CONN

VOLUME 386. Imaging in Biological Research (Part B)
Edited by P. MICHAEL CONN

VOLUME 419. Adult Stem Cells
Edited by IRINA KLIMANSKAYA AND ROBERT LANZA

VOLUME 420. Stem Cell Tools and Other Experimental Protocols
Edited by IRINA KLIMANSKAYA AND ROBERT LANZA

VOLUME 421. Advanced Bacterial Genetics: Use of Transposons and Phage for Genomic Engineering
Edited by KELLY T. HUGHES

VOLUME 422. Two-Component Signaling Systems, Part A
Edited by MELVIN I. SIMON, BRIAN R. CRANE, AND ALEXANDRINE CRANE

VOLUME 423. Two-Component Signaling Systems, Part B
Edited by MELVIN I. SIMON, BRIAN R. CRANE, AND ALEXANDRINE CRANE

VOLUME 424. RNA Editing
Edited by JONATHA M. GOTT

VOLUME 425. RNA Modification
Edited by JONATHA M. GOTT

VOLUME 426. Integrins
Edited by DAVID CHERESH

VOLUME 427. MicroRNA Methods
Edited by JOHN J. ROSSI

VOLUME 428. Osmosensing and Osmosignaling
Edited by HELMUT SIES AND DIETER HAUSSINGER

VOLUME 429. Translation Initiation: Extract Systems and Molecular Genetics
Edited by JON LORSCH

VOLUME 430. Translation Initiation: Reconstituted Systems and Biophysical Methods
Edited by JON LORSCH

VOLUME 431. Translation Initiation: Cell Biology, High-Throughput and Chemical-Based Approaches
Edited by JON LORSCH

VOLUME 432. Lipidomics and Bioactive Lipids: Mass-Spectrometry–Based Lipid Analysis
Edited by H. ALEX BROWN

VOLUME 489. The Unfolded Protein Response and Cellular Stress, Part A
Edited by P. MICHAEL CONN

VOLUME 490. The Unfolded Protein Response and Cellular Stress, Part B
Edited by P. MICHAEL CONN

VOLUME 491. The Unfolded Protein Response and Cellular Stress, Part C
Edited by P. MICHAEL CONN

VOLUME 492. Biothermodynamics, Part D
Edited by MICHAEL L. JOHNSON, JO M. HOLT, AND GARY K. ACKERS

VOLUME 493. Fragment-Based Drug Design Tools,
Practical Approaches, and Examples
Edited by LAWRENCE C. KUO

VOLUME 494. Methods in Methane Metabolism, Part A
Methanogenesis
Edited by AMY C. ROSENZWEIG AND STEPHEN W. RAGSDALE

VOLUME 495. Methods in Methane Metabolism, Part B
Methanotrophy
Edited by AMY C. ROSENZWEIG AND STEPHEN W. RAGSDALE

VOLUME 496. Research on Nitrification and Related Processes, Part B
Edited by MARTIN G. KLOTZ AND LISA Y. STEIN

VOLUME 497. Synthetic Biology, Part A
Methods for Part/Device Characterization and Chassis Engineering
Edited by CHRISTOPHER VOIGT

VOLUME 498. Synthetic Biology, Part B
Computer Aided Design and DNA Assembly
Edited by CHRISTOPHER VOIGT

VOLUME 499. Biology of Serpins
Edited by JAMES C. WHISSTOCK AND PHILLIP I. BIRD

VOLUME 500. Methods in Systems Biology
Edited by DANIEL JAMESON, MALKHEY VERMA, AND HANS V. WESTERHOFF

VOLUME 501. Serpin Structure and Evolution
Edited by JAMES C. WHISSTOCK AND PHILLIP I. BIRD

VOLUME 502. Protein Engineering for Therapeutics, Part A
Edited by K. DANE WITTRUP AND GREGORY L. VERDINE

VOLUME 503. Protein Engineering for Therapeutics, Part B
Edited by K. DANE WITTRUP AND GREGORY L. VERDINE

VOLUME 504. Imaging and Spectroscopic Analysis of Living Cells
Optical and Spectroscopic Techniques
Edited by P. MICHAEL CONN

VOLUME 505. Imaging and Spectroscopic Analysis of Living Cells
Live Cell Imaging of Cellular Elements and Functions
Edited by P. MICHAEL CONN

VOLUME 506. Imaging and Spectroscopic Analysis of Living Cells
Imaging Live Cells in Health and Disease
Edited by P. MICHAEL CONN

VOLUME 507. Gene Transfer Vectors for Clinical Application
Edited by THEODORE FRIEDMANN

VOLUME 508. Nanomedicine
Cancer, Diabetes, and Cardiovascular, Central Nervous System, Pulmonary and Inflammatory Diseases
Edited by NEJAT DÜZGÜNEŞ

VOLUME 509. Nanomedicine
Infectious Diseases, Immunotherapy, Diagnostics, Antifibrotics, Toxicology and Gene Medicine
Edited by NEJAT DÜZGÜNEŞ

VOLUME 510. Cellulases
Edited by HARRY J. GILBERT

VOLUME 511. RNA Helicases
Edited by ECKHARD JANKOWSKY

VOLUME 512. Nucleosomes, Histones & Chromatin, Part A
Edited by CARL WU AND C. DAVID ALLIS

VOLUME 513. Nucleosomes, Histones & Chromatin, Part B
Edited by CARL WU AND C. DAVID ALLIS

VOLUME 514. Ghrelin
Edited by MASAYASU KOJIMA AND KENJI KANGAWA

VOLUME 515. Natural Product Biosynthesis by Microorganisms and Plants, Part A
Edited by DAVID A. HOPWOOD

VOLUME 516. Natural Product Biosynthesis by Microorganisms and Plants, Part B
Edited by DAVID A. HOPWOOD

VOLUME 517. Natural Product Biosynthesis by Microorganisms and Plants, Part C
Edited by DAVID A. HOPWOOD

VOLUME 518. Fluorescence Fluctuation Spectroscopy (FFS), Part A
Edited by SERGEY Y. TETIN

40 Years of FCS: How It All Began

Elliot L. Elson[1]
Department of Biochemistry and Molecular Biophysics, Washington University School of Medicine, Campus Box 8231, St. Louis, Missouri, USA
[1]Corresponding author: e-mail address: elson@wustl.edu

Abstract

Fluorescence correlation spectroscopy (FCS) determines rates of molecular transport and chemical reactions from measurements of spontaneous concentration fluctuations in small open sub-volumes of systems in equilibrium or non-equilibrium steady state. The concentration fluctuations are monitored via fluorescence. Although difficult at first, FCS measurements are now in routine use in areas of chemistry, physics and biology. The initial approach has given rise to a wide range of extensions and variations that yield information about the dynamics of molecular processes in simple systems and living cells. FCS provides a window on a mesoscopic world in which molecular fluctuations are both detectable and (for biological cells) possibly functionally important. Moreover, FCS is a precursor to single molecule measurements.

Fluorescence correlation spectroscopy (FCS) has evolved from a conceptually elegant method that was difficult to use in practice to one that is now routine in the laboratory and has given rise to a host of successor methods. This evolution illustrates how efforts focused on a very specific research problem can ramify beyond the wildest expectations of the original investigators. FCS was introduced 40 years ago (Magde, Elson, & Webb, 1972). Two laboratories independently developed different aspects of the approach, one at Cornell University in Ithaca, New York, and the other at the Karolinska Institute in Stockholm, Sweden. A primary motivation for both groups was to develop a method that could measure the kinetics of chemical reactions in systems in equilibrium and thereby to evade the need to perturb the state of the system, for example, by temperature jump, as is required for conventional measurements of chemical kinetics. For the Cornell group, the problem was to investigate the kinetics of interior loop formation during DNA untwisting (Elson, 2004, 2011). The aim was to dissect the kinetic contributions of the minimum number of loops at any stage of the untwisting and so to discriminate among states that were closely spaced in free energy. The minimum number of loops would be opened by the minimum free energy change, that is, no free energy

Methods in Enzymology, Volume 518
ISSN 0076-6879
http://dx.doi.org/10.1016/B978-0-12-388422-0.00001-7

change at all, and so the intention was to measure spontaneous fluctuations of helicity within the helix-random coil transition range. For the Stockholm group, the initial motivation was to measure kinetics relating to the acetylcholine receptor. The preferred method of temperature jump was inapplicable due to the high concentration of detergent in the receptor preparation (Rigler, 2009). The Cornell group developed their approach using a simpler experimental system, the interaction of ethidium bromide with DNA, which was favorable due to the large increase in fluorescence upon binding of the dye to DNA and because the rate of the process was in a convenient range. This leads to a demonstration of the ability of FCS to measure both diffusion and chemical reaction kinetics (Elson & Magde, 1974; Magde et al., 1972; Magde, Webb, & Elson, 1978). The Stockholm group focused on rotational motion. They demonstrated that for molecules for which the fluorescence lifetime was much shorter than the rotational relaxation time, it was possible using continuous excitation to measure rotational diffusion that was much slower than fluorescence de-excitation, thereby overcoming a limit to methods based on pulsed excitation (Ehrenberg & Rigler, 1974; Ehrenberg & Rigler, 1976; Kask, Piksarv, Mets, Pooga, & Lippmaa, 1987). At about this time, Feher and Weissman demonstrated that the kinetics of dissociation of $BeSO_4$ could be measured at equilibrium via conductivity fluctuations (Feher & Weissman, 1973) and also carried out a fluorescence fluctuation experiment to measure the molecular weight of DNA (Weissman, Schindler, & Feher, 1976).

The plan to study unwinding fluctuations in DNA depended on the absorbance hypochromic effect as an indicator of helicity (Cantor & Schimmel, 1980). Because the absorbance change due to base unstacking is much weaker than the fluorescence change due to binding of ethidium to DNA, the experience gained developing FCS argued convincingly that measurement of helicity fluctuations by hypochromicity was impractical. At that time, however, there was great interest in the mobility of biological membrane proteins. This was sparked by the fluid mosaic model, a proposal that membrane proteins could diffuse freely in a sea of fluid bilayer lipid (Singer & Nicolson, 1972), and by the demonstration by Frye and Edidin that cell surface proteins could rapidly intermix between cells that had been fused using Sendai virus (Frye & Edidin, 1970). It therefore seemed promising to use FCS to measure the mobility of cell surface proteins. Although FCS could be used to measure lipid mobility in model membranes even at this stage of its technical development (Fahey et al., 1977), the small fluctuations signals and the long data accumulations required then made its application to cells impractical. Therefore, the Cornell group developed a fluorescence photobleaching recovery method (FPR also called fluorescence recovery after photobleaching, FRAP) (Axelrod, Koppel, Schlessinger, Elson, & Webb, 1976; Koppel, Axelrod, Schlessinger, Elson, & Webb, 1976) during the same period that other photobleaching methods were introduced independently (Edidin, Zagyansky, & Lardner, 1976; Peters, Peters, Tews, & Bahr, 1974). FPR revealed that membrane proteins diffused more slowly than expected from the fluid mosaic model, for example, Schlessinger et al. (1976), and led to intense study of this phenomenon by many groups. Single particle tracking was also developed to investigate protein diffusion (Gross & Webb, 1988; Qian, 2000; Sheetz, Turney, Qian, & Elson, 1989) and led to a model in which the membrane was partitioned into submicron compartments that impede free diffusion over long distances (Kusumi et al., 2005).

The Stockholm focused on improving the sensitivity of FCS measurements by exploiting confocal detection to minimize the observation volume. This correspondingly minimized background fluorescence and enabled the detection of single molecules in FCS measurements (Rigler, 2009). The measurement of small numbers of molecules per detection volume increased the amplitude of the correlation function. Minimizing the dimensions of the observation volume also diminished the diffusion correlation times and correspondingly increased the rate of measuring diffusion fluctuations. This and other advances in fluorescence microscopy and on-line calculation of correlation functions made FCS measurements faster and more robust and so opened the way to its adoption as a routine measurement method with applicability to a wide range of subjects from polymer physics (Lumma, Keller, Vilgis, & Radler, 2003) and photophysics (Schwille, Kummer, Heikal, Moerner, & Webb, 2000; Widengren, Mets, & Rigler, 1995) to measurements on cells (Kim, Heinze, & Schwille, 2007).

Since the time of its introduction, FCS has given rise to a host of variants including those based on cross-correlation, scanning, imaging, and other approaches. Table 1.1 lists some of these methods and Fig. 1.1 shows a corresponding family tree.

It seems likely that the innovative impulse that has characterized fluorescence fluctuation measurements will continue to produce new and useful approaches and that these approaches will be applied to an increasingly wide range of subjects. One of these could be to study nonequilibrium steady states (NESSs), especially metabolic and signaling NESSs in cells (Qian &

Table 1.1 Acronyms of some FCS-related methods

Acronym	Name	Reference
FCS	Fluorescence correlation spectroscopy	Magde et al. (1972)
rFCS	Rotational FCS	Ehrenberg and Rigler (1974)
HOA	High-order autocorrelation	Palmer and Thompson (1987)
HMA	High moment analysis	Qian and Elson (1990)
PCH	Photon count histogram	Chen, Müller, So, and Gratton (1999)
dcPCH	Dual color PCH	Chen et al. (2005)
FIDA	Fluorescence intensity distribution analysis	Kask, Palo, Ullmann, and Gall (1999)
FIMDA	Fluorescence intensity multiple distribution analysis	Palo, Mets, Jager, Kask, and Gall (2000)
2dFIDA	Two-dimensional FIDA	Kask et al. (2000)
FILDA	Fluorescence intensity and lifetime distribution analysis	Palo et al. (2002)
fFCS	Flow FCS	Magde et al. (1978)
sFCS	Scanning FCS	Petersen (1986)
ICS	Image correlation spectroscopy	Petersen, Hoddelius, Wiseman, Seger, and Magnusson (1993)
TICS	Temporal ICS	Srivastava and Petersen (1996)
STICS	Space-time ICS	Hebert, Costantino, and Wiseman (2005)
kICS	*k*-space ICS	Kolin, Ronis, and Wiseman (2006)
ICCS	Image cross-correlation spectroscopy	Wiseman, Squier, Ellisman, and Wilson (2000)
TICCS	Temporal ICCS	Kolin and Wiseman (2007)
STICCS	Space-time ICCS	Kolin and Wiseman (2007)
SpIDA	Spatial intensity distribution analysis	Godin et al. (2011)
RICS	Raster image correlation spectroscopy	Digman et al. (2005)

Table 1.1 Acronyms of some FCS-related methods—cont'd

Acronym	Name	Reference
N&B	Number and brightness analysis	Digman, Dalal, Horwitz, and Gratton (2008)
PCF	Pair correlation function	Digman and Gratton (2009)
FCCS	Fluorescence cross-correlation spectroscopy	Koltermann, Kettling, Bieschke, Winkler, and Eigen (1998)
FRET-FCS	Forster resonance energy transfer-FCS	Torres and Levitus (2007)
iFCS	Inverse FCS	Wennmalm, Thyberg, Xu, and Widengren (2009)
flFCS	Fluorescence lifetime FCS	Kapusta, Wahl, Benda, Hof, and Enderlein (2007)
FPR	Fluorescence photobleaching recovery	Axelrod, Koppel, Schlessinger, Elson, and Webb (1976)
FRAP	Fluorescence recovery after photobleaching	Edidin et al. (1976)
FM	Fluorescence microphotolysis	Peters et al. (1974)
CFM	Continuous fluorescence microphotolysis	Peters, Brunger, and Schulten (1981)
FLIP	Fluorescence loss in photobleaching	Lippincott-Schwartz, Altan-Bonnet, and Patterson (2003)
FLAP	Fluorescence localization after photobleaching	
PA	Photoactivation	
TIRF-FCS	Total Internal reflection-FCS	Thompson, Burghardt, and Axelrod (1981)
STED-FCS	Stimulated emission depletion-FCS	Eggeling et al. (2009)
NSOM-FCS	Near-field scanning optical microscopy-FCS	Vobornik et al. (2008)
SPIM-FCS	Single plane illumination FCS	Wohland, Shi, Sankaran, and Stelzer (2010)
ZMW-FCS	Zero-mode waveguide-FCS	Edel, Wu, Baird, and Craighead (2005)

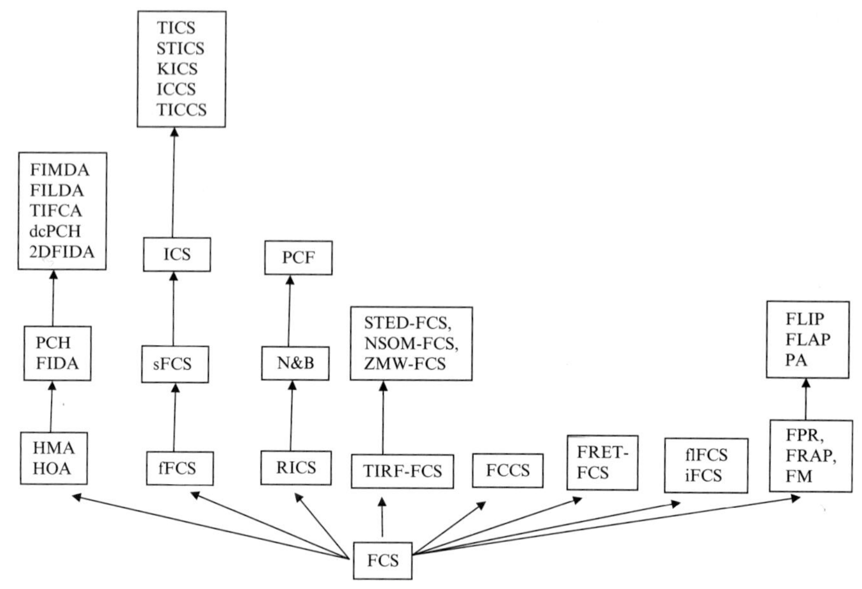

Figure 1.1 *Family tree of methods derived from or related to FCS.* An overall term for FCS and derived methods is fluorescence fluctuation spectroscopy (FFS). The most commonly used version of FPR/FRAP was developed as an alternative to FCS for studies of cells and is closely related in theory and instrumentation to FCS (Axelrod et al., 1976; Elson, 1985; Koppel, Axelrod, Schlessinger, Elson, & Webb, 1976). The method in different forms was also developed independently of FCS at about the same time (Edidin et al., 1976; Peters et al., 1974). Similarly, the successor methods, FLAP, FLIP, and PA, were developed independent of FCS and are included for completeness. The acronyms are decoded in Table 1.1.

Elson, 2004, 2009; Qian, Saffarian, & Elson, 2002). Another is to carry out FCS measurements using observation volumes smaller than allowed by the classical diffraction limit (Wenger & Rigneault, 2010). This will enable the extension of FCS measurements to higher concentrations and also to measure diffusion in smaller volumes. Both of these possibilities should open up a wider range of studies of dynamics in living cells including small cells such as yeast and bacteria.

REFERENCES

Axelrod, D., Koppel, D. E., Schlessinger, J., Elson, E., & Webb, W. W. (1976). Mobility measurement by analysis of fluorescence photobleaching recovery kinetics. *Biophysical Journal*, *16*, 1055–1069.

Cantor, C. R., & Schimmel, P. R. (1980). *Biophysical chemistry, part II, techniques for the study of biological structure and function*. San Francisco: W.H. Freeman.

Chen, Y., Müller, J. D., So, P. T., & Gratton, E. (1999). The photon counting histogram in fluorescence fluctuation spectroscopy. *Biophysical Journal*, 77, 553–567.

Chen, Y., Tekmen, M., Hillesheim, L., Skinner, J., Wu, B., & Müller, J. D. (2005). Dual-color photon-counting histogram. *Biophysical Journal*, *88*, 2177–2192.

Digman, M. A., Brown, C. M., Sengupta, P., Wiseman, P. W., Horwitz, A. R., & Gratton, E. (2005). Measuring fast dynamics in solutions and cells with a laser scanning microscope. *Biophysical Journal*, *89*, 1317–1327.

Digman, M. A., Dalal, R., Horwitz, A. F., & Gratton, E. (2008). Mapping the number of molecules and brightness in the laser scanning microscope. *Biophysical Journal*, *94*, 2320–2332.

Digman, M. A., & Gratton, E. (2009). Imaging barriers to diffusion by pair correlation functions. *Biophysical Journal*, *97*, 665–673.

Edel, J. B., Wu, M., Baird, B., & Craighead, H. G. (2005). High spatial resolution observation of single-molecule dynamics in living cell membranes. *Biophysical Journal*, *88*, L43–L45.

Edidin, M., Zagyansky, Y., & Lardner, T. J. (1976). Measurement of membrane protein lateral diffusion in single cells. *Science*, *191*, 466–468.

Eggeling, C., Ringemann, C., Medda, R., Schwarzmann, G., Sandhoff, K., Polyakova, S., et al. (2009). Direct observation of the nanoscale dynamics of membrane lipids in a living cell. *Nature*, *457*, 1159–1162.

Ehrenberg, M., & Rigler, R. (1974). Rotational Brownian-motion and fluorescence intensity fluctuations. *Chemical Physics*, *4*, 390–401.

Ehrenberg, M., & Rigler, R. (1976). Fluorescence correlation spectroscopy applied to rotational diffusion of macromolecules. *Quarterly Reviews of Biophysics*, *9*, 69–81.

Elson, E. L. (1985). Fluorescence correlation spectroscopy and photobleaching recovery. *Annual Review of Physical Chemistry*, *36*, 379–406.

Elson, E. L. (2004). Quick tour of fluorescence correlation spectroscopy from its inception. *Journal of Biomedical Optics*, *9*, 857–864.

Elson, E. L. (2011). Fluorescence correlation spectroscopy: Past, present, future. *Biophysical Journal*, *101*, 2855–2870.

Elson, E. L., & Magde, D. (1974). Fluorescence correlation spectroscopy. I. Conceptual basis and theory. *Biopolymers*, *13*, 1–27.

Fahey, P. F., Koppel, D. E., Barak, L. S., Wolf, D. E., Elson, E. L., & Webb, W. W. (1977). Lateral diffusion in planar lipid bilayers. *Science*, *195*, 305–306.

Feher, G., & Weissman, M. (1973). Fluctuation spectroscopy: Determination of chemical reaction kinetics from the frequency spectrum of fluctuations. *Proceedings of the National Academy of Sciences of the United States of America*, *70*, 870–875.

Frye, L. D., & Edidin, M. (1970). The rapid intermixing of cell surface antigens after formation of mouse-human heterokaryons. *Journal of Cell Science*, 7, 319–335.

Godin, A. G., Costantino, S., Lorenzo, L.-E., Swift, J. L., Sergeev, M., Ribeiro-da-Silva, A., et al. (2011). Revealing protein oligomerization and densities in situ using spatial intensity distribution analysis. *Proceedings of the National Academy of Sciences*, *108*, 7010–7015.

Gross, D. J., & Webb, W. W. (Eds.), (1988). *Cell surface clustering and mobility of the ligand LDL receptor measured by digital fluorescence microscopy*. Boca Raton: CRC Press.

Hebert, B., Costantino, S., & Wiseman, P. W. (2005). Spatiotemporal image correlation spectroscopy (STICS) theory, verification, and application to protein velocity mapping in living CHO cells. *Biophysical Journal*, *88*, 3601–3614.

Kapusta, P., Wahl, M., Benda, A., Hof, M., & Enderlein, J. (2007). Fluorescence lifetime correlation spectroscopy. *Journal of Fluorescence*, *17*, 43–48.

Kask, P., Palo, K., Fay, N., Brand, L., Mets, U., Ullmann, D., et al. (2000). Two-dimensional fluorescence intensity distribution analysis: Theory and applications. *Biophysical Journal*, *78*, 1703–1713.

Kask, P., Palo, K., Ullmann, D., & Gall, K. (1999). Fluorescence-intensity distribution analysis and its application in biomolecular detection technology. *Proceedings of the National Academy of Sciences of the United States of America*, *96*, 13756–13761.

Kask, P., Piksarv, P., Mets, U., Pooga, M., & Lippmaa, E. (1987). Fluorescence correlation spectroscopy in the nanosecond time range: Rotational diffusion of bovine carbonic anhydrase B. *European Biophysics Journal, 14*, 257–261.

Kim, S. A., Heinze, K. G., & Schwille, P. (2007). Fluorescence correlation spectroscopy in living cells. *Nature Methods, 4*, 963–973.

Kolin, D. L., Ronis, D., & Wiseman, P. W. (2006). k-Space image correlation spectroscopy: A method for accurate transport measurements independent of fluorophore photophysics. *Biophysical Journal, 91*, 3061–3075.

Kolin, D. L., & Wiseman, P. W. (2007). Advances in image correlation spectroscopy: Measuring number densities, aggregation states, and dynamics of fluorescently labeled macromolecules in cells. *Cell Biochemistry and Biophysics, 49*, 141–164.

Koltermann, A., Kettling, U., Bieschke, J., Winkler, T., & Eigen, M. (1998). Rapid assay processing by integration of dual-color fluorescence cross-correlation spectroscopy: High throughput screening for enzyme activity. *Proceedings of the National Academy of Sciences of the United States of America, 95*, 1421–1426.

Koppel, D. E., Axelrod, D., Schlessinger, J., Elson, E. L., & Webb, W. W. (1976). Dynamics of fluorescence marker concentration as a probe of mobility. *Biophysical Journal, 16*, 1315–1329.

Kusumi, A., Nakada, C., Ritchie, K., Murase, K., Suzuki, K., Murakoshi, H., et al. (2005). Paradigm shift of the plasma membrane concept from the two-dimensional continuum fluid to the partitioned fluid: High-speed single-molecule tracking of membrane molecules. *Annual Review of Biophysics and Biomolecular Structure, 34*, 351–378.

Lippincott-Schwartz, J., Altan-Bonnet, N., & Patterson, G. H. (2003). Photobleaching and photoactivation: Following protein dynamics in living cells. *Nature Cell Biology, 5* (Suppl.), S7–S14.

Lumma, D., Keller, S., Vilgis, T., & Radler, J. O. (2003). Dynamics of large semiflexible chains probed by fluorescence correlation spectroscopy. *Physical Review Letters, 90*, 218301.

Magde, D., Elson, E. L., & Webb, W. W. (1972). Thermodynamic fluctuations in a reacting system—Measurement by fluorescence correlation spectroscopy. *Physical Review Letters, 29*, 705–708.

Magde, D., Webb, W. W., & Elson, E. L. (1978). Fluorescence correlation spectroscopy. III. Uniform translation and laminar flow. *Biopolymers, 17*, 361–376.

Palmer, A. G., & Thompson, N. L. (1987). Molecular aggregation characterized by high order autocorrelation in fluorescence correlation spectroscopy. *Biophysical Journal, 52*, 257–270.

Palo, K., Brand, L., Eggeling, C., Jager, S., Kask, P., & Gall, K. (2002). Fluorescence intensity and lifetime distribution analysis: Toward higher accuracy in fluorescence fluctuation spectroscopy. *Biophysical Journal, 83*, 605–618.

Palo, K., Mets, U., Jager, S., Kask, P., & Gall, K. (2000). Fluorescence intensity multiple distributions analysis: Concurrent determination of diffusion times and molecular brightness. *Biophysical Journal, 79*, 2858–2866.

Peters, R., Brunger, A., & Schulten, K. (1981). Continuous fluorescence microphotolysis: A sensitive method for study of diffusion process in single cells. *Proceedings of the National Academy of Sciences of the United States of America, 78*, 962–966.

Peters, R., Peters, J., Tews, K. H., & Bahr, W. (1974). A microfluorimetric study of translational diffusion in erythrocyte membranes. *Biochimica et Biophysica Acta, 367*, 282–294.

Petersen, N. O. (1986). Scanning fluorescence correlation spectroscopy. I. Theory and simulation of aggregation measurements. *Biophysical Journal, 49*, 809–815.

Petersen, N. O., Hoddelius, P. L., Wiseman, P. W., Seger, O., & Magnusson, K. E. (1993). Quantitation of membrane receptor distributions by image correlation spectroscopy: Concept and application. *Biophysical Journal, 65*, 1135–1146.

Qian, H. (2000). Single-particle tracking: Brownian dynamics of viscoelastic materials. *Biophysical Journal, 79*, 137–143.

Qian, H., & Elson, E. L. (1990). On the analysis of high order moments of fluorescence fluctuations. *Biophysical Journal, 57*, 375–380.

Qian, H., & Elson, E. L. (2004). Fluorescence correlation spectroscopy with high-order and dual-color correlation to probe nonequilibrium steady states. *Proceedings of the National Academy of Sciences of the United States of America, 101*, 2828–2833.

Qian, H., & Elson, E. (2009). Chemical fluxes in cellular steady states measured by fluorescence correlation spectroscopy. In A. Graslund, R. Rigler & J. Widengren (Eds.), *Single molecule spectroscopy in chemistry, physics and biology* (pp. 119–138). Heidelberg: Springer.

Qian, H., Saffarian, S., & Elson, E. L. (2002). Concentration fluctuations in a mesoscopic oscillating chemical reaction system. *Proceedings of the National Academy of Sciences of the United States of America, 99*, 10376–10381.

Rigler, R. (2009). FCS and single molecule spectroscopy. In A. Graslund, R. Rigler & J. Widengren (Eds.), *Single molecule spectroscopy in chemistry, physics, and biology* (pp. 77–103). Heidelberg: Springer-Verlag.

Schlessinger, J., Koppel, D. E., Axelrod, D., Jacobson, K., Webb, W. W., & Elson, E. L. (1976). Lateral transport on cell membranes: Mobility of concanavalin A receptors on myoblasts. *Proceedings of the National Academy of Sciences of the United States of America, 73*, 2409–2413.

Schwille, P., Kummer, S., Heikal, A. A., Moerner, W. E., & Webb, W. W. (2000). Fluorescence correlation spectroscopy reveals fast optical excitation- driven intramolecular dynamics of yellow fluorescent proteins. *Proceedings of the National Academy of Sciences of the United States of America, 97*, 151–156.

Sheetz, M. P., Turney, S., Qian, H., & Elson, E. L. (1989). Nanometre-level analysis demonstrates that lipid flow does not drive membrane glycoprotein movements [see comments]. *Nature, 340*, 284–288.

Singer, S. J., & Nicolson, G. L. (1972). The fluid mosaic model of the structure of cell membranes. *Science, 175*, 720–731.

Srivastava, M., & Petersen, N. O. (1996). Image cross-correlation spectroscopy: A new experimental biophysical approach to measurement of slow diffusion of fluorescent molecules. *Methods in Cell Science, 18*, 47–54.

Thompson, N. L., Burghardt, T. P., & Axelrod, D. (1981). Measuring surface dynamics of biomolecules by total internal reflection fluorescence with photobleaching recovery or correlation spectroscopy. *Biophysical Journal, 33*, 435–454.

Torres, T., & Levitus, M. (2007). Measuring conformational dynamics: A new FCS-FRET approach. *The Journal of Physical Chemistry. B, 111*, 7392–7400.

Vobornik, D., Banks, D. S., Lu, Z. F., Fradin, C., Taylor, R., & Johnston, L. J. (2008). Fluorescence correlation spectroscopy with sub-diffraction-limited resolution using near-field optical probes. *Applied Physics Letters, 93*, 163904.

Weissman, M., Schindler, H., & Feher, G. (1976). Determination of molecular weights by fluctuation spectroscopy: Application to DNA. *Proceedings of the National Academy of Sciences of the United States of America, 73*, 2776–2780.

Wenger, J., & Rigneault, H. (2010). Photonic methods to enhance fluorescence correlation spectroscopy and single molecule fluorescence detection. *International Journal of Molecular Sciences, 11*, 206–221.

Wennmalm, S., Thyberg, P., Xu, L., & Widengren, J. (2009). Inverse-fluorescence correlation spectroscopy. *Analytical Chemistry, 81*, 9209–9215.

Widengren, J., Mets, U., & Rigler, R. (1995). Fluorescence correlation spectroscopy of triplet states in solution: A theoretical and experimental study. *The Journal of Physical Chemistry, 99*, 13368–13379.

Wiseman, P. W., Squier, J. A., Ellisman, M. H., & Wilson, K. R. (2000). Two-photon image correlation spectroscopy and image cross-correlation spectroscopy. *Journal of Microscopy, 200*, 14–25.

Wohland, T., Shi, X., Sankaran, J., & Stelzer, E. H. (2010). Single plane illumination fluorescence correlation spectroscopy (SPIM-FCS) probes inhomogeneous three-dimensional environments. *Optics Express, 18*, 10627–10641.

CHAPTER TWO

Brief Introduction to Fluorescence Correlation Spectroscopy

Elliot L. Elson[1]
Department of Biochemistry and Molecular Biophysics, Washington University School of Medicine, Campus Box 8231, St. Louis, Missouri, USA
[1]Corresponding author: e-mail address: elson@wustl.edu

Contents

Abstract

Fluorescence correlation spectroscopy (FCS) measures rates of transport (diffusion coefficients, convection velocities) and chemical reactions (rate constants) in equilibrium or nonequilibrium steady-state systems without the need for a perturbation of the state of the system. The rates are extracted from a record of fluorescence fluctuations observed from an open laser-illuminated observation volume. In addition, FCS provides the number of measured fluorescent molecules in the observation volume and, therefore, their molecular brightness. This enables sensitive measurements of aggregation of the fluorescent system components. FCS is now used as a routine tool for studying systems in physics,

Methods in Enzymology, Volume 518
ISSN 0076-6879
http://dx.doi.org/10.1016/B978-0-12-388422-0.00002-9

chemistry, and biology. Recent extensions of the fluorescence fluctuation approach have included methods based on imaging and cross correlation. Because of the sensitivity and chemical specificity of fluorescence detection and also the ability to form a very small diffraction-limited detection volume, FCS is especially useful for studies of dynamic molecular processes in living biological cells. FCS provides a window on mesoscopic systems and is closely related to fluorescence methods for studying single molecules.

1. INTRODUCTION

Fluorescence correlation spectroscopy (FCS) measures rates of molecular processes in systems in equilibrium or nonequilibrium steady states (NESSs). Among others, these processes include transport, for example, diffusion and convection, chemical reaction, and photophysical processes such as triplet decay. FCS can also supply the steady-state mean number of fluorescent molecules within an observation region and from that, molecular brightness, which is a sensitive measure of molecular aggregation. As spontaneous fluctuations supply the information about molecular parameters, FCS is applied to mesoscopic systems, that is, systems in which, due to the small number of molecules present, fluctuations are relatively large and so can be measured. FCS accomplishes this by measuring the fluorescence fluctuations that arise from molecules either moving across an observation region formed by a focused laser beam or undergoing changes of state, for example, via chemical reactions, that cause changes in fluorescence. In the simplest version of FCS, single spot FCS, a focused laser beam defines the open observation region from which the fluorescence fluctuations are registered, typically through a confocal fluorescence microscope. The fluorescence fluctuations are directly related to fluctuations of the number of observed molecules. As the molecular events that give rise to the fluorescence fluctuations are stochastic, they must be analyzed statistically. According to the Onsager hypothesis (Onsager, 1931), the statistically averaged time course of the fluctuations yields the same phenomenological rate coefficients, for example, diffusion coefficients and chemical rate constants, as are obtained using conventional methods that observe the relaxation of macroscopic concentration gradients. Thus, the same diffusion coefficients and chemical rate constants are obtained by FCS as by measuring the spread of a concentration boundary (Tanford, 1961) or the relaxation of a reaction system displaced from chemical equilibrium, for example, by a temperature jump or rapid mixing (Eigen & De Maeyer, 1963).

A crucial distinction between macroscopic measurements and those based on spontaneous concentration fluctuations in a mesoscopic system is that the former yield the desired phenomenological parameters from the measurement of a single concentration transient to an accuracy determined by the accuracy of the concentration measurement. In contrast, no matter how accurate the measurement of any individual concentration fluctuation, due to their stochastic nature, many fluctuations must be measured and statistically analyzed to obtain meaningful parameter estimates. Hence, FCS typically requires longer measurement times than macroscopic measurements. Fluorescence photobleaching recovery (FPR, also commonly called fluorescence recovery after photobleaching) provides a macroscopic analogy to FCS (see below).

Advantageous properties of FCS include the following:

1. Steady-state measurement. Dynamic parameters are obtained without the need to establish initial concentration gradients or displacement of chemical reaction mixtures from equilibrium. This is especially important for measurements on cells in which the establishment of initial concentration gradients may be difficult or impossible.

2. Fluorescence detection.

  A. Spectroscopic selectivity. This allows the detection of a specific fluorescent compound mixed with other compounds that are not fluorescent in the same spectral range (for excitation and/or emission). This property is crucial for measurements of both complex systems such as biological cells and chemical reaction kinetics, the progress of which can be sensitively indicated by fluorescence changes. In this way, FCS contrasts with a precursor method, dynamic light scattering (Berne & Pecora, 1976). Light scattering can detect neither compounds at very low concentration nor a minority component of a complex mixture and also is typically a very poor indicator of chemical reaction progress (Elson, 2011).

  B. High sensitivity. Because of the high sensitivity of fluorescence detection that extends to the single molecule regime, FCS measurements are typically performed at the nanomolar concentration level. Hence, FCS is very suitable for characterizing systems with very low concentrations of specific fluorescence molecules as, for example, in cells.

3. Microscopic detection volume.

  A. Resolution. Using confocal microscopy, FCS detection volumes are in the femtoliter regime with diffraction-limited linear dimensions in the submicrometer range. This is very useful for measuring distinct regions of biological cells and makes possible measurements in very small cells such as yeast (Slaughter, Huff, Wiegraebe, Schwartz, & Li, 2008; Slaughter & Li, 2010; Slaughter, Schwartz, & Li, 2007; Wood et al., 2011).

B. Conservation of materials. The combination of small detection volumes and high sensitivity of detection allows measurements on rare or expensive materials and is very useful for the economic application of FCS to high-throughput screening for pharmaceuticals (Auer et al., 1998) and as a tool for single-nucleotide-polymorphism genotyping (Bannai et al., 2004).

4. Wide dynamic range. The lower limit of the time range suitable for FCS is set by the rate of photon count detection. Measurements in the nanosecond domain are possible, for example, in using FCS to measure molecular rotational diffusion via fluorescence anisotropy (Ehrenberg & Rigler, 1974, 1976). The upper time limit is set only by the stability of the experimental system and the patience of the experimentalist. A rough but useful rule of thumb is that the precision of an FCS measurement varies as $1/\sqrt{N}$, where N is the number of fluctuations detected. Hence, for $\sim 1\%$ precision, the duration of a measurement would be $\sim 10^4 \tau_c$, where τ_c is the characteristic lifetime of a fluctuation. FCS measurements become impractical except for very stable systems or are at least very tedious for $\tau_c > \sim 1$–10 s.

Finally, it is worth noting that FCS was a precursor to current single molecule measurements (Rigler, 2009) and provides a window on the mesoscopic world in which molecular fluctuations not only provide a basis for measurements but also can have important effects on the overall behavior of a system such as a living cell (Elson, 2011; Laughlin, Pines, Schmalian, Stojkovic, & Wolynes, 2000; Raj & van Oudenaarden, 2008).

2. CONCEPTUAL BASIS AND THEORY OF FCS

The basic theory and conceptual basis of FCS have been described in some detail (e.g., Elson, 2011; Elson & Magde, 1974; Krichevsky & Bonnet, 2002), and so I will present only an outline here. The main experimental task in an FCS measurement is to acquire a record of fluorescence fluctuations over an extended period of time. The main theoretical task is to interpret these fluctuations to yield phenomenological coefficients for diffusion, chemical reaction, and other dynamic processes.

2.1. Relationship between fluorescence and molecular number fluctuations

The relationship between the fluorescence, $F_j(\mathbf{r},t)$, detected from position $\mathbf{r}$ at time t and the concentration, $c_j(\mathbf{r},t)$, of the jth component is weighted by the laser intensity profile, $I(\mathbf{r})$, that defines the observation region. Hence, $F_j(\mathbf{r},t) = Q_j I(\mathbf{r}) c_j(\mathbf{r},t)$ and the fluctuation $\delta F_j(\mathbf{r},t) = F_j(\mathbf{r},t) - \langle F_j \rangle$ is correspondingly related to the concentration fluctuation: $\delta F_j(\mathbf{r},t) = Q_j I(\mathbf{r}) \delta c_j(\mathbf{r},t)$, where $\delta c_j(\mathbf{r},t) = c_j(\mathbf{r},t) - \langle c_j \rangle$; Q_j accounts for the extinction coefficient and quantum yield at the relevant excitation wavelength as well as instrumental factors such as optical losses; and $\langle F_j \rangle$ and $\langle c_j \rangle$ are the steady-state means of F_j and c_j. The measurement detects the total fluorescence emitted from the observation region: $F_j(t) = Q_j \int I(\mathbf{r}) c_j(\mathbf{r},t) d\mathbf{r}$ and so $\delta F_j(t) = Q_j \int I(\mathbf{r}) \delta c_j(\mathbf{r},t) d\mathbf{r}$, and the integration is carried out over the entire observation region. A Gaussian laser profile is used for most spot FCS measurements. The detection of the emitted fluorescence depends on the confocal pinhole of the fluorescence microscope system used for the measurement (Koppel, Axelrod, Schlessinger, Elson, & Webb, 1976). A relatively simple analysis is available at the level of geometric optics (Qian & Elson, 1991). It is often useful to minimize the size of the observation volume by focusing on a diffraction-limited spot. This minimizes background fluorescence and decreases the duration of fluorescence fluctuations due to transport, thereby allowing the acquisition of more fluctuations per time interval. Under these conditions, however, the distortion of the observation volume by diffraction fringes can lead to difficulties in the interpretation of the measurements (Hess & Webb, 2002). A commonly used approximation is to take the combination of excitation profile and detection efficiency as represented by an ellipsoidal Gaussian function: $I(x,y,z,) = I(\mathrm{r},z) = I_0 \exp(-2\mathrm{r}^2/w^2) \exp(-2z^2/w_z^2)$, where $\mathrm{r} = x^2 + y^2$ is the coordinate in the focal plane and z is the coordinate along the optical axis perpendicular to the focal plane (Rigler, Mets, Widengren, & Kask, 1993). A fluorescent molecule at the point $\mathrm{r} = 0$, $z = 0$, at the center of the laser intensity profile yields the maximum of the detected fluorescence intensity.

2.2. Statistical analysis of fluorescence fluctuations

To obtain a statistically averaged time course of fluorescence fluctuations, it is conventional to compute a fluorescence fluctuation autocorrelation function normalized by $\langle F \rangle^2$, $G(\tau) = \frac{\langle \delta F(t) \delta F(t+\tau) \rangle}{\langle F(t) \rangle^2}$. (Dividing by $\langle F(t) \rangle^2$ renders $G(\tau)$ independent of the absorption and quantum yield of the fluorophore, the detection efficiency, and the laser power.) In $\langle \delta F(t) \delta F(t+\tau) \rangle$, the product of δF at some time t and of δF at a time $t + \tau$ later is calculated for many times t over the fluctuation record. As the delay time τ increases, the fluctuations become uncorrelated, and so the average of fluctuation products decreases for larger values of τ (Elson & Webb, 1975). Hence, $G(\tau)$ is a decreasing function of τ and the rate of correlation decay is related to the phenomenological parameters that describe the dynamic processes in the system. (Note that both $\langle \delta F(t) \delta F(t+\tau) \rangle$ and $\langle F(t) \rangle$ are averaged over the time t, and so neither is a function of t.) Operationally, the experimentally determined $\hat{G}(\tau)$ is obtained from the measured fluorescence fluctuations as

$$\hat{G}(\tau) = \frac{\lim_{N\to\infty} \left(\frac{1}{N}\right) \sum_{i=1}^{N} \delta\hat{F}(i\mathrm{d}t)\delta\hat{F}(i\mathrm{d}t+\tau)}{\langle \hat{F} \rangle^2}$$

with $\langle \hat{F} \rangle = \lim_{N\to\infty} \left(\frac{1}{N}\right) \sum_{i=1}^{N} \hat{F}(i\mathrm{d}t)$ and $\delta\hat{F}(i\mathrm{d}t) = \hat{F}(i\mathrm{d}t) - \langle \hat{F} \rangle$.The "hat" on $\hat{F}$ denotes the experimentally measured value of F. Note that the correlation function is sometimes defined as $G'(\tau) = \frac{\langle F(t)F(t+\tau)\rangle}{\langle F(t)\rangle^2}$. It is straightforward to show that $G(\tau) = G'(\tau) - 1$. Alternative to using the autocorrelation function is to calculate the power spectrum from the Fourier components of the fluctuation record. The correlation function and power spectrum are related by Fourier transformation as provided by the Wiener–Khinchin theorem (Kittel, 1958).

In a single component system, the correlation function

$$G(\tau) = \frac{\langle \delta F(t) F(t+\tau) \rangle}{\langle F \rangle^2} = \frac{\int I(r) I(r') \langle \delta c(r,t) \delta c(r', t+\tau) \rangle \mathrm{d}r \mathrm{d}\mathbf{r}'}{\langle c \rangle^2 P^2}$$

where $P = \int \exp\left(\frac{-2r^2}{w^2}\right) \mathrm{d}r$, is the incident laser power. For a two-dimensional system, $P = \pi I_0 w^2/2$ and $G(0) = 1/\pi w^2 \langle c \rangle = 1/\langle N \rangle$, where $\langle N \rangle$ is the average number of fluorescent molecules in the effective beam area, πw^2. For a multicomponent system with M components,

$$G(\tau) = \frac{\left\langle \sum_{j=1}^{M} \delta F_j(t) \sum_{l=1}^{M} \delta F_l(t+\tau) \right\rangle}{\left\langle \sum_{j=1}^{M} F_j(t) \right\rangle^2} = \frac{\sum_{j=1}^{M} \sum_{l=1}^{M} (2 - \delta_{jl}) G'_{jl}(\tau)}{\left\langle \sum_{j=1}^{M} F_j(t) \right\rangle^2} \quad [2.1]$$

where $G'_{jl}(\tau) = \langle \delta F_j(t) \delta F_l(t+\tau) \rangle$ (Elson & Magde, 1974).

2.3. Partial differential equations describe the dynamics of the molecular system

The next task is to relate the measured $\hat{G}(\tau)$ to the phenomenological rate parameters. This requires the formulation of a model for the experimental system. For a simple diffusion system, the model is embodied in the diffusion equation: $\frac{\partial \delta c_i(r,t)}{\partial t} = D_i \nabla^2 \delta c_i(r,t)$, where D_i is the desired diffusion coefficient and the Laplacian operator

$$\nabla^2 = \frac{\partial^2}{\partial x^2} + \frac{\partial^2}{\partial y^2} + \frac{\partial^2}{\partial z^2}$$

Molecular transformations, for example, as in chemical reactions, require a more extensive treatment. As for conventional chemical kinetics measurements in closed systems, it is necessary to posit a model for the chemical reaction. For measurements in an open observation volume, the coupling of diffusion to reaction substantially increases the complexity of the analysis. For example, in a closed system, the simple reaction $A + B \underset{k_b}{\overset{k_f}{\rightleftarrows}} C$ is subject to the conservation conditions: $\delta c_A = \delta c_B = -\delta c_C$. Hence, the kinetics can be described in a single ordinary differential equation, which, supposing that the deviations of the concentrations from their equilibrium values are small, takes the form $\frac{\mathrm{d}\delta c_C}{\mathrm{d}t} = -R\delta c_C$ yielding $c_C(t) = c_C(0)\exp(-Rt)$, where the chemical relaxation rate is $R = k_f(\langle c_A \rangle + \langle c_B \rangle) + k_b$. In an FCS measurement, however, A, B, and C can independently diffuse into and out of the open observation region. Hence, the system must be described in terms of three equations, one for each component, for example, $\frac{\partial \delta c_A(r,t)}{\partial t} = D_A \nabla^2 \delta c_A(r,t) - k_f \langle c_B \rangle \delta c_A(r,t) - k_f \langle c_A \rangle \delta c_B(r,t) + k_b \delta c_C(r,t)$ with similar equations for components B and C. Equations that account for the variation over time and position of the ith component of a more general reaction system containing M reacting components have the form:

$$\frac{\partial \delta c_i(r,t)}{\partial t} = D_i \nabla^2 \delta c_i(r,t) - V_i \frac{\partial c_i(r,t)}{\partial x} + \sum_{j=1}^{M} T_{ij} \delta c_j(r,t) \qquad [2.2]$$

The first term on the right accounts for diffusion; the second, for uniform convection along the x-axis with constant velocity V and the third, for chemical reactions. The coefficient T_{ij} accounts for reactions that transform component j into component i, and is composed of kinetic rate constants and, for second and higher order reactions, equilibrium concentrations of reaction participants. A general method for solving these kinds of equation systems and then expressing $G(\tau)$ in terms of the diffusion coefficients, chemical kinetic rate constants, and other parameters has been developed using Fourier transforms (Elson & Magde, 1974). The solutions that couple diffusion and chemical reaction can be quite complex. A significant simplification is available, however, if the diffusion coefficients of all the system components are identical. Then the reaction and diffusion contributions to $G(\tau)$ can be expressed as the product of two factors, one that describes the contribution of diffusion and the other that accounts for chemical relaxation (Palmer & Thompson, 1987b).

2.4. Different fluctuation relaxation mechanisms yield different forms of $G(\tau)$

Diffusion, convection, and chemical reaction yield different correlation functions that are summarized in Table 2.1, where a two-dimensional geometry, for example, a planar membrane, is assumed for simplicity (Elson, 2011). Although mathematically distinct, these correlation functions may nevertheless be difficult to distinguish experimentally unless data of very high quality are available. The relative contributions of diffusion and chemical reaction can be tested by varying, w, the radius of the observation volume. Rates of chemical reaction are independent of w while the characteristic diffusion time is $\tau_D = w^2/4D$. Active transport in cells can be distinguished from diffusion by using biological inhibitors, for example, cytoskeletal disruptors or energy poisons to suppress active cellular mechanical functions. For chemical reactions of second or higher order, the chemical relaxation times are functions of the mean concentrations of the chemical components. For example, the mechanism for the simple reaction, $A + B \underset{k_b}{\overset{k_f}{\rightleftarrows}} C$, can be tested by changing $\langle c_A \rangle$ or $\langle c_B \rangle$ to vary the relaxation rate, $R = k_f(\langle c_A \rangle + \langle c_B \rangle) + k_b$ (Magde et al., 1974). As for any study of chemical reaction kinetics, it is important to test the proposed reaction mechanism, if at all possible, by independent kinds of measurements, for example, Saffarian, Collier, Marmer, Elson, and Goldberg (2004).

2.5. Extraction of the rate parameters from the measured correlation function

Standard methods based on least squares are available for determining the rate parameters by optimizing the fit of the experimental $\hat{G}(\tau)$ to the $G(\tau)$ derived from the theoretical model (Bevington & Robinson, 2003). There remains the task of assessing the accuracy of these best fit parameters.

Table 2.1 Forms of the correlation function, $G(\tau)$ for different relaxation mechanisms

Process	$\frac{\partial \delta c_i}{\partial t}$	$\frac{G(\tau)}{G(0)}$	Parameter	Reference
Diffusion	$D_i \nabla^2 \delta c_i$	$\left(1+\frac{\tau}{\tau_{Di}}\right)^{-1}$	$\tau_{Di}=\frac{w^2}{4D_i}$	Magde, Elson, and Webb (1974)
Convection	$-V_i \nabla \delta c_i$	$\exp\left[-\left(\frac{\tau}{\tau_{V_i}}\right)^2\right]$	$\tau_{V_i}=\frac{w}{V_i}$	Magde, Webb, and Elson (1978)
Diffusion + Convection	$D_i \nabla^2 \delta c_i - V_i \nabla \delta c_i$	$\frac{\exp\left[\frac{-\left(\frac{\tau}{\tau_{V_i}}\right)^2}{1+\frac{\tau}{\tau_{Di}}}\right]}{1+\frac{\tau}{\tau_{Di}}}$	As above	Magde et al. (1978)
Chemical reaction (general)	$\sum_{j=1}^{M} T_{ij}\delta c_j$	$\sum_{s=1}^{N} A_s \exp\left[-\frac{\tau}{\tau_s}\right]$	T_{ij}, A_s, and τ_s are functions of the rate constants and equilibrium concentrations	Magde et al. (1974)

We have assumed a two-dimensional system for simplicity. There are M chemical components and N chemical reactions. For the ith component, the diffusion coefficient and convection velocity are D_i and V_i, respectively. The coupled diffusion and chemical reaction systems that arise due to the open observation volume in FCS yield complicated functions that mix contributions of transport and reaction (Elson, 2011).

As for any experimental measurement, it is important to minimize systematic errors. For example, misalignment of the optics could distort the size and shape of the observation volume and thereby induce errors in diffusion measurements. Calibration against a molecule of known diffusion coefficient can check for this potential problem. In addition to random errors, due to limited measurement precision that affect both macroscopic and mesoscopic systems, measurements on mesoscopic systems are subject to additional sources of randomness. Because of the small number of photon counts registered from a sample containing only a few fluorescent molecules, shot noise can limit the precision with which a fluctuation is measured. Moreover, as pointed out above, even a highly accurate measurement of the time course of a single stochastic fluctuation would not yield an accurate measurement of the diffusion coefficient. Statistical analysis of many fluctuation measurements is needed for an accurate result.

Central to the analysis of the random errors in FCS is to calculate the variance of the correlation function, which is a gauge of the breadth of the distribution of the measurements. This analysis serves two purposes. One is the optimization of the experimental method to maximize the signal to noise ratio (SNR). For an FCS measurement, it is appropriate to define the "signal" as the correlation function, $G(\tau)$. Then, the SNR is $G(\tau)/[\mathrm{var}(G(\tau))]^{1/2}$ (Koppel, 1974; Qian, 1990). In addition, to determine the best fitted parameters from experimental measurements, it is often useful to weight the individual measured values of the correlation function by the inverse of their variance so that those values with the least uncertainty have the greatest weight in the fitting. Wohland, Rigler, and Vogel (2001) have presented an empirical approach based on averaging over a number of measurements. This very direct method requires multiple measurements to obtain the variance. Analytical calculation of the variance evades this requirement. Early in the development of the FCS method, Koppel (1974) presented an approximate analytical approach that has since been extended by several authors (Kask, Gunther, & Axhausen, 1997; Qian, 1990; Saffarian & Elson, 2003).

3. MEASUREMENTS BASED ON FLUCTUATION AMPLITUDES

3.1. General concepts

The size of a fluorescence fluctuation is proportional to the size of the corresponding fluctuation of the number of fluorescent molecules in the observation volume. To determine the average number of fluorescent molecules at concentration $\langle c \rangle$ that are in this volume, we define the confocal volume as

$$V_{\text{conf}} = \int I(x,y,z)\text{d}x\text{d}y\text{d}z = 2\pi\int_0^{\infty} \exp\left(-\frac{2r^2}{w^2}\right) r\text{d}r \int_{-\infty}^{\infty} \exp\left(-\frac{2z^2}{w_z^2}\right)\text{d}z$$
$$= \left(\frac{\pi}{2}\right)^{\frac{3}{2}} w^2 w_z$$

Then the effective observation volume is defined (somewhat arbitrarily) as $V_{\text{obs}} = 2^{3/2} V_{\text{conf}} = \pi^{3/2} w^2 w_z$ (Ruttinger, 2006). Hence, the average number of fluorescent molecules in the observation volume at time t is $\langle n(t) \rangle = \langle c(t) \rangle V_{\text{obs}}$. Typically, the concentrations of fluorescent molecules in FCS measurements are very low (in the nanomolar range), and so the molecular systems are effectively ideal. A Poisson distribution describes the number of fluorescent molecules in the observation volume, and the probability that $n(t_j)$ particles are within V_{obs} at time t_j is $P\left(n\left(t_j\right)\right) = \frac{\left\langle n\left(t_j\right)\right\rangle^{n\left(t_j\right)} \exp\left(-\left\langle n\left(t_j\right)k\right\rangle\right)}{n\left(t_j\right)!}$, where $\langle n(t_j) \rangle$ is the average of $n(t_j)$ over all t_j. (In these equations, we have included the arguments, t_j, to emphasize that averages are taken over a series of measurement times.) For the Poisson distribution, the variance $\left\langle \delta n\left(t_j\right)\right\rangle^2 = \left\langle n\left(t_j\right)\right\rangle$. Suppose, for simplicity, that instead of a Gaussian profile, the intensity of the excitation laser is uniform over the observation volume. Then, the fluorescence measured at t_j is $F(t_j) = qn(t_j)$ and $\delta F(t_j) = q\delta n(t_j)$, where q is the "brightness" of the molecules, that is, the number of photon counts emitted per second by a molecule in the observation region. Then, $G(0) = \langle \delta F(0) \rangle^2 / \langle F \rangle^2 = q^2 \left\langle \left(\delta n\left(t_j\right)\right)\right\rangle^2 / \left(q\left\langle n\left(t_j\right)\right\rangle\right)^2 = \left\langle n\left(t_j\right)\right\rangle^{-1}$. Hence, G(0) yields the average number of the molecules of the fluorophore in the observation region. Dividing by the volume of the observation region yields their concentration. The mean fluorescence, $\langle F \rangle$, provides the brightness: $q = \langle F \rangle / \langle n \rangle = G(0)\langle F \rangle$. (For a system with N species of fluorescent molecules, the result is more complex: $G(0) = \frac{1}{\pi w^2} \frac{\sum_{i=1}^{N} \langle c_i \rangle q_i^2}{\left(\sum_{i=1}^{N} \langle c_i \rangle q_i\right)^2}$; Elson, 2011.) Adapting this concept to laser scanning microscopy provides the basis of the useful "number and brightness" (N&B) method for characterizing fluorophores on cell surfaces (Digman, Dalal, Horwitz, & Gratton, 2008). Often, systems, for example, aggregation systems, contain particles with various brightness values at various concentrations $\{c_i, q_i\}$. Methods are available that, in principle, provide information about $\{c_i, q_i\}$.

3.2. Higher moments and PCH

One approach is to evaluate the higher moments of the fluorescence fluctuations (Qian & Elson, 1990a,1990b). An alternative is to evaluate the zero-time amplitudes of high-order correlation functions, for example, $G_{m,n}(\tau) = \frac{\langle \delta F^m(0)\delta F^n(\tau)\rangle - \langle \delta F^m\rangle\langle \delta F^n\rangle}{\langle F\rangle^{m+n}}$ (Palmer & Thompson, 1987a, 1989). A simplified and extended version has been proposed but not yet tested experimentally (Melnykov & Hall, 2009).

Efforts to determine $\{c_j, q_j\}$ have mostly centered on the photon count histogram (PCH) (Kask, Palo, Ullmann, & Gall, 1999; Müller, Chen, & Gratton, 2000). The data record for an FCS measurement is a sequence of photon counts, n, collected during time bins of brief duration, T. The interval T should be enough shorter than the diffusion times of the system components, so that there is no significant motion during T but, otherwise, as long as possible to minimize shot noise. The PCH is the probability, $P(n)$, that a time bin contains n photons (including $n=0$). To determine the values of the $\{n_j, q_j\}$, one first postulates an analytical model that specifies the values of n_j and q_j. A good fit, for example, using nonlinear least squares, of the measured PCH to that derived from the model supports the adequacy of the model and yields values of $\{n_i, q_j\}$. To illustrate, for a single component $P(n) = \sum_{m=0}^{\infty} P_m(m)P_n(n\,|\,m)$, where $P_m(m)$ is the probability that m molecules are in the laser-illuminated observation volume and $P_n(n\,|\,m)$ is the conditional probability that n photons are emitted from the m molecules (Kask et al., 1999). Both P_m and P_n are the Poisson probability distributions. The PCH for a multicomponent system can be expressed either as convolutions of the PCHs of the individual system components (Müller et al., 2000) or as a product of the generating functions, $H(\xi)$ (where $H(\xi) = \sum_{n=1}^{\infty} P(n)\xi^n$), for the PCHs of each component (Kask et al., 1997).

In theory, this approach can supply the values of $\{n_i, q_i\}$ for an arbitrary multicomponent system. In practice, to determine n and q even for only a few components can require an impractically extensive record of fluctuation data. One source of ambiguity is the fact that a dim particle near the center of the Gaussian laser excitation profile appears similar to a bright particle farther from the beam center (Pryse, Rong, et al., 2012). The ability of the PCH approach to distinguish between models has been explored (Müller et al., 2000) and the approach has been extended to the time domain and the use of several fluorescence colors, for example, Kask et al. (2000), Palo et al. (2002), Palo, Mets, Jager, Kask, and Gall (2000), Wu, Chen, and Müller (2006), Wu and Müller (2005).

4. EXPERIMENTAL MEASUREMENTS

The basic requirements for conventional spot FCS measurements include one or more laser excitation sources, an optical system to convey the excitation light into the sample and the emitted fluorescence from the sample to a detector, the detector, and the means to compute the correlation function from the record of fluorescence fluctuations (Krichevsky & Bonnet, 2002). Most often the heart of the optical system is a confocal fluorescence microscope that is especially useful to localize measurements on cells but is also convenient for any FCS measurements (Koppel et al., 1976). It is desirable to have several laser wavelengths to be able to use of different fluorophores and also to have two detectors for cross correlation measurements (see below). Detectors are either photomultipliers or avalanche photodiodes. Special-purpose correlation function cards are available to register fluorescence fluctuation data and to compute continuously updating correlation functions during the data acquisition. Excellent FCS instruments are available from several commercial sources.

Two-photon laser excitation (Berland, So, & Gratton, 1995; Schwille, Haupts, Maiti, & Webb, 1999) has a number of advantages over conventional one-photon excitation including greater depth of penetration into a sample, avoidance of diffraction effects (incurred by the confocal pinhole required for one-photon excitation) that can distort the shape of the observation volume (Hess & Webb, 2002), and a broader excitation wavelength range that is useful for two-color cross correlation measurements (Heinze, Koltermann, & Schwille, 2000).

In conventional FCS measurements, diffraction limits the minimum observation volume to characteristic distances, d, set by the wavelength of the excitation light, λ, and the numerical aperture, NA, of the objective lens that focuses the laser beam into the sample and receives the emitted fluorescence. In the focal plane, $w = d_{xy} \approx 0.61\lambda_0/\mathrm{NA}$; along the optical axis, $w_z = d_z \approx n\lambda_0/\mathrm{NA}^2$, where n is the index of refraction of the medium containing the fluorescent molecules (Toomre & Bewersdorf, 2010). For high NA objectives, $d_{xy} \sim 200$–250 nm and $d_z \sim 500$–700 nm. Access to smaller observation volumes would be useful for several reasons: to enable diffusion measurements in small compartments, for example, endocytic vesicles, to diminish background fluorescence, and to extend FCS measurements to higher concentrations of fluorescent molecules. Recently, several approaches to shrink the observation volume have emerged (Toomre & Bewersdorf, 2010) that are applicable to FCS (Wenger & Rigneault, 2010).

Among these is stimulated emission depletion (STED) microscopy, a method that shrinks the excitation laser diameter by using a second, ring-shaped laser intensity profile to deplete the peripheral regions of the excitation beam (Hell & Wichmann, 1994). STED FCS measurements have shown that small (<20 nm) cholesterol-rich nanodomains transiently trap glycophosphatidylinositol-anchored cell membrane proteins (Eggeling et al., 2009). Methods based on evanescent radiation can also reduce the observation volume. Total internal reflection microscopy (TIRFM), in which an evanescent field illuminates a thin layer of the sample above its interface with a glass substratum, is the best known of these. Applications of TIRFM to FCS and to photobleaching recovery date back 30 years (Thompson, Burghardt, & Axelrod, 1981) and continue to the present (Lieto, Cush, & Thompson, 2003; Ohsugi, Saito, Tamura, & Kinjo, 2006; Vobornik et al., 2008). Although TIRFM strongly reduces the illuminated dimension parallel to the focal plane, further measures are needed to obtain a small spot in the x- and y-plane, for example, using a parabolic mirror objective (Ruckstuhl & Seeger, 2004). Using this approach with a detection volume as small as 5 al ($<5 \times 10^{-18}$ l), it was possible to perform FCS measurements at concentrations as high as 0.2 μM (Ruckstuhl & Seeger, 2004). Zero-mode waveguides or nanometric apertures allow FCS to be performed with even smaller detection volumes in the zeptoliter range (10^{-21} l) at which samples are in the single molecule regime at concentrations as high as 200 μM (Levene et al., 2003). FCS with zero-mode waveguides can measure single molecule motion on cell membranes (Edel, Wu, Baird, & Craighead, 2005). Another way to reduce observation volume is via near field optical microscopy (Lewis et al., 2003). An interesting FCS application is a study of transport through nuclear pores (Herrmann et al., 2009).

5. CLOSE RELATIONSHIP BETWEEN FCS AND FLUORESCENCE PHOTOBLEACHING

In the early days, after its introduction, FCS required long periods of data acquisition and very stable samples mainly because the measurements were carried out with relatively large observation volumes and relatively large numbers of molecules within the observation volumes. The former imposed long diffusion correlation times and the latter, small correlation function amplitudes. Hence, it was difficult to use FCS to study living cells.

This was one motivation for the development of fluorescence photobleaching to study diffusion on live cells (Axelrod, Koppel, Schlessinger, Elson, & Webb, 1976; Edidin, Zagyansky, & Lardner, 1976; Elson, 1985; Jacobson, Ishihara, & Inman, 1987; Jacobson, Wu, & Poste, 1976; Peters, Peters, Tews, & Bahr, 1974). In a photobleaching measurement, a brief intense pulse of excitation intensity generates a localized macroscopic concentration gradient by irreversibly photolyzing the fluorophore in the laser-illuminated observation volume. Then, using reduced excitation intensity, the relaxation of the concentration gradient is recorded. The macroscopic concentration gradient supplies a larger signal than that due to spontaneous concentration fluctuations. Furthermore, the magnitude of the signal increases with the concentration of the fluorophore, and so photobleaching is complementary to FCS, which is best performed at low concentrations of fluorophore. Equally important, since the relaxation of a macroscopic concentration gradient is not stochastic, measurement of a single concentration relaxation transient yields the desired diffusion coefficients to the limit of accuracy of the measurement. As might be expected from the fact that a concentration gradient generated by photobleaching approaches a spontaneous fluctuation as the magnitude of the bleach pulse approaches zero, there are close similarities in theory between photobleaching and FCS measurements although the two differ in the quantitative weighting of diffusion and chemical kinetic components (Elson, 1985). Also, FCS and photobleaching measurements are similarly implemented on a fluorescence microscope, the main differences being the necessity to provide a means of rapidly switching between high and low fluorescence excitation intensity for photobleaching and to compute fluctuation correlation functions for FCS (Koppel et al., 1976). The 1990s brought substantial technological improvements in FCS measurements. One of the main advances was to diminish the size of the observation volume to the diffraction limit. This had a number of beneficial effects, including the reduction of diffusion correlation times and background fluorescence. As a result it was possible to increase correlation function amplitudes by carrying out measurements with very few fluorophores, even down to the single molecule level (Rigler, 2009). Due to the larger fluctuation amplitudes and shorter integration times, FCS is now commonly used for measurements on cells (Vukojevic et al., 2005). Nevertheless, photobleaching methods are still both usefully complementary to FCS and continue to be used for studies on cells (Day & Schaufele, 2005; Lippincott-Schwartz, Altan-Bonnet, & Patterson, 2003; Sprague & McNally, 2005).

6. SCANNING AND IMAGING APPROACHES TO FCS

Measurement of brightness, q, provides a sensitive way to measure aggregation or polymerization of fluorescent particles. Suppose that the brightness of a fluorescent monomer is q_{m}. Then, in the absence of electronic interactions among the monomers (fluorescence quenching or enhancement), a polymer or aggregate is composed of n monomers with a brightness of nq_{m} (Magde et al., 1978). This concept can be applied to determine the degree of aggregation of diffusing particles. FCS measurements are difficult, however, on slowly diffusing particles (large τ_D), because the fluorescence fluctuations have long durations that necessitate long periods of data acquisition to acquire an accurate fluctuation autocorrelation function. One way to accelerate data acquisition is to flow the sample at velocity V through the observation volume (Magde et al., 1978). This approach is effective if the correlation time for flow, $\tau_V = w/V$, is shorter than that for diffusion, $\tau_D = w^2/4D$ (Table 2.1), but it is not useful for studies on cells. A simpler but similar approach, scanning FCS (sFCS), readily applicable to cells and to samples containing both slowly moving and immobile aggregates, is to translate the laser beam over the sample (Petersen, 1986; Skinner, Chen, & Müller, 2005). Using sFCS one computes an autocorrelation function from a record of fluorescence fluctuations from sequential positions along the scan line: $G(\xi) = \frac{\langle \delta F(x)\delta F(x+\xi)\rangle}{\langle F(x)\rangle^2}$. Both FCS and sFCS detect series of independent fluorescence fluctuations. Hence, for sFCS as well as for FCS, $G(0) = 1/\langle N\rangle$, where $\langle N\rangle$ is the mean number of fluorescent molecules in the observation region, and so both methods can yield average particle brightness and therefore the extent of aggregation.

This approach, based on spatial rather than temporal autocorrelation, can be further generalized to fluorescence fluctuations in two-dimensional images obtained by scanning laser microscopy, a method termed, "Image Correlation Spectroscopy" (ICS) (Petersen, Hoddelius, Wiseman, Seger, & Magnusson, 1993). In an ICS measurement, the photon counts in the pixels that make up the image are used to compute a two-dimensional spatial fluctuation autocorrelation function $G_{\mathrm{ICS}}(\xi,\eta)$(Petersen et al., 1993). The (zero-lag) amplitude is $G_{\mathrm{ICS}}(0,0) = 1/\langle N\rangle$. Again, $\langle N\rangle$ is the average number of fluorescent "particles" in the beam area ($=\pi w^2$). As for spot FCS, the brightness of the fluorescent particles is $q = G_{\mathrm{ICS}}(0,0)\langle F\rangle$ and the aggregate size from $G_{\mathrm{ICS}}(0,0)\langle F\rangle/q_{\mathrm{m}}$, where $\langle F\rangle$ is the mean fluorescence over the measured surface. In contrast to FCS that measures fluctuations from particles that diffuse into and out of the sample volume, for ICS and sFCS, the sampling process is independent of the dynamics of the particles in the system. Indeed, ICS is often used on surfaces of fixed cells on which proteins are immobilized. Hence, FCS but not ICS or sFCS detects aggregates that are dynamically linked. ICS like other image correlation methods has the important advantage that a large quantity of fluorescence fluctuation data is rapidly acquired in individual images enhancing the statistical significance of the results.

ICS has given rise to several image correlation methods (Kolin & Wiseman, 2007) that also are used mainly to measure aggregation of cell surface proteins. For example, temporal ICS measures fluctuations of fluorescence from pixels that are correlated over a time sequence of images (Wiseman, Squier, Ellisman, & Wilson, 2000). This method can supply spatially resolved information about a number of dynamic mechanisms including diffusion, convective flow, or chemical reaction. Similarly, spatiotemporal ICS includes both spatial and temporal correlation of image fluorescence and reveals the direction and velocity of systematic motion, for example, convective flow, of fluorescent particles (Hebert, Costantino, & Wiseman, 2005).

A scanning microscope sequentially scans the object field line by line to form a rectangular grid of pixels that compose the image. Raster ICS takes advantage of the three distinct time scales on which a scanned image is acquired (Digman et al., 2005). First, there is a rapid scan along a line with a dwell time on the order of a few microseconds per pixel. Scanning the entire line of pixels requires a few milliseconds. Finally, acquiring the full complement of lines needed for the entire image requires on the order of a second. Hence, correlating fluorescence fluctuations along a single line, across a range of lines, and from image to image yields information about dynamic processes that occur on the microsecond, millisecond, and second time scales. Advantages of this approach include the wide time range over which dynamics, for example, diffusion, can be measured, the ability to relate the dynamic information to the structural information in the image, and that readily available commercial scanning confocal microscopes suffice for the measurements.

7. A BRIEF AND PARTIAL SURVEY OF APPLICATIONS

FCS has been used to study so wide a range of subjects that a comprehensive survey would be far beyond the scope of this chapter. Rather, a brief account of a few selected subjects provides an idea of the breadth of its applications.

7.1. Polymer conformational fluctuations

The segments of a random coil polymer molecule, which on average conform approximately to a Gaussian spatial distribution, are in constant Brownian motion relative to each other. A simple but useful model takes the molecule to be a series of beads connected by entropic springs (Doi & Edwards, 1988). The beads encounter viscous resistance to their motion through the solvent and also experience hydrodynamic interactions among one another. One fundamental theory of the viscoelastic properties of dilute solutions of polymers includes the former viscous force but not the latter (Rouse, 1953), while the theory derived by Zimm (1956) includes both kinds of viscous interactions. FCS can measure the segmental motion of polymer molecules to which fluorescent tags are attached and that are large compared to the observation volume (Lumma, Keller, Vilgis, & Radler, 2003). This is because the diffusion of the center of mass of the large DNA molecule is much slower than the segmental motions. Therefore, conformational fluctuations large enough to traverse a significant fraction of the Gaussian laser excitation profile cause measurable fluorescence fluctuations that have a shorter lifetime than those caused by diffusion of the entire molecule across the observation volume. An FCS study of the segmental motion of single- and double-stranded DNA has shown that the Zimm model explains the former and the Rouse model the latter (Shusterman, Alon, Gavrinyov, & Krichevsky, 2004). The difference presumably arises from the fact that single-stranded DNA is relatively compact but double-stranded DNA is a worm-like coil with a large persistence length. As a result the segments of the double-stranded DNA are relatively distant from one another diminishing their hydrodynamic interactions (Shusterman et al., 2004).

Conformational fluctuations of proteins have also been observed, for example, for both the native globular and unfolded disordered states of the intestinal fatty-acid-binding protein (IFABP). IFABP has the form of a clamshell composed of two five-stranded β-sheets. A 35-μs conformational fluctuation detected by FCS in the native protein was attributed to a modulation of the interaction between a fluorescein and a trp residue within the cavity (Chattopadhyay, Saffarian, Elson, & Frieden, 2002). (A different IFABP derivative with the fluorescein located outside the cavity did not display the 35-μs fluctuation nor did IFABP molecules when unfolded at low pH.) If we suppose that this conformational motion can be modeled as Brownian motion of a harmonic spring that connects the two sheets of the clamshell, then the fluctuation correlation time is $\tau_c = \zeta/k$, where ζ is the frictional coefficient for the viscous resistance to the motion and k is the spring constant (Doi & Edwards, 1988). Hence, supposing a value for ζ allows an estimation of the potential energy that governs the relative motions of the clam shell β-sheets. The initial state for protein folding, at least *in vitro*, is composed of multiple continuously fluctuating disordered conformations. To understand the folding process, it is important to measure the structural and dynamic properties of this initial state. Conformational fluctuations were measured for guanidinium-unfolded IFABP molecules to which two rhodamine flurophores had been attached at different locations along the polypeptide chain (Chattopadhyay, Elson, & Frieden, 2005). As the conformation of the unfolded protein fluctuated, the rhodamines would occasionally come into close proximity and quench, yielding measurable fluorescence fluctuations that occurred with a correlation time ~1.6 μs as measured by FCS. This was distinct from a 2.5-μs correlation time measured for the protein in a condensed molten globule state. The rate at which these conformational fluctuations occur depends on the internal mobility of the polypeptide chain and the size of the molecule. Despite the fact that the amino acids were confined to a smaller volume in the molten globule state, the correlation time was shorter for the more extended unfolded form, presumably due to greater frictional resistance to the motion of the protein segments in the more compact form. In another study, FCS measured the relaxation times for conformational fluctuations of the yeast prion protein SUP35 in the 20–300 ns time range (Mukhopadhyay, Krishnan, Lemke, Lindquist, & Deniz, 2007). Together with single molecule FRET, this indicated that one domain of the molecule experiences an ensemble of collapsed and rapidly fluctuating conformations.

7.2. Molecular interactions

Molecular interaction's important for biology range from simple bimolecular associations ($A + B \underset{k_b}{\overset{k_f}{\rightleftarrows}} C$) to the formation of large polymers, aggregates, and other supra-molecular structures, for example, ribosomes. FCS can be used in a variety of formats to probe these processes. One approach is to measure the change in diffusion coefficient of a small rapidly diffusing fluorescent molecule when it binds to a larger more slowly diffusing molecule. This can work well if the difference in diffusion coefficients is large enough. To distinguish between two components by FCS, for good data (high count rate), the diffusion times of two components with comparable brightness must differ by at least 1.6-fold (Meseth, Wohland, Rigler, & Vogel, 1999). Unfortunately, translational diffusion coefficients depend weakly on the size of the diffusing particle. For spheres, $D \sim M^{-1/3}$ where M is the molecular mass of the particle. Hence, the effective molecular mass of the complex must be grater than four-fold that of the fluorescent reaction partner to distinguish the change in diffusion rate. Evidently, this is not a favorable approach for measuring limited molecular aggregation processes. In contrast, brightness scales linearly with the number of fluorescent particles in an aggregate so long as there are no electronic interactions among the fluorphores that quench or enhance fluorescence. Dimerization yields a twofold change in brightness but only $\sim 2^{1/3} = 1.24$-fold change in diffusion rate. Often, aggregation processes include many components that contain different numbers of monomers. In principle, PCH can provide the distribution of the number and brightnesses of all the components $\{n_i, q_i\}$ in the system. As we have pointed out above, however, the accuracy and sensitivity of this approach are limited by requirements for very long fluctuation data records. If one could assume that the aggregation process was dominated by a narrow range of aggregate sizes, then an estimate from the amplitude of the correlation function might suffice, $q = \langle F \rangle / \langle n \rangle = G(0)\langle F \rangle$, as pointed out above. For example the "N&B" method provides values for the number of fluorescent particles and their brightness from fluorescence microscopy image data (Digman et al., 2008). Advantages of this approach are that it supplies a large quantity of data sampled over the many pixels in an image and it allows localized measurements of N&B to be related to the morphology of the region of the image from which they are acquired. A recent publication provides an informative application of this approach to measure clustering of ErbB receptors as well as a survey of other approaches to this subject such as ICS and related methods (Nagy, Claus, Jovin, & Arndt-Jovin, 2010). In addition, PCH has been used to study clustering of cell surface proteins (Chen, Wei, & Müller, 2003) including EGF (ErbB-1) receptors (Saffarian, Li, Elson, & Pike, 2007).

Fluorescence cross correlation spectroscopy (FCCS) provides a very specific way to detect the association of two molecules, say a and b, that emit fluorescence at different wavelengths. If molecules a and b move independently of one another, their concentration fluctuations in the FCS observation volume are uncorrelated. Then, the fluorescence cross correlation function $G'_{ab}(\tau) = \langle \delta F_a(t) \delta F_b(t+\tau) \rangle = 0$. If, however, the two molecules are linked so that they move together into and out of the observation volume, their concentration fluctuations are correlated and $\langle \delta F_a(t) \delta F_b(t+\tau) \rangle \neq 0$. For stable aggregates, the amplitude of the cross correlation function $G_{ab}(0)$ measures the extent of linkage of the two species of fluorescent molecules. This approach together with applications has been extensively reviewed (Bacia, Kim, & Schwille, 2006; Bacia & Schwille, 2007; Heinze, Jahnz, & Schwille, 2004; Kim, Heinze, Bacia, Waxham, & Schwille, 2005).

7.3. Chemical reaction kinetics

One of the main motivations for developing FCS was to provide a method for measuring the kinetics of chemical reaction systems in equilibrium (Elson, 2004, 2011). Yet there have been far fewer applications to chemical kinetics than to transport rates and brightness. For reactions that are slow enough, correlation functions could be measured at a series of time points during the relaxation of a reaction system to equilibrium. Then changes in diffusion coefficients or brightness of the reactants would indicate reaction progress. Of greater interest, reaction kinetics can be measured by their direct effect on the relaxation of concentration fluctuations. For this to happen, the reaction must change the fluorescence of the reaction system. The larger the fluorescence change, the better the kinetics can be measured. Also, the chemical relaxation time must be short compared to the diffusion relaxation time. Consider a simple reaction $A+B\underset{k_b}{\overset{k_f}{\rightleftarrows}}C$. If the chemical relaxation time, $\tau_{\text{chem}}=\left[k_f(\langle c_A\rangle+\langle c_B\rangle)+k_b\right]^{-1}$, is long compared to the diffusion relaxation time, $\tau_D=w^2/(4D_B)$, where B is the fastest diffusing species, then the reactants rapidly diffuse through the observation volume. The reaction will occur rarely before concentration fluctuations have relaxed due to diffusion, and so little data on the kinetics will be registered. If $\tau_{\text{chem}}\ll\tau_D$, however, then the reaction will occur many times during a diffusion correlation time, and so sufficient data on the reaction kinetics will be contained in the correlation function.

The kinetics of conformational change has been studied in several systems described above as well as others (Kim et al., 2002; Michalet, Weiss, & Jager, 2006). FCS measurements have been applied fairly extensively to study the kinetics of formation of DNA hairpins (Orden & Jung, 2008). The first application of FCS to this subject used oligonucleotides with a fluorophore and a quencher attached to the 5′ and 3′ ends, respectively (Bonnet, Krichevsky, & Libchaber, 1998). The nucleotide sequences at the two ends were complementary so that they could form a helical stem. Intervening loop nucleotides consisted of varying lengths of T or A. In the hairpin helix conformation the fluorescence was quenched. When the stem helix melted, the fluorophore and quencher were no longer held together and the fluorescence increased. At temperatures near the helix opening transition FCS measurements reflected both diffusion ($\tau_D \sim 150\,\mu s$) and the opening and closing of the hairpin helix ($\tau_R = 5\,\mu s$–$1\,ms$). Assuming that the diffusion coefficient of the oligonucleotide was the same for the open and closed states, the correlation function could be expressed as $G(\tau) = G_D(\tau)G_R(\tau)$, and $G_D(\tau)$ could be measured with oligonucleotides that did not produce fluorescence fluctuations due to hairpin formation. Then $G_R(\tau) = G(\tau)/G_D(\tau)$ provides τ_R. An important assumption is that the transition between open (O) and closed (C) conformations is a single step process: $O \underset{k_-}{\overset{k_+}{\rightleftarrows}} C$, with rate constants k_+ and k_- and the equilibrium constant $K = k_-/k_+$. Then, $G_R(\tau) = 1 + \frac{1-p}{p}\exp(-\tau/\tau_R)$, where p is the fraction of the oligonucleotides in the open conformation and $\frac{1}{\tau_R} = k_+ + k_-$. Measurements of the temperature dependence of the hairpin opening/closing reaction independently determined $K(T)$. Then $k_+ = \frac{1/\tau_R}{(1+K)}$ and $k_- = k_+K$. Thus, the rate constants are obtained from the FCS results and the activation energies for opening and closing can be determined from the dependence of k_+ and k_- on temperature (Bonnet et al., 1998).

This study and its successors (Bonnet, Tyagi, Libchaber, & Kramer, 1999; Goddard, Bonnet, Krichevsky, & Libchaber, 2000) elegantly demonstrated how FCS can be used to dissect a chemical reaction system. Nevertheless, an important flaw in this analysis appears to be its assumption of a one-step opening–closing transition. Additional studies using both FCS and laser temperature-jump for kinetics and PCH to determine the relative concentrations of open and closed states revealed that the one-step model was insufficient to describe the data (Orden & Jung, 2008). A three-state mechanism was proposed in which the unfolded oligonucleotide forms an intermediate with closed loop and quenched fluorescence but incomplete helical stem in the time range of tens of microseconds. Formation of the complete hairpin helix is a slower process requiring milliseconds or longer.

Studies of molecular association by FCCS as discussed above typically suppose that the links between binding partners are stable on the time scale of the measurement. Cross correlation can, however, supply important information about the kinetics of interaction. For example, consider the cross correlation function, $G_{BC}(\tau)$, for the reaction $A + B \underset{k_b}{\overset{k_f}{\rightleftarrows}} C$ in Fig. 2 of Elson and Magde (1974) supposing that $\tau_{chem} \ll \tau_D$. Given the presumed ideality

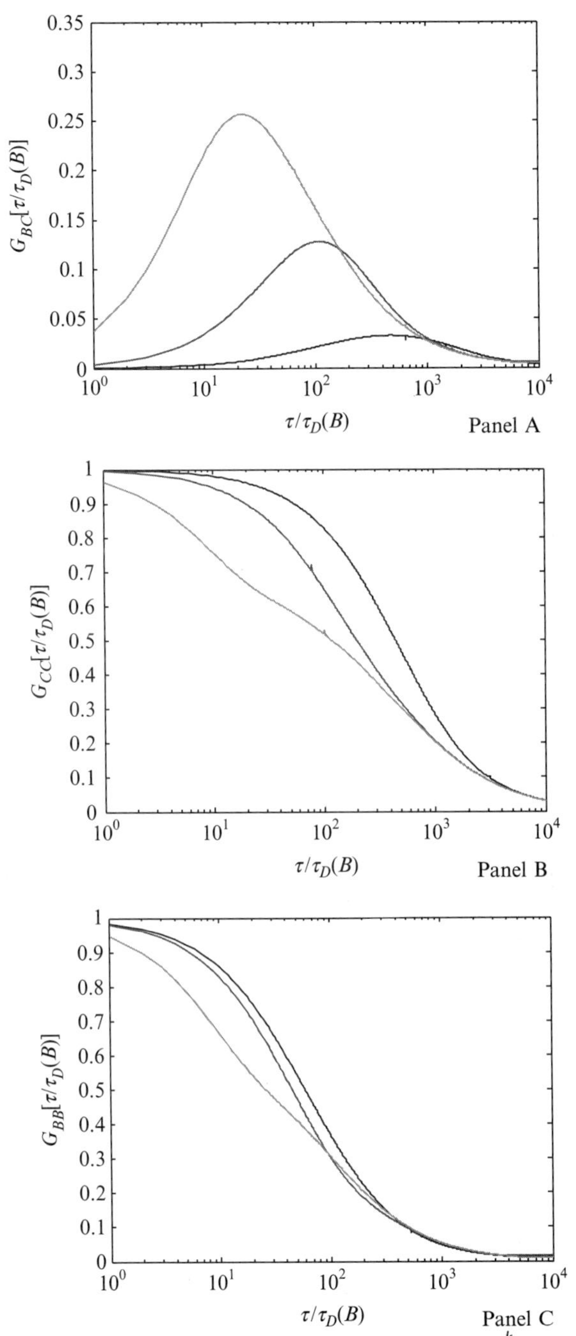

Figure 2.1 Correlation functions for the chemical reaction $A + B \underset{k_b}{\overset{k_f}{\rightleftarrows}} C$. The fluorescence fluctuation autocorrelation functions $G_{CC}(\tau)$ and $G_{BB}(\tau)$ (Panels B and C, respectively) and

of the reaction system $\langle \delta c_B(0) \delta c_C(0) \rangle = 0$, and so $G_{BC}(0) = 0$. Fluctuations in B, however, induce corresponding fluctuations in C (and vice versa) due to the chemical reaction. Therefore, $G_{BC}(\tau)$ increases over a time period of the order of τ_{chem} (Fig. 2.1). At later times the correlation decays to zero due to diffusion. In addition to providing kinetic data about reaction systems in equilibrium, cross correlation functions can also supply unique information about reaction fluxes in NESS systems (Qian & Elson, 2004, 2009). In equilibrium systems, due to microscopic reversibility, $\langle \delta c_X(t) \delta c_Y(t+\tau) \rangle = \langle \delta c_Y(t) \delta c_X(t+\tau) \rangle$ and so $G_{XY}(\tau) = G_{YX}(\tau)$. In contrast, in NESS systems, these two cross correlation functions are not equal. Indeed, $G_{XY}(\tau)$ and $G_{YX}(\tau)$ supply the flux $X \to Y$ and $Y \to X$, respectively (Qian & Elson, 2004). In principle, this could be especially useful to study reaction kinetics in steady-state metabolic and signaling systems in cells (Qian & Elson, 2009). How might one measure the needed fluorescence cross correlation functions? One possibility is to use FRET FCS. For example, if X were a donor and Y an acceptor, $G_{XY}(\tau)$ could be obtained by measuring the fluorescence of X and Y at their distinct emission wavelengths. Although this is a promising method, in principle, it requires a large change in energy transfer efficiency, that is, the efficiency is practically 0 on one side of the reaction and 1 on the

---

the cross correlation function $G_{BC}(\tau)$ (Panel A) were calculated for a two-dimensional system with the following parameters:

Diffusion coefficients: $D_A = D_C = 10^{-9}$ cm^2 s^{-1}; $D_B = 10^{-8}$ cm^2 s^{-1}

Gaussian laser beam radius: $w = 10^{-4}$ cm so that the diffusion correlation time for component B is $\tau_D(B) = w^2/(4D_B) = 0.25$ s.

Concentrations: $\langle C_A \rangle = \langle C_B \rangle = \langle C_C \rangle = 10^{-9} M$.

Rate constants and chemical relaxation times ($R = k_f(\langle C_A \rangle + \langle C_B \rangle) + k_b$; $\tau_{\text{chem}} = R^{-1}$)

For the blue curves: $k_f = 10^8 M^{-1}$ s^{-1}; $k_b = 0.1$ s^{-1}; $\tau_{\text{chem}}/\tau_D(B) = 13.2$;

For the red curves: $k_f = 10^9 M^{-1}$ s^{-1}; $k_b = 1.0$ s^{-1}; $\tau_{\text{chem}}/\tau_D(B) = 1.32$

For the green curves: $k_f = 10^{10} M^{-1}$ s^{-1}; $k_b = 10$ s^{-1}; $\tau_{\text{chem}}/\tau_D(B) = 0.132$

The cross-correlation function, $G_{BC}(\tau)$, vanishes at $\tau = 0$, increases due to chemical correlations and then decreases due to diffusion. The autocorrelation function $G_{CC}(\tau)$ displays distinct chemical kinetic and diffusion phases for the smallest values of $\tau_{\text{chem}}/\tau_D$. The autocorrelation function $G_{BB}(\tau)$ also shows a separation between the diffusion and chemical kinetic components but is less sensitive to the presence of the chemical reaction. Supposing that $Q_A = 0$, the total correlation function that would be experimentally measured is $G(\tau) = \frac{G'_{BB}(\tau) + G'_{CC}(\tau) + 2G'_{BC}(\tau)}{(\langle F_B \rangle + \langle F_C \rangle)^2}$ (cf., Eq. 2.1). For the reaction kinetics to be directly detectable, it is necessary that $Q_B \neq Q_C$. For example, in the favorable case that $Q_B = 0$, $G(\tau) = G_{CC}(\tau)$ (cf., Elson & Magde, 1974). (For color version of this figure, the reader is referred to the online version of this chapter.)

other. If this condition does not hold, the donor–acceptor fluorescence cross correlation function for FRET FCS includes autocorrelation function contributions from donor and acceptor that can obscure the contribution of the desired cross correlation. A useful analysis of this approach has been supplied by Torres and Levitus (2007). The challenge for this interesting application, then, is to find a suitable FRET system that can be incorporated into the reaction system of interest.

8. SUMMARY

FCS, a method that is elegant in principle, has now become a practical tool for studies of a wide range of phenomena including molecular transport, chemical reaction kinetics, photophysics, and molecular interactions among others. It provides a way to investigate mesoscopic systems like cells in which effects of random fluctuations due to the smallness of the number of molecular participants can lead to significant stochastic biological consequences (Eldar & Elowitz, 2010; Raj & van Oudenaarden, 2008, 2009). The availability of commercial instruments and the robustness of the results obtained by current technology make FCS available for routine laboratory applications. Even with the wide elaboration of methods based on FCS including, for example, imaging and cross-correlation approaches, as well as other recently introduced methods (Kapusta, Wahl, Benda, Hof, & Enderlein, 2007; Wennmalm, Thyberg, Xu, & Widengren, 2009; Wennmalm & Widengren, 2010), it seems likely that new kinds of applications are still possible. One promising field would be to study NESSs both *in vitro* and *in vivo* (Elson, 2011; Qian & Elson, 2009, 2010; Qian, Saffarian, & Elson, 2002).

ACKNOWLEDGMENTS

The development and application of FCS at Cornell and beyond resulted from the work of many people. I can list only a few whose contributions seem to me notably conspicuous. All of the work at Cornell could not have flourished without the wisdom, enthusiasm, and experimental expertise of Watt Webb. The initial demonstration of FCS was entirely dependent on the persistence, skill, and technical know-how of Douglas Magde. Dennis Koppel and Daniel Axelrod established the microscope-based versions of FCS and developed our version of FPR. Joseph Schlessinger was the prime mover in the early applications of these methods to cells. In later work, Hong Qian provided a number of important contributions including, among others, analyses of the FCS optical system, measurement noise, and the high moment approach to measuring molecule number and brightness. Saveez Saffarian extended the analysis of measurement noise and carried out studies of the aggregation of EGF receptors and of a matrix metalloproteinase Brownian ratchet. I would like to thank Michael Edidin for his advice and Ken Jacobson for his early participation in our measurements on cells. I am also grateful to Rudolf Rigler for the work that he and his colleagues have done that led to the establishment of FCS as a widely accessible measurement method as well as for the many interesting and wide ranging applications that they have provided. Finally, I thank NIH (R01-GM084200) and NSF (CMMI 0826518) for financial support.

REFERENCES

Auer, M., Moore, K. J., Meyer-Almes, F. J., Guenther, R., Pope, A. J., & Stoeckli, J. A. (1998). Fluorescence correlation spectroscopy: Lead discovery by miniaturized HTS. *Drug Discovery Today, 3*, 457–465.

Axelrod, D., Koppel, D. E., Schlessinger, J., Elson, E., & Webb, W. W. (1976). Mobility measurement by analysis of fluorescence photobleaching recovery kinetics. *Biophysical Journal, 16*, 1055–1069.

Bacia, K., Kim, S. A., & Schwille, P. (2006). Fluorescence cross-correlation spectroscopy in living cells. *Nature Methods, 3*, 83–89.

Bacia, K., & Schwille, P. (2007). Practical guidelines for dual-color fluorescence cross-correlation spectroscopy. *Nature Protocols, 2*, 2842–2856.

Bannai, M., Higuchi, K., Akesaka, T., Furukawa, M., Yamaoka, M., Sato, K., et al. (2004). Single-nucleotide-polymorphism genotyping for whole-genome-amplified samples using automated fluorescence correlation spectroscopy. *Analytical Biochemistry, 327*, 215–221.

Berland, K. M., So, P. T., & Gratton, E. (1995). Two-photon fluorescence correlation spectroscopy: Method and application to the intracellular environment. *Biophysical Journal, 68*, 694–701.

Berne, B. J., & Pecora, R. (1976). *Dynamic light scattering with applications to chemistry, biology, and physics*. New York: John Wiley & Sons.

Bevington, D., & Robinson, D. K. (2003). *Data reduction and error analysis for the physical sciences*. McGraw-Hill.

Bonnet, G., Krichevsky, O., & Libchaber, A. (1998). Kinetics of conformational fluctuations in DNA hairpin-loops. *Proceedings of the National Academy of Sciences of the United States of America, 95*, 8602–8606.

Bonnet, G., Tyagi, S., Libchaber, A., & Kramer, F. R. (1999). Thermodynamic basis of the enhanced specificity of structured DNA probes. *Proceedings of the National Academy of Sciences of the United States of America, 96*, 6171–6176.

Chattopadhyay, K., Elson, E. L., & Frieden, C. (2005). The kinetics of conformational fluctuations in an unfolded protein measured by fluorescence methods. *Proceedings of the National Academy of Sciences of the United States of America, 102*, 2385–2389.

Chattopadhyay, K., Saffarian, S., Elson, E. L., & Frieden, C. (2002). Measurement of microsecond dynamic motion in the intestinal fatty acid binding protein by using fluorescence correlation spectroscopy. *Proceedings of the National Academy of Sciences of the United States of America, 99*, 14171–14176.

Chen, Y., Wei, L. N., & Müller, J. D. (2003). Probing protein oligomerization in living cells with fluorescence fluctuation spectroscopy. *Proceedings of the National Academy of Sciences of the United States of America, 100*, 15492–15497.

Day, R. N., & Schaufele, F. (2005). Imaging molecular interactions in living cells. *Molecular Endocrinology, 19*, 1675–1686.

Digman, M. A., Brown, C. M., Sengupta, P., Wiseman, P. W., Horwitz, A. R., & Gratton, E. (2005). Measuring fast dynamics in solutions and cells with a laser scanning microscope. *Biophysical Journal, 89*, 1317–1327.

Digman, M. A., Dalal, R., Horwitz, A. F., & Gratton, E. (2008). Mapping the number of molecules and brightness in the laser scanning microscope. *Biophysical Journal, 94*, 2320–2332.

Doi, M., & Edwards, S. F. (1988). *The theory of polymer dynamics*. Oxford: Clarendon Press.

Edel, J. B., Wu, M., Baird, B., & Craighead, H. G. (2005). High spatial resolution observation of single-molecule dynamics in living cell membranes. *Biophysical Journal, 88*, L43–L45.

Edidin, M., Zagyansky, Y., & Lardner, T. J. (1976). Measurement of membrane protein lateral diffusion in single cells. *Science, 191*, 466–468.

Eggeling, C., Ringemann, C., Medda, R., Schwarzmann, G., Sandhoff, K., Polyakova, S., et al. (2009). Direct observation of the nanoscale dynamics of membrane lipids in a living cell. *Nature, 457*, 1159–1162.

Ehrenberg, M., & Rigler, R. (1974). Rotational Brownian-motion and fluorescence intensity fluctuations. *Chemical Physics, 4*, 390–401.

Ehrenberg, M., & Rigler, R. (1976). Fluorescence correlation spectroscopy applied to rotational diffusion of macromolecules. *Quarterly Reviews of Biophysics, 9*, 69–81.

Eigen, M., & De Maeyer, L. C. (1963). S. L. Freiss, E. S. Lewis & A. Weissberger (Eds.), *Techniques of organic chemistry*, Vol. 8, Part II, (p. 895). New York: Interscience.

Eldar, A., & Elowitz, M. B. (2010). Functional roles for noise in genetic circuits. *Nature, 467*, 167–173.

Elson, E. L. (1985). Fluorescence correlation spectroscopy and photobleaching recovery. *Annual Review of Physical Chemistry, 36*, 379–406.

Elson, E. L. (2004). Quick tour of fluorescence correlation spectroscopy from its inception. *Journal of Biomedical Optics, 9*, 857–864.

Elson, Elliot L. (2011). Fluorescence correlation spectroscopy: Past, present, future. *Biophysical Journal, 101*, 2855–2870.

Elson, E., & Magde, D. (1974). Fluorescence correlation spectroscopy. I. Conceptual basis and theory. *Biopolymers, 13*, 1–27.

Elson, E. L., & Webb, W. W. (1975). Concentration correlation spectroscopy: A new biophysical probe based on occupation number fluctuations. *Annual Review of Biophysics and Bioengineering, 4*, 311–334.

Goddard, N. L., Bonnet, G., Krichevsky, O., & Libchaber, A. (2000). Sequence dependent rigidity of single stranded DNA [In Process Citation]. *Physical Review Letters, 85*, 2400–2403.

Hebert, B., Costantino, S., & Wiseman, P. W. (2005). Spatiotemporal image correlation spectroscopy (STICS) theory, verification, and application to protein velocity mapping in living CHO cells. *Biophysical Journal, 88*, 3601–3614.

Heinze, K. G., Jahnz, M., & Schwille, P. (2004). Triple-color coincidence analysis: One step further in following higher order molecular complex formation. *Biophysical Journal, 86*, 506–516.

Heinze, K. G., Koltermann, A., & Schwille, P. (2000). Simultaneous two-photon excitation of distinct labels for dual-color fluorescence crosscorrelation analysis. *Proceedings of the National Academy of Sciences of the United States of America, 97*, 10377–10382.

Hell, S. W., & Wichmann, J. (1994). Breaking the diffraction resolution limit by stimulated emission: Stimulated-emission-depletion fluorescence microscopy. *Optics Letters, 19*, 780–782.

Herrmann, M., Neuberth, N., Wissler, J., Perez, J., Gradl, D., & Naber, A. (2009). Near-field optical study of protein transport kinetics at a single nuclear pore. *Nano Letters, 9*, 3330–3336.

Hess, S. T., & Webb, W. W. (2002). Focal volume optics and experimental artifacts in confocal fluorescence correlation spectroscopy. *Biophysical Journal, 83*, 2300–2317.

Jacobson, K., Ishihara, A., & Inman, R. (1987). Lateral diffusion of proteins in membranes. *Annual Review of Physiology, 49*, 163–175.

Jacobson, K., Wu, E., & Poste, G. (1976). Measurement of the translational mobility of concanavalin A in glycerol-saline solutions and on the cell surface by fluorescence recovery after photobleaching. *Biochimica et Biophysica Acta, 433*, 215–222.

Kapusta, P., Wahl, M., Benda, A., Hof, M., & Enderlein, J. (2007). Fluorescence lifetime correlation spectroscopy. *Journal of Fluorescence, 17*, 43–48.

Kask, P., Gunther, R., & Axhausen, P. (1997). Statistical accuracy in fluorescence fluctuations experiments. *European Biophysics Journal with Biophysics Letters, 25*, 163–169.

Kask, P., Palo, K., Fay, N., Brand, L., Mets, U., Ullmann, D., et al. (2000). Two-dimensional fluorescence intensity distribution analysis: Theory and applications. *Biophysical Journal, 78*, 1703–1713.

Kask, P., Palo, K., Ullmann, D., & Gall, K. (1999). Fluorescence-intensity distribution analysis and its application in biomolecular detection technology. *Proceedings of the National Academy of Sciences of the United States of America, 96*, 13756–13761.

Kim, S. A., Heinze, K. G., Bacia, K., Waxham, M. N., & Schwille, P. (2005). Two-photon cross-correlation analysis of intracellular reactions with variable stoichiometry. *Biophysical Journal, 88*, 4319–4336.

Kim, H. D., Nienhaus, G. U., Ha, T., Orr, J. W., Williamson, J. R., & Chu, S. (2002). Mg2+-dependent conformational change of RNA studied by fluorescence correlation and FRET on immobilized single molecules. *Proceedings of the National Academy of Sciences of the United States of America, 99*, 4284–4289.

Kittel, C. (1958). *Elementary statistical physics*. New York: John Wiley & Sons.

Kolin, D. L., & Wiseman, P. W. (2007). Advances in image correlation spectroscopy: Measuring number densities, aggregation states, and dynamics of fluorescently labeled macromolecules in cells. *Cell Biochemistry and Biophysics, 49*, 141–164.

Koppel, D. (1974). Statistical accuracy in fluorescence correlation spectroscopy. *Physical Review A, 10*, 1938–1945.

Koppel, D. E., Axelrod, D., Schlessinger, J., Elson, E. L., & Webb, W. W. (1976). Dynamics of fluorescence marker concentration as a probe of mobility. *Biophysical Journal, 16*, 1315–1329.

Krichevsky, O., & Bonnet, G. (2002). Fluorescence correlation spectroscopy: The technique and its applications. *Reports on Progress in Physics, 65*, 251–297.

Laughlin, R. B., Pines, D., Schmalian, J., Stojkovic, B. P., & Wolynes, P. (2000). The middle way. *Proceedings of the National Academy of Sciences of the United States of America, 97*, 32–37.

Levene, M. J., Korlach, J., Turner, S. W., Foquet, M., Craighead, H. G., & Webb, W. W. (2003). Zero-mode waveguides for single-molecule analysis at high concentrations. *Science, 299*, 682–686.

Lewis, A., Taha, H., Strinkovski, A., Manevitch, A., Khatchatouriants, A., Dekhter, R., et al. (2003). Near-field optics: From subwavelength illumination to nanometric shadowing. *Nature Biotechnology, 21*, 1378–1386.

Lieto, A. M., Cush, R. C., & Thompson, N. L. (2003). Ligand–receptor kinetics measured by total internal reflection with fluorescence correlation spectroscopy. *Biophysical Journal, 85*, 3294–3302.

Lippincott-Schwartz, J., Altan-Bonnet, N., & Patterson, G. H. (2003). Photobleaching and photoactivation: Following protein dynamics in living cells. *Nature Cell Biology*, S7–S14 Suppl.

Lumma, D., Keller, S., Vilgis, T., & Radler, J. O. (2003). Dynamics of large semiflexible chains probed by fluorescence correlation spectroscopy. *Physical Review Letters, 90*, 218301.

Magde, D., Elson, E. L., & Webb, W. W. (1974). Fluorescence correlation spectroscopy. II. An experimental realization. *Biopolymers*, *13*, 29–61.

Magde, D., Webb, W. W., & Elson, E. L. (1978). Fluorescence correlation spectroscopy. III. Uniform translation and laminar flow. *Biopolymers*, *17*, 361–376.

Melnykov, A. V., & Hall, K. B. (2009). Revival of high-order fluorescence correlation analysis: Generalized theory and biochemical applications. *The Journal of Physical Chemistry. B*, *113*, 15629–15638.

Meseth, U., Wohland, T., Rigler, R., & Vogel, H. (1999). Resolution of fluorescence correlation measurements. *Biophysical Journal*, *76*, 1619–1631.

Michalet, X., Weiss, S., & Jager, M. (2006). Single-molecule fluorescence studies of protein folding and conformational dynamics. *Chemistry Review*, *106*, 1785–1813.

Mukhopadhyay, S., Krishnan, R., Lemke, E. A., Lindquist, S., & Deniz, A. A. (2007). A natively unfolded yeast prion monomer adopts an ensemble of collapsed and rapidly fluctuating structures. *Proceedings of the National Academy of Sciences of the United States of America*, *104*, 2649–2654.

Müller, J. D., Chen, Y., & Gratton, E. (2000). Resolving heterogeneity on the single molecular level with the photon-counting histogram. *Biophysical Journal*, *78*, 474–486.

Nagy, P., Claus, J., Jovin, T. M., & Arndt-Jovin, D. J. (2010). Distribution of resting and ligand-bound ErbB1 and ErbB2 receptor tyrosine kinases in living cells using number and brightness analysis. *Proceedings of the National Academy of Sciences of the United States of America*, *107*, 16524–16529.

Ohsugi, Y., Saito, K., Tamura, M., & Kinjo, M. (2006). Lateral mobility of membrane-binding proteins in living cells measured by total internal reflection fluorescence correlation spectroscopy. *Biophysical Journal*, *91*, 3456–3464.

Onsager, L. (1931). Reciprocal relations in irreversible processes. I. *Physical Review*, *37*, 405–426.

Orden, A. V., & Jung, J. (2008). Review fluorescence correlation spectroscopy for probing the kinetics and mechanisms of DNA hairpin formation. *Biopolymers*, *89*, 1–16.

Palmer, A. G., & Thompson, N. L. (1987a). Molecular aggregation characterized by high order autocorrelation in fluorescence correlation spectroscopy. *Biophysical Journal*, *52*, 257–270.

Palmer, A. G., 3rd, & Thompson, N. L. (1987b). Theory of sample translation in fluorescence correlation spectroscopy. *Biophysical Journal*, *51*, 339–343.

Palmer, A. G., 3rd, & Thompson, N. L. (1989). High-order fluorescence fluctuation analysis of model protein clusters. *Proceedings of the National Academy of Sciences of the United States of America*, *86*, 6148–6152.

Palo, K., Brand, L., Eggeling, C., Jager, S., Kask, P., & Gall, K. (2002). Fluorescence intensity and lifetime distribution analysis: Toward higher accuracy in fluorescence fluctuation spectroscopy. *Biophysical Journal*, *83*, 605–618.

Palo, K., Mets, U., Jager, S., Kask, P., & Gall, K. (2000). Fluorescence intensity multiple distributions analysis: Concurrent determination of diffusion times and molecular brightness. *Biophysical Journal*, *79*, 2858–2866.

Peters, R., Peters, J., Tews, K. H., & Bahr, W. (1974). A microfluorimetric study of translational diffusion in erythrocyte membranes. *Biochimica et Biophysica Acta*, *367*, 282–294.

Petersen, N. O. (1986). Scanning fluorescence correlation spectroscopy. I. Theory and simulation of aggregation measurements. *Biophysical Journal*, *49*, 809–815.

Petersen, N. O., Hoddelius, P. L., Wiseman, P. W., Seger, O., & Magnusson, K. E. (1993). Quantitation of membrane receptor distributions by image correlation spectroscopy: Concept and application. *Biophysical Journal*, *65*, 1135–1146.

Pryse, K. M., Rong, X., Whisler, J. A., McConnaughey, W. B., Jiang, Y. F., Melnykov, A. V., Elson, E. L., & Genin, G. M. (2012). Confidence intervals for

concentration and brightness from fluorescence fluctuation measurements. *Biophysical Journal, 103*, 898–906.
Qian, H. (1990). On the statistics of fluorescence correlation spectroscopy. *Biophysical Chemistry, 38*, 49–57.
Qian, H., & Elson, E. L. (1990a). Distribution of molecular aggregation by analysis of fluctuation moments. *Proceedings of the National Academy of Sciences of the United States of America, 87*, 5479–5483.
Qian, H., & Elson, E. L. (1990b). On the analysis of high order moments of fluorescence fluctuations. *Biophysical Journal, 57*, 375–380.
Qian, H., & Elson, E. L. (1991). Analysis of confocal laser-microscope optics for 3-D fluorescence correlation spectroscopy. *Applied Optics, 30*, 1185–1195.
Qian, H., & Elson, E. L. (2004). Fluorescence correlation spectroscopy with high-order and dual-color correlation to probe nonequilibrium steady states. *Proceedings of the National Academy of Sciences of the United States of America, 101*, 2828–2833.
Qian, H., & Elson, E. (2009). Chemical fluxes in cellular steady states measured by fluorescence correlation spectroscopy. In A. Graslund, R. Rigler & J. Widengren (Eds.), *Single molecule spectroscopy in chemistry, physics and biology* (pp. 119–138). Heidelberg: Springer.
Qian, H., & Elson, E. L. (2010). Chemical fluxes in cellular steady states measured by fluorescence correlation spectroscopy. In: A. Graslund, R. Rigler & J. Widengren (Eds.), *Single molecule spectroscopy in chemical physics and biology*, Vol. 96, (pp. 119–137). New York: Springer.
Qian, H., Saffarian, S., & Elson, E. L. (2002). Concentration fluctuations in a mesoscopic oscillating chemical reaction system. *Proceedings of the National Academy of Sciences of the United States of America, 99*, 10376–10381.
Raj, A., & van Oudenaarden, A. (2008). Nature, nurture, or chance: Stochastic gene expression and its consequences. *Cell, 135*, 216–226.
Raj, A., & van Oudenaarden, A. (2009). Single-molecule approaches to stochastic gene expression. *Annual Review of Biophysics, 38*, 255–270.
Rigler, R. (2009). FCS and single molecule spectroscopy. In A. Graslund, R. Rigler & J. Widengren (Eds.), *Single molecule spectroscopy in chemistry, physics, and biology* (pp. 77–103). Heidelberg: Springer-Verlag.
Rigler, R., Mets, U., Widengren, J., & Kask, P. (1993). Fluorescence correlation spectroscopy with high count rate and low background: Analysis of translational diffusion. *European Biophysics Journal, 22*, 169–175.
Rouse, P. E. (1953). A theory of the linear viscoelastic properties of dilute solutions of coiling polymers. *The Journal of Chemical Physics, 21*, 1272–1280.
Ruckstuhl, T., & Seeger, S. (2004). Attoliter detection volumes by confocal total-internal-reflection fluorescencemicroscopy. *Optics Letters, 29*, 569–571.
Ruttinger, S. (2006). Confocal microscopy and quantitative single molecule techniques for metrology in molecular medicine. Mathematik und Naturwissenschaften, Vol. Dr. rer. nat. Technischen Universitat Berlin, Berlin. p. 141.
Saffarian, S., Collier, I. E., Marmer, B. L., Elson, E. L., & Goldberg, G. (2004). Interstitial collagenase is a Brownian ratchet driven by proteolysis of collagen. *Science, 306*, 108–111.
Saffarian, S., & Elson, E. L. (2003). Statistical analysis of fluorescence correlation spectroscopy: The standard deviation and bias. *Biophysical Journal, 84*, 2030–2042.
Saffarian, S., Li, Y., Elson, E. L., & Pike, L. J. (2007). Oligomerization of the EGF receptor investigated by live cell fluorescence intensity distribution analysis. *Biophysical Journal, 93*, 1021–1031.
Schwille, P., Haupts, U., Maiti, S., & Webb, W. W. (1999). Molecular dynamics in living cells observed by fluorescence correlation spectroscopy with one- and two-photon excitation. *Biophysical Journal, 77*, 2251–2265.

Shusterman, R., Alon, S., Gavrinyov, T., & Krichevsky, O. (2004). Monomer dynamics in double- and single-stranded DNA polymers. *Physical Review Letters*, *92*, 048303.
Skinner, J. P., Chen, Y., & Müller, J. D. (2005). Position-sensitive scanning fluorescence correlation spectroscopy. *Biophysical Journal*, *89*, 1288–1301.
Slaughter, B. D., Huff, J. M., Wiegraebe, W., Schwartz, J. W., & Li, R. (2008). SAM domain-based protein oligomerization observed by live-cell fluorescence fluctuation spectroscopy. *PloS One*, *3*, e1931.
Slaughter, B. D., & Li, R. (2010). Toward quantitative "*in vivo* biochemistry" with fluorescence fluctuation spectroscopy. *Molecular Biology of the Cell*, *21*, 4306–4311.
Slaughter, B. D., Schwartz, J. W., & Li, R. (2007). Mapping dynamic protein interactions in MAP kinase signaling using live-cell fluorescence fluctuation spectroscopy and imaging. *Proceedings of the National Academy of Sciences of the United States of America*, *104*, 20320–20325.
Sprague, B. L., & McNally, J. G. (2005). FRAP analysis of binding: Proper and fitting. *Trends in Cell Biology*, *15*, 84–91.
Tanford, C. (1961). *Physical chemistry of macromolecules*. New York: John Wiley & Sons.
Thompson, N. L., Burghardt, T. P., & Axelrod, D. (1981). Measuring surface dynamics of biomolecules by total internal reflection fluorescence with photobleaching recovery or correlation spectroscopy. *Biophysical Journal*, *33*, 435–454.
Toomre, D., & Bewersdorf, J. (2010). A new wave of cellular imaging. *Annual Review of Cell and Developmental Biology*, *26*, 285–314.
Torres, T., & Levitus, M. (2007). Measuring conformational dynamics: A new FCS-FRET approach. *The Journal of Physical Chemistry. B*, *111*, 7392–7400.
Vobornik, D., Banks, D. S., Lu, Z. F., Fradin, C., Taylor, R., & Johnston, L. J. (2008). Fluorescence correlation spectroscopy with sub-diffraction-limited resolution using near-field optical probes. *Applied Physics Letters*, *93*, 163904-1–163904-3.
Vukojevic, V., Pramanik, A., Yakovleva, T., Rigler, R., Terenius, L., & Bakalkin, G. (2005). Study of molecular events in cells by fluorescence correlation spectroscopy. *Cellular and Molecular Life Sciences*, *62*, 535–550.
Wenger, J., & Rigneault, H. (2010). Photonic methods to enhance fluorescence correlation spectroscopy and single molecule fluorescence detection. *International Journal of Molecular Sciences*, *11*, 206–221.
Wennmalm, S., Thyberg, P., Xu, L., & Widengren, J. (2009). Inverse-fluorescence correlation spectroscopy. *Analytical Chemistry*, *81*, 9209–9215.
Wennmalm, S., & Widengren, J. (2010). Inverse-fluorescence cross-correlation spectroscopy. *Analytical Chemistry*, *82*, 5646–5651.
Wiseman, P. W., Squier, J. A., Ellisman, M. H., & Wilson, K. R. (2000). Two-photon image correlation spectroscopy and image cross-correlation spectroscopy. *Journal of Microscopy*, *200*, 14–25.
Wohland, T., Rigler, R., & Vogel, H. (2001). The standard deviation in fluorescence correlation spectroscopy. *Biophysical Journal*, *80*, 2987–2999.
Wood, C., Huff, J., Marshall, W., Yu, Q., Unruh, J., Slaughter, B., et al. (2011). Fluorescence correlation spectroscopy as tool for high-content-screening in yeast (HCS-FCS). *Proceedings of SPIE*, *7905*, 14.
Wu, B., Chen, Y., & Müller, J. D. (2006). Dual-color time-integrated fluorescence cumulant analysis. *Biophysical Journal*, *91*, 2687–2698.
Wu, B., & Müller, J. D. (2005). Time-integrated fluorescence cumulant analysis in fluorescence fluctuation spectroscopy. *Biophysical Journal*, *89*, 2721–2735.
Zimm, B. H. (1956). Dynamics of polymer molecules in dilute solution: Viscoelasticity, flow birefringence, and dielectric loss. *The Journal of Chemical Physics*, *24*, 269.

CHAPTER THREE

Dual-Color Fluorescence Cross-Correlation Spectroscopy with Continuous Laser Excitation in a Confocal Setup

Thomas Weidemann[1,2], Petra Schwille[1,2]
Biophysics/BIOTEC, Technische Universität Dresden, Tatzberg 47-51, Dresden, Germany
[1]Current address: Max Planck Institute of Biochemistry, Am Klopferspitz 18, 82152 Martinsried, Germany
[2]Corresponding authors: e-mail address: weidemann@biochem.mpg.de; schwille@biochem.mpg.de

Contents

Abstract

Fluorescence correlation spectroscopy evaluates local signal fluctuations arising from stochastic movements of fluorescent particles in solution. The measured fluctuating signal is correlated in time and analyzed with appropriate model functions containing the parameters that describe the underlying molecular behavior. The dual-color extension, fluorescence cross-correlation spectroscopy (FCCS) allows for a comparison between spectrally well-separated channels to extract codiffusion events that reflect interactions between differently labeled molecules. In addition to solution measurements, FCCS can be applied with subcellular resolution and is therefore a very promising approach for a quantitative biochemical assessment of molecular networks in living cells. To derive thermodynamic and kinetic reaction parameters, the influence of a number of other factors like background noise, illumination intensity profiles, photophysical processes, and cross talk between the channels have to be treated. Here, we provide a roadmap to derive binding reaction data with dual-color FCCS using continuous wave laser excitation, as it is now accessible with many state-of-the-art confocal microscopes.

Methods in Enzymology, Volume 518
ISSN 0076-6879
http://dx.doi.org/10.1016/B978-0-12-388422-0.00003-0

1. INTRODUCTION

Biochemists and cell biologists aim for a quantitative characterization of molecular interaction networks, and this endeavor is now at the heart of the emerging field of systems biology. During the past decades, fluorescence-based microscopic techniques have revolutionized our understanding of cellular processes because they provided means to selectively observe tagged molecules among tens of thousands of non- or weakly fluorescent cellular components. Hand in hand with the use of increasingly versatile fluorescent-labeling chemistry and genetically encoded fluorescent proteins, confocal laser scanning microscopy (CLSM) is now a standard technique to dissect reaction pathways in cell biology. CLSM imaging provides snapshots of spatial distributions of fluorescent molecules in cells but cannot tell much about the underlying molecular behavior. Pairwise interactions can be studied with the colocalization approach, measuring the degree of spatial overlap of differently colored molecules. However, this approach fails for homogeneous distributions, since these interactions are not assessed on a molecular level. Thus, colocalization cannot distinguish between a molecular complex and the mere presence in the same cellular compartment. Accordingly, confocal imaging does not provide access to relevant thermodynamic parameters like mobility, concentrations, and affinity constants.

A promising approach to extend the analysis to a molecular level has emerged with the development of fluorescence correlation spectroscopy (FCS; Elson & Magde, 1974; Magde, Elson, & Webb, 1974; Rigler, Mets, Widengren, & Kask, 1993). FCS employs the same optical setup as a conventional CLSM. However, while confocal imaging measures the fluorescence distribution in space, FCS is based on a time-dependent analysis. With FCS, molecular properties are determined from recording a fluctuating signal with submicrosecond resolution at a particular spot. The signal is generated by fluorescent molecules diffusing stochastically through a small, illuminated open volume in solution (see Chapter 1). A statistical analysis of the time-dependent signal, a correlation analysis, directly yields the average dwell times (called *diffusion time*) of the molecules. In addition, the relative magnitude of the signal fluctuations (called *correlation amplitude*) reflects the average number of observed molecules. Thus, after a proper calibration of the optical setup, concentrations and diffusion coefficients of up to three differently sized molecular species can be quantified and thereby any biochemical reaction causing changes in these parameters (Eigen & Rigler, 1994). Since a CLSM provides subfemtoliter observation volumes, positioning and measuring *in situ* in living cells are straightforward (Schwille, Haupts, Maiti, & Webb, 1999). Different cellular compartments like the nucleus, the cytoplasm, or the plasma membrane can be selectively probed. Importantly, FCS data always reflect the dynamical molecular behavior in conjunction with environmental properties. If the photophysical and diffusion properties of the fluorescent molecules are known, these molecules can in turn serve as sensors to characterize the local environment. For example, ion concentrations, pH, or viscosity were determined. Thus, FCS is regarded as a versatile and attractive technique to study interactions and dynamics of fluorescent molecules in free solution as well as in cellular systems (Kim, Heinze, & Schwille, 2007).

The use of single-color FCS to monitor binding reactions bears certain limitations. (i) The diffusion times of the observed complexes are quite insensitive with respect to molecular size. According to the Stokes–Einstein equation, the diffusion time relates linearly to the molecular diameter and therefore only to the cubic root of the molecular weight. Therefore, a simple dimerization is hardly detectable with a conventional confocal setup. (ii) The size and shape of a confocal detection volume is prone to distortions depending on local changes in refractive index within the sample. This is a major source of error for amplitude-based measurements, especially within cells and tissues. (iii) Even if the changes in diffusion times of a free and bound molecular species are significant, changes of the molecular brightness upon binding may complicate the analysis. This problem is even aggravated in the case of statistical labeling of multiple sites leading to broad distributions of differently bright molecules. While this problem has been noticed long ago, a work around was achieved in only a few simple cases. (iv) Finally, in complex environments, like cells, many different processes take place on different time scales: for example, Brownian motion, vesicular trafficking, motor-mediated directed transport, as well as slow changes in cellular shape. All of these processes, when linked to fluorescence, appear in the correlation curves. Because collective movement of fluorescent aggregates distorts the curves, it is not always possible to derive true binding-related concentrations.

These limitations can largely be overcome by discriminating the interaction partners based on color instead of mobility (Bacia, Kim, & Schwille, 2006; Bacia & Schwille, 2007; Hwang & Wohland, 2007; Schwille, Meyer-Almes, & Rigler, 1997; Weidemann, Wachsmuth, Tewes, Rippe, & Langowski, 2002). Dual-color fluorescence cross-correlation (FCCS) works with two differently labeled spectrally distinguishable molecules and extends the FCS setup to simultaneous recording of both detection channels; autocorrelation in the individual color channels and additionally the cross-correlation between the two channels. The cross-correlation amplitude is directly proportional to the fraction of codiffusing particles carrying both dyes simultaneously through the detection volume. Thus, a positive cross-correlation amplitude is indicative of molecular interactions above a background of noninteracting particles, even if they show similar mobility (Heinze, Rarbach, Jahnz, & Schwille, 2002; Kohl, Haustein, & Schwille, 2005; Kohl, Heinze, Kuhlemann, Koltermann, & Schwille, 2002; Koltermann, Kettling, Bieschke, Winkler, & Eigen, 1998; Rigler et al., 1998; Rippe, 2000; Schwille et al., 1997). In this chapter, we describe the theoretical framework and experimental procedure of how to detect and quantify binding reactions with FCCS in free solution or in cells using a conventional multichannel confocal microscope and continuous wave laser excitation.

2. FCS IN A NUTSHELL

From a technical perspective, FCS developed its full potential through the use of fiber-coupled avalanche photodiodes (APDs) as single-photon counting detectors in combination with high-numerical aperture objectives in a confocal setup (Rigler et al., 1993). The main features are sketched in Fig. 3.1. Laser sources are combined and reflected by a dichroic mirror (DM1) onto the back aperture of the objective and focused in solution. A software-controlled scanning unit in between allows for prior imaging and positioning of the focus within the sample (Wachsmuth et al., 2003). Usually, water immersion objectives are used to minimize focal distortions for larger penetration depths in aqueous media (10–200 μm). The red-shifted fluorescence is collected by the same optics, passes the first dichroic, and is fed by a second dichroic mirror (DM2) into two subsequent color channels. Lenses project the beam onto confocal pinholes as well as the sensitive detection area of the APDs. The main working principle of a confocal setup is that the pinholes reject out-of-focus fluorescence along the optical axis z. Therefore, proper pinhole alignment is crucial to reaching single-molecule sensitivity with FCS.

For measurements, the fluorescence signal $F(t)$ is computed by either a hardware correlator in real time or a correlator software postmeasurement to derive the normalized second-order autocorrelation function (ACF; Fig. 3.2A and B)

$$G(\tau) = \frac{\langle F(t)\cdot F(t+\tau)\rangle}{\langle F\rangle^2} - 1 = \frac{\langle \delta F(t)\cdot \delta F(t+\tau)\rangle}{\langle F\rangle^2} \qquad [3.1]$$

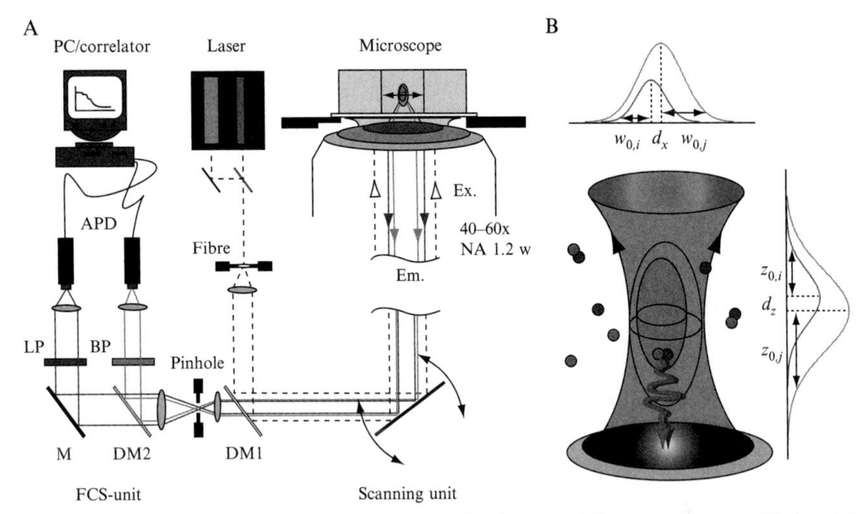

Figure 3.1 Optical setup as used in this chapter for the eGFP/Alexa647 dye pair. (A) Coaxial laser beams (488, 633 nm) are reflected by a dichroic mirror (DM1, 488/633 nm), guided to the back aperture of the objective and focused into the sample. From there the emitted fluorescence is collected, passes DM1 and the detection pinhole (70 μm) after which a second dichroic (DM2, transmission >635 nm) distributes the light into the two APD terminated detection channels. A band pass (BP, transmission 505–560 nm) and a long pass (LP, transmission >655 nm) filter were used to further narrow the spectral range of the fluorescence signal. (B) Gaussian-shaped intensity distributions of the superimposed foci. The centers exhibit a certain width and displacement for each color. (See Color Insert.)

The second equation can be easily verified by expressing the signal as temporal deviations from the mean $F(t) = \langle F(t) \rangle + \delta F$. Angular brackets indicate time averaging. $G(\tau)$ is an empirical, dimensionless function which contains no information about the underlying physical nature of the signal fluctuations. Therefore, an analytical expression is needed, in which the expected parameters are modeled and implemented. For example, 3D diffusion with triplet-blinking is described by the widely used function (Widengren, Mets, & Rigler, 1995)

$$\begin{cases} G(\tau) = \dfrac{1}{cV_{\text{eff}}} G^{\text{T}}(\tau) G^{\text{D}}(\tau), \\ G^{\text{T}}(\tau) = \left(\dfrac{1 - f_{\text{nf}} + f_{\text{nf}} \mathrm{e}^{-\tau/\tau_{\text{nf}}}}{1 - f_{\text{nf}}} \right), \\ G^{\text{D}}(\tau) = \sum_i f_i \left(1 + \dfrac{\tau}{\tau_{\text{diff},i}} \right)^{-1} \left(1 + \dfrac{\tau}{S^2 \tau_{\text{diff},i}} \right)^{-1/2}, \end{cases} \quad [3.2]$$

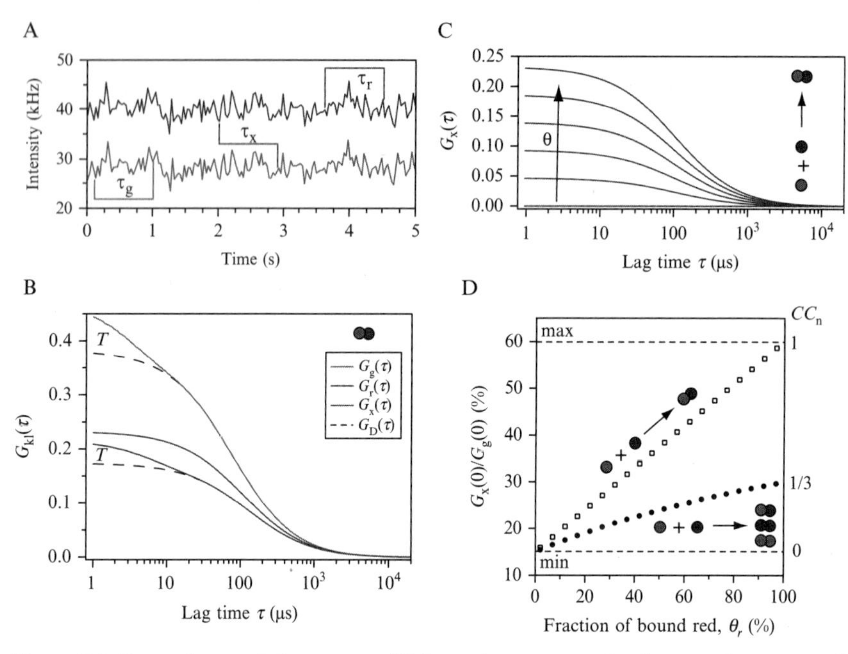

Figure 3.2 Example correlation curves. (A) Intensity traces as simultaneously measured in the green and the red channel. For correlation, pairs of values separated by a lag time τ are multiplied and normalized with the square of the mean intensity. (B) Numerical correlation functions for a hypothetical double-labeled standard ($D = 200\ \mu m^2/s$, $c = 10\ nM$, $\tau_{trip,g} = 3\ \mu s$, and $\tau_{trip,r} = 6\ \mu s$, $f_{nf} = 20\%$) with a typical focal geometry ($w_{0,g} = 0.25\ \mu m$, $w_{0,r} = 0.325\ \mu m$, $S = 5$). The CCF contains displacement of the centers by 20% of the 488 nm $1/e^2$ radii in all dimensions. Dashed curves show the purely diffusion related decay. (C) Increase of the cross-correlation amplitude with increasing fractions of ligand bound θ (heterotypic binding, no cross talk). (D) Progression of the ratio between cross- and autocorrelations for a heterotypic (open squares) and homotypic (dots) dimerization. In the latter case the curve is nonlinear and depressed due to double-labeled, single-colored species. (See Color Insert.)

where f_{nf} and τ_{nf} denote the fraction and time constant of the triplet-dependent decay, c the total concentration, $f_i = c_i/c$ the molar fractions, and $\tau_{diff} = w_0^2/4D$ the average dwell times of the fluorescent particles of type i (<3) within the effective observation volume V_{eff}. To derive Eq. (3.2), the excitation intensity distribution was assumed to be 3D Gaussian and stretched along the optical axis by the structure parameter $S = z_0/\omega_0$ (Fig. 3.1B). Due to the diffraction limit, S cannot be smaller than 3; however, values between 4 and 6 are assumed to be realistic. Because the structure parameter S is only weakly defined in correlation functions, and trades with diffusion times when fitting, it should be kept fixed during the experimental session.

Numerical examples for ACFs are shown in Fig. 3.2B. The decaying correlation function $G(\tau)$ is usually plotted versus the lag time in logarithmic scale which leads to a sigmoid shape. According to Eq. (3.2), the intercept

$$G(0) = \frac{1}{(1-f_{nf})cV_{eff}} = \frac{1}{(1-f_{nf})N} \quad [3.3]$$

is apart from a blinking-related increase $T = 1/(1-f_{nf})$, the inverse average number of observed molecules N. Although the exact diffusion times are returned by the fit, visual inspection allows to estimate their time scale on the x-axis at the position where the decay has decreased to 50% (mathematically exact only for infinite S). As illustrated in Fig. 3.1B, the illuminated open volume can be sketched as an ellipsoid with half axes corresponding to the $1/e^2$ radii of the Gaussian intensity profiles. Note that the focal volume contains no finite boundaries; the $1/e^2$ radii w_0 and z_0 are a direct result of integrating the Gaussian intensity profile to infinity. Diffusion coefficients and detection volume must be calibrated with a standard in separate measurements under the same conditions, for example, a small hydrophilic dye for which the diffusion coefficients are known. Then N_{ob} and τ_{diff} can be converted into absolute concentrations and diffusion coefficients:

$$\begin{cases} c = \dfrac{N_{ob}}{V_{eff}} = \dfrac{N_{ob}}{S\left(4\pi D_{st}\tau_{diff,st}\right)^{3/2}}, \\ D = \dfrac{D_{st}\tau_{diff,\,st}}{\tau_{diff}}. \end{cases} \quad [3.4]$$

Note that the standard (index "st") must be measured at the same conditions (temperature, solvent) for which the diffusion coefficient has been reported for, while the microscopic setup (type of cover glass, optical alignment, and the structure parameter used for fitting) should closely match the experimental conditions.

3. THE DUAL-COLOR FCCS EXTENSION

The formalism for FCCS is analogous (Schwille et al., 1997). The correlator device computes the normalized correlation function from the fluorescence intensities $F_j(t)$ recorded at the individual detectors

$$\begin{cases} G_j(\tau) = \dfrac{\langle \delta F_j(t)\delta F_j(t+\tau)\rangle}{\langle F_j\rangle^2}, \\ G_\times(\tau) = \dfrac{\langle \delta F_g(t)\delta F_r(t+\tau)\rangle}{\langle F_g\rangle\langle F_r\rangle}. \end{cases} \quad [3.5]$$

The simultaneously measured signals are autocorrelated in each channel ($j=$g for green or $j=$r for red) or cross-correlated between the two channels ($\times$) (Fig. 3.2A). For perfectly separated detection channels, the numerator in Eq. (3.5) will be a positive finite number only when both fluorophores diffuse collectively through the observation volume. Therefore, increasing fractions of double-labeled complexes in solution will lead to a successive increase of the cross-correlation amplitude (Fig. 3.2C). Like ACFs, the amplitude of the cross-correlation function (CCF) is inversely proportional to the absolute number of dual-colored particles (see Eqs. 3.6 and 3.8). However, in binding reactions, where also monomers are present, this can be ambiguous. For instance, the cross-correlation amplitude may decrease because of dissociation of the complexes into monomers or because of an increase in the absolute concentration of complexes—based on the CCF alone, this cannot be distinguished.

For this reason, the ratio of the cross- and the autocorrelation amplitude was introduced as a useful observable to monitor binding reactions (Rippe, 2000). To see how this works, let us assume an identical observation volume ($V_{\mathrm{eff,g}} = V_{\mathrm{eff,r}} = V_{\mathrm{eff},\times}$) for all three channels and write the correlation amplitude as the sum of the individual molecular contributions (Rippe, 2000; Ruan & Tetin, 2008; Weidemann et al., 2002).

$$\begin{cases} \hat{G}_j^{\mathrm{D}}(0) = \dfrac{\sum_i N_i \eta_{i,j}^2}{\left(\sum_i N_i \eta_{i,j}\right)^2}, \\ \hat{G}_\times(0) = \dfrac{\sum_i N_i \eta_{i,\mathrm{g}} \eta_{i,\mathrm{r}}}{\left(\sum_i N_i \eta_{i,\mathrm{g}}\right)\left(\sum_i N_i \eta_{i,\mathrm{r}}\right)}. \end{cases} \quad [3.6]$$

The background and cross talk free correlation amplitudes (circumflexed) depend on the number of contributing particles N weighted with their characteristic molecular brightness η_j in the respective color channel (j=g or r). Considering a mixture of single (N_R and N_G) and double-labeled (N_{GR}) particles (i=R, G, or GR) and assuming identical molecular brightness (which could be adjusted by varying the excitation power for the individual channels), the autocorrelation in the red channel is governed solely by red particles

$$\hat{G}_r^D(0) = \frac{N_R + N_{GR}}{(N_R + N_{GR})^2} = \frac{1}{(N_R + N_{GR})} = \frac{1}{N_{R,\mathrm{total}}} \quad [3.7]$$

while the cross-correlation amplitude will depend on the abundance of all three species

$$\hat{G}_\times(0) = \frac{N_{GR}}{(N_R + N_{GR})(N_G + N_{GR})}. \quad [3.8]$$

Calculating the ratio

$$\frac{\hat{G}_\times(0)}{\hat{G}_r^D(0)} = \frac{N_{GR}(N_R + N_{GR})}{(N_R + N_{GR})(N_G + N_{GR})} = \frac{N_{GR}}{(N_G + N_{GR})} = \frac{N_{GR}}{N_{G,\mathrm{total}}}, \quad [3.9]$$

then extracts the fraction of double-labeled species in the mixture. However, the ratio signifies the fraction of bound particles with respect to the other color, corresponding to the labeled species not used for normalization (in this case green). Equation (3.6) is the basis to calculate the possible outcome for any kind of mixture and thus the starting point to discriminate different binding models. For example, the maximum achievable cross-correlation for a homotypic dimerization is only 1/3 as compared to the heterotypic case (Fig. 3.2D; Eq. 3.19).

In experimental practice, applying Eqs. (3.6)–(3.9) is not straightforward. A complication in dual-color FCCS is that the effective observation volumes for the individual colors as well as for the cross-correlation channel are different in shape and size, and must each be calibrated with an appropriate standard. The size of the focal volumes scales with the excitation wavelength. Moreover, the centers of the two observation volumes are usually displaced due to chromatic aberrations of the objective. For a solution of 100% double-labeled dimers, the molecular brightness in Eq. (3.6) cancel and the amplitude is governed by the displacement d leading to an additional exponential factor in the model function for the cross-correlation curve (Weidemann et al., 2002)

$$\begin{cases} G_\times(\tau) = G_{\text{eff}}^{\text{D}}(\tau)\exp\left(-\dfrac{d_x^2+d_y^2}{4D\tau+\omega_{0,\text{eff}}^2}-\dfrac{d_z^2}{4D\tau+z_{0,\text{eff}}^2}\right) \\ G_\times(0) = \dfrac{1}{cV_{\text{eff},\times}}\exp\left(-\dfrac{d_x^2+d_y^2}{\omega_{0,\text{eff}}^2}-\dfrac{d_z^2}{z_{0,\text{eff}}^2}\right) = \dfrac{d}{cV_{\text{eff},\times}} \end{cases} \quad [3.10]$$

with $\omega_{0,\text{eff}}^2 = (\omega_{0,\text{g}}^2+\omega_{0,\text{r}}^2)/2$, $z_{0,\text{eff}}^2 = (z_{0,\text{g}}^2+z_{0,\text{r}}^2)/2$, and c denoting their concentration. Because dark states of individual fluorophores are statistically independent, the CCF contains no triplet or blinking fractions. With a larger displacement the maximum achievable cross-correlation amplitude decreases. In addition, the diffusion times of the double-labeled species appear slightly increased in cross-correlation as compared to the autocorrelation functions. Apart from a recent publication (Stromqvist et al., 2011), Eq. (3.10) was rarely used for fitting. Instead, the effective maximum ratio is usually assessed experimentally with a 100% dual-colored positive control (Fig. 3.2D) and Eq. (3.2) is applied to both auto and cross-correlation.

4. DATA ANALYSIS

Fitting correlation curves with Eq. (3.2) yields inverse correlation amplitudes, and therefore apparent particle numbers, which under ideal conditions could be directly converted into true concentrations. However, the detectors, the optical setup, and the medium in which the molecules are observed pose inherent limits. In addition, one has to separate photophysical from binding-related contributions. Therefore, evaluation of FCCS measurements requires careful postprocessing of the data. Instrumental nonidealities such as background intensity, cross talk into the "wrong" color channel, and quantum yields have significant impact on the data and should be treated before concluding on different binding models and stoichiometries.

4.1. Noncorrelated background

The signal recorded at the detector is a sum of fluorescent and nonfluorescent contributions. Instrumental properties like shot noise of the APDs, excitation stray light, and insufficiently shaded ambient light lead to background signal at each detector. In addition, solutions and cells may contain scattering or weakly fluorescent components. Cells in particular synthesize a large number of aromatic metabolites. Some enzymes contain fluorescent prosthetic groups like, for example, porphyrins in hemoglobin or flavin in flavoproteins. Flavoproteins were even suggested to produce correlating fluorescence in the wavelength range between 550 and 600 nm (Brock, Hink, & Jovin, 1998). However, in most of the experimental cases, cells exhibit an increased, noncorrelating background intensity which extenuates for red-shifted excitation in the visible range.

Background intensity, if determined in separate measurements, can be used to correct amplitudes. Considering the total signal I_{tot} to be composed of correlated signal fluctuations F and uncorrelated background I_{bg}, we can extend Eq. (3.2) with a correction factor (Schwille et al., 1997; Weidemann et al., 2002)

$$\begin{cases} G_j^{\mathrm{D}}(0) = \dfrac{1}{cV_{\mathrm{eff},j}} \left(\dfrac{\langle I_{\mathrm{tot},j} \rangle}{\langle I_{\mathrm{tot},j} \rangle - \langle I_{\mathrm{bg},j} \rangle} \right)^{-2}, \\ G_{\times}(0) = \dfrac{d}{cV_{\mathrm{eff},\times}} \left(\dfrac{\langle I_{\mathrm{tot,g}} \rangle}{\langle I_{\mathrm{tot,g}} \rangle - \langle I_{\mathrm{bg,g}} \rangle} \right)^{-1} \left(\dfrac{\langle I_{\mathrm{tot,r}} \rangle}{\langle I_{\mathrm{tot,r}} \rangle - \langle I_{\mathrm{bg,r}} \rangle} \right)^{-1}. \end{cases} \quad [3.11]$$

Noncorrelating background contributions rescale the correlation amplitudes and therefore particle numbers derived from ACF and CCF functions. An example is shown in Fig. 3.3. A stock solution of Alexa488 (A488) in 10 mM Tris (pH 8) was used to produce a fourfold dilution series (1-fold, 4-fold, 16-fold, and 64-fold). The four A488 solutions plus buffer were measured at constant excitation laser power producing about 10 kHz counts per particle (CPP) in the absence and presence of ambient daylight. The intensity trace and correlation curves for the 16-fold dilution are shown in Fig. 3.3A and B, respectively. The gap between total signal and background (300 Hz or 22.4 kHz) is maintained; however, the fluctuation amplitude, that is, the relative changes of the intensity, is decreased due to the significant background contributions. The corresponding correlation curves show that background leads to a dramatic depression of the intercept and thus particle numbers are overestimated. However, applying Eq. (3.11) perfectly restores the hyperbolic progression of the dilution series for both cases (Fig. 3.3C). In summary, a 2-min measurement at a signal-to-noise ratio of only 1/25 revealed that the 64-fold dilution contained in average about 0.13 particles in the observation volume ($V_{\mathrm{eff}} = 0.33$ femtoliter → 1.6 nM).

4.2. Spectral cross talk

Cross talk into the "wrong" detection channel produces a false-positive CCF amplitude even in the absence of interactions and reduces the accessible measurement window in binding experiments (Fig. 3.2D). In particular,

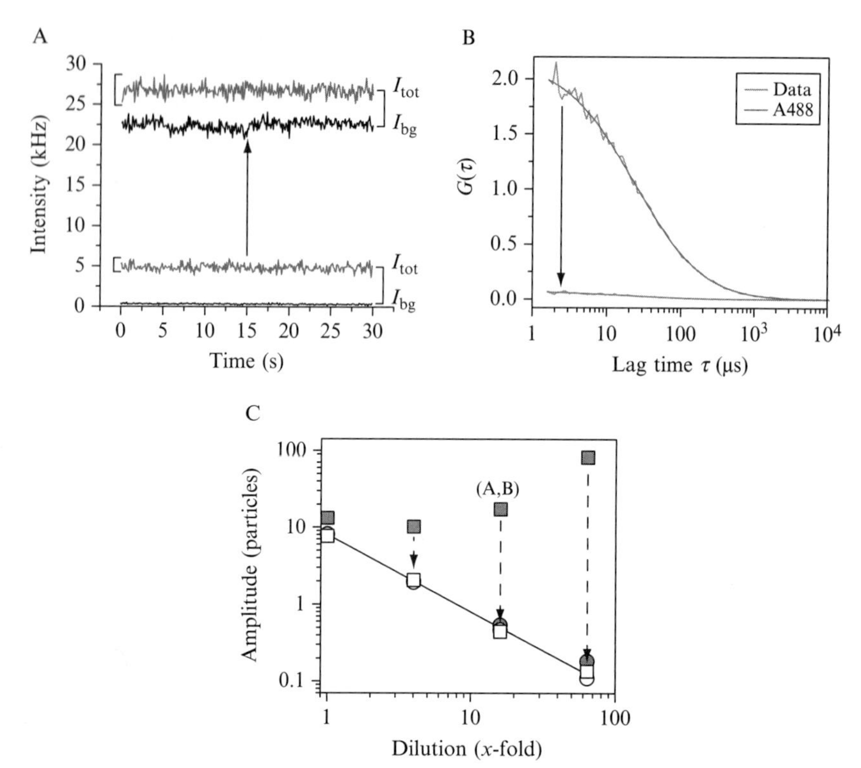

Figure 3.3 Noncorrelating background. (A) Intensity traces obtained in buffer (black) and a Alexa488 solution (green) with the detectors either shaded or exposed to ambient day light (arrow). Note that background slightly adds the absolute fluctuation amplitude. However, due to the overall higher mean intensity (I_{tot}), the relative amplitude decreases dramatically (open brackets). (B) Corresponding autocorrelation functions. The lower relative amplitude corresponds to a smaller intercept of the ACF (arrow) (C) Apparent and corrected particle numbers (Eq. 3.11) for a fourfold dilution series (1/1, 1/4, 1/16, and 1/64). The data points corresponding to (A) and (B) are indicated. (For interpretation of the references to color in this figure legend, the reader is referred to the online version of this chapter.)

cross talk from the green into the red channel was one of the major issues in FCCS for life cell applications. Therefore, cross talk considerations are an important starting point for experimental planning to select fluorophores matching the instrument settings.

One successful strategy to prevent cross talk is to use organic dyes emitting in the far red (>650 nm), like Cy5 or Alexa647 (A647) in combination with enhanced green fluorescent protein (eGFP; Ohrt et al., 2011; Weidemann et al., 2011). However, this approach seems inefficient for large scale studies of intracellular protein–protein interaction networks, since the recombinant protein must be produced, site specifically labeled, and purified. Additionally, life cell applications rely on efficient delivery strategies to bring the labeled protein back into the cell. Cell-penetrating techniques like injection with glass capillaries or detergent-mediated membrane permeabilization are rather harsh, nonphysiological treatments. From this point of view, using genetically encoded dual-color tagging strategies seem advantageous. However, GFPs exhibit a pronounced red-shifted tail in the emission spectrum. For many years, suitable red fluorescent proteins (RFPs) for FCCS applications did not exist. Although a handful of useful candidates like mRFP or mCherry are now in use, most of these red-shifted variants show unfortunate properties like low-quantum yield, strong blinking, photo-instability, slow-maturation kinetics, and a tendency to oligomerize (Baudendistel, Müller, Waldeck, Angel, & Langowski, 2005; Foo, Naredi-Rainer, Lamb, Ahmed, & Wohland, 2012; Maeder et al., 2007; Slaughter, Schwartz, & Li, 2007; Worch, Bökel, Höfinger, Schwille, & Weidemann, 2010).

However, cross talk can be rigorously treated. In the very first paper on FCCS, Riçka and Binkert developed a mathematical formalism for the dual-color FCCS extension and showed that the fluorescence signal as well as the correlation functions (ACF and CCF) can be described as linear combinations of the true, cross talk free contributions weighted with factors describing the degree of cross talk between them (Riçka & Binkert, 1989). The weighing factors can be noted as elements in so-called color matrices (Riçka & Binkert, 1989; Weidemann et al., 2002). Recently, Bacia et al. evaluated these matrices very thoroughly for a comprehensive back-correction of cross talk (Bacia, Petrášek, & Schwille, 2012). Using these results, we show how cross talk affects the measured ratio between CCF and ACF amplitudes. To translate this formalism into a realistic scenario, we first have to relate the particle numbers from Eqs. (3.7)–(3.9) to true concentrations by introducing effective focal volumes for each individual channel

$$\begin{cases} \hat{G}_{g}^{D}(0) = \dfrac{1}{c_{G} V_{\mathrm{eff},g}}, \\ \hat{G}_{r}^{D}(0) = \dfrac{1}{c_{R} V_{\mathrm{eff},r}}, \\ \hat{G}_{\times}(0) = \dfrac{c_{GR}}{c_{G} c_{R}} \dfrac{d}{V_{\mathrm{eff},\times}}. \end{cases} \quad [3.12]$$

The circumflexed correlation functions imply that it is already corrected for background and cross talk, and therefore reflects the true number of particles in the respective detection volume V_{eff}. The concentrations of diffusing particles are now indexed according to color also to include single-labeled species as one would encounter with a dynamic heterotypic binding reaction (Eq. 3.6). We further denote the ratio between CCF and ACF amplitudes by CC_j (j = g or r), indexed by the color channel of the ACF. In addition, cross talk from the green into the red channel κ was varied while the opposite cross talk from the red into the green channel was neglected. To link the FCCS formalism to the language of biochemistry, let us name the green-labeled particles "ligands" ($c_G = L_0$) and the red particles "receptors" ($c_R = R_0$); the choice of color is without loss of generality. Finally, evaluating the color matrix, one can see with

$$\begin{cases} CC_g \equiv \dfrac{G_\times(0)}{G_g^D(0)} = \kappa f + (1-\kappa f)\dfrac{V_{\text{eff,g}}}{V_{\text{eff},\times}}\dfrac{d\,L_b}{R_0} \\ f = \dfrac{F_g}{F_r} = \dfrac{\eta_g L_0}{\eta_r R_0 + \kappa \eta_g L_0} \end{cases} \quad [3.13]$$

that CC_g depends still linearly on the fraction of bound green ligands ($c_{GR} = L_b$) with respect to the total concentration of red receptors R_0, that is, the "fraction of receptor occupied." Here, f denotes the ratio of the measured background-corrected intensities of the green and red detection channels, each comprised of the ligand and receptor concentrations scaled with their molecular brightness η_j plus cross talk.

Some particular examples are shown in Fig. 3.4. Since the molecular brightness could be adjusted by varying the excitation power, we assumed identical molecular brightness ($\eta_g = \eta_r = 1$). When using the 488 and 633 nm laser lines, detection volume in the red channel is about 1.5-fold enlarged ($V_{\text{eff,r}}/V_{\text{eff,g}} = 1.5$), and the effective overlap for cross-correlation is reduced to 50% (Fig. 3.4A). Note that cross talk in combination with chromatic mismatch affects not only the minimum but also the maximum reached with a perfectly double-labeled control (Fig. 3.4B). However, normalizing by the ACF amplitude of the red, cross talk affected channel creates a more complicated dependence since these contributions do not fully cancel.

$$CC_r \equiv \frac{G_\times(0)}{G_r^D(0)} = \frac{\kappa f \frac{V_{\text{eff,r}}}{V_{\text{eff,g}}}\frac{R_0}{L_0} + (1-\kappa f)\frac{V_{\text{eff,r}}}{V_{\text{eff},\times}}\frac{d\,L_b}{L_0}}{(\kappa f)^2 \frac{V_{\text{eff,r}}}{V_{\text{eff,g}}}\frac{R_0}{L_0} + (1-\kappa f)^2 + 2\kappa f(1-\kappa f)\frac{V_{\text{eff,r}}}{V_{\text{eff},\times}}\frac{d\,L_b}{L_0}}. \quad [3.14]$$

The cross talk-induced shifts are now more pronounced, and for larger cross talk values, even the nonlinear shape becomes apparent (Fig. 3.4C).

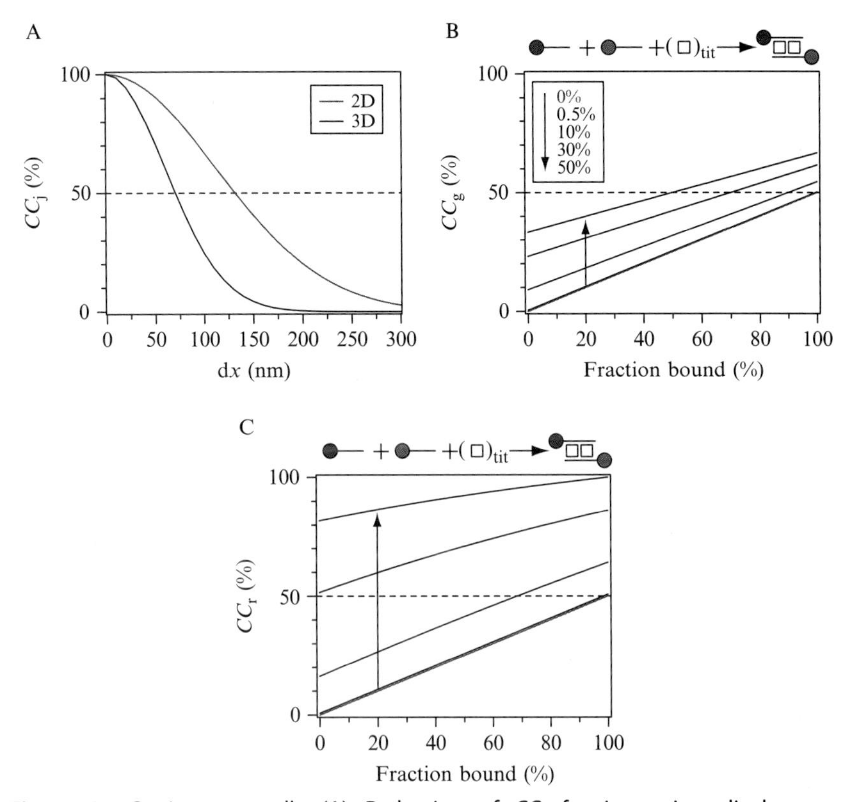

Figure 3.4 Static cross talk. (A) Reduction of CC_j for increasing displacement (dx) (Eq. 3.10) and a typical observation volume with $w_0 = 224$ nm for the two-dimensional (e.g., diffusion in a horizontal membrane, red) and the three-dimensional case (e.g., diffusion in the cytoplasm, black). Dimensions and displacement along the optical axis z were scaled with the structure parameter ($S = 5$). (B) Effect of successive cross talk from the green into the red channel (legend, arrows) for a static binding experiment in which the concentrations c_i and molecular brightness η_j of labeled ligands are constant (here $c_g = c_r$ and $\eta_g = \eta_r = 1$), while a nonfluorescent reaction partner (open square) is titrated. The examples assumes a chromatic displacement of 50%. (C) Same plot as in (B) but now the red, cross talk affected channel was used for normalization. The predicted progression of CC_j with 0% cross talk is shown for reference ((B) and (C), red). (See Color Insert.)

Also, significant cross talk overrides the chromatic mismatch which tends to lower the CCF amplitudes. Therefore, if one wants to keep the FCCS readout linear, CC_g should be consistently used and the labels chosen in such a way that CC_g reflects the desired fraction. This approach was suggested some time ago and successfully followed up in biochemical studies, where the minimum and the maximum were determined experimentally with a mixture of noninteracting and a solution of perfectly complexed binding partners (Rippe, 2000; Weidemann et al., 2002). Note that in the case of homotypic interactions also, double green and double red particles are formed which change the shape of the binding curve (Fig. 3.2D).

Titrating a nonlabeled subunit of a ternary complex is a convenient approach for FCCS analysis, however, from the perspective of biochemistry, a quite-specific situation (Fig. 3.4, reaction schemes). The simplest binding reaction imaginable is a heterotypic dimerization (Fig. 3.5, reaction schemes). A common strategy to determine the affinity between two binding partners is to titrate one of them and measure the "fractions bound." In FCCS, such titrations can only span a limited concentration range because a low abundance of one of the two colors renders the FCCS data very uncertain. This problem is especially aggravated in the presence of cross talk. To study this issue, we plotted binding isotherms for the case in which the ligand is titrated across a fixed receptor concentration for different dissociation constants K_d. We chose $K_d = 10$ nM (e.g., typical for antibodies) and a receptor at 50 nM, a concentration well suited for FCS analysis (Fig. 3.5A, red curves). Since the ligand is titrated and the receptor concentration is fixed, binding curves look differently depending on which color combination was assigned to the ligand and the receptor and the two ratio (CC_g and CC_r) now have different meaning. As can be seen in the absence of cross talk (Fig. 3.5B and C, red curves), the "fraction receptor occupied" is reflected by CC_g when titrating green ligands (Fig. 3.5B, left panel) and CC_r when titrating red ligands (Fig. 3.5C, right panel). However, cross talk affects these situations differently. It can be also seen that the "fraction of ligand bound" shows an overall weaker dependence on the total ligand concentration and is therefore a less well-suited variable to determine the K_d by fitting (Fig. 3.5B, right panel, and C, left panel). In summary, a scenario in which a green-labeled ligand is titrated and CC_g is used for the analysis seems to be the most stable experimental setup.

Finally, we want to note the equations which allow the experimenter to correct measured correlation amplitudes and intensities to obtain true fractions of bound particles in the case of green ligands and red receptors (in the opposite case simply be swap $L \leftrightarrow R$):

$$\frac{L_b}{R_0} = \frac{V_{\mathrm{eff},\times}}{V_{\mathrm{eff},g} d} \frac{CC_g - \kappa f}{(1 - \kappa f)} \qquad [3.15]$$

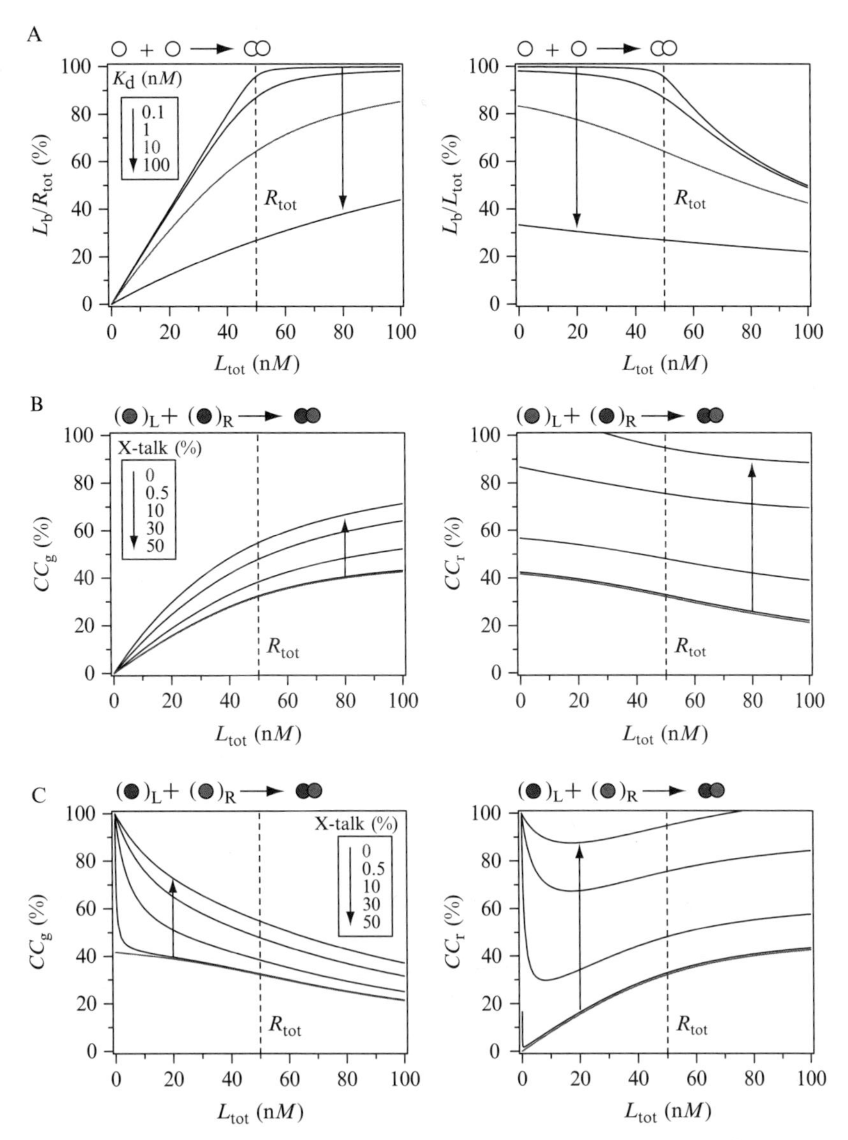

Figure 3.5 Titration-dependent cross talk. (A) Binding isotherms calculated for a heterotypic reaction. The "fraction receptor occupied" (left panel) and the "fraction ligand bound" (right panel) are plotted for increasing K_d-values (legend, arrows) and a fixed receptor concentration ($R_0 = 50$ nM, dashed vertical line). The 10 nM curve (red) was used for (B) and (C). (B) Titration with a green-labeled ligand. Effect of successive cross talk from the green into the red channel (legend, arrows) assuming a constant molecular brightness η_j of the labeled ligands ($\eta_g = \eta_r = 1$), a chromatic displacement of 50%, and a 1.5-fold enlarged detection volume in the red channel. The predicted progression of CC_j with 0% cross talk is shown for reference (red). (C) Same as in (B), but titration with a red labeled ligand. (See Color Insert.)

and

$$\frac{L_b}{L_0} = \frac{V_{\mathrm{eff},\times}}{V_{\mathrm{eff},r}d} \frac{\mathrm{CC_r} - \kappa f(1-\kappa f)\frac{G_g^D(0)}{G_r^D(0)}}{1+(\kappa f)^2\frac{G_g^D(0)}{G_r^D(0)} - 2\kappa f \mathrm{CC_r}} \quad [3.16]$$

which on the right hand side contain only experimentally accessible quantities. Equations (3.15) and (3.16) also reveal how the ratio of the detection volumes can be determined in a separate control measurement with a 100% cross-correlating dimer, where $L_b/R_0 = L_b/L_0 = 1$.

4.3. Multiple binding sites

The correlation functions build up from individual molecular fractions weighted with their molecular brightness (Eq. 3.6). Therefore, characterizing a heterogeneous mixture of reactants can be a daunting task. Even in a simple heterotypic dimerization, the quantum yield of the fluorophore may be changed in the bound state. While such a complication can be treated by characterizing the individual molecular species in separate control measurements, this approach is not always feasible (Hwang & Wohland, 2005). Stochastic chemical labeling often involves multiple binding sites, for example, reactive amine groups or cysteines, which may be located in different chemical environments of the protein. Thus, such a labeled protein constitutes an ensemble of chemical species exhibiting different molecular brightness, which cannot be easily purified. In particular, when labeling proteins, a quantitative removal of nonconjugated dye, exhibiting different molecular brightness, may be difficult to achieve. A similar situation is encountered in dynamic binding reactions with higher-order stoichiometries. When multiple copies of red and green monomers dynamically form a complex, the molecular brightness is multiplied, and the mixture also contains reaction intermediates of variable stoichiometry, which are not accessible in isolation.

Measuring heterogeneous mixtures with FCS, the $1/N$ dependence of the ACF is changed. Since a broadening of molecular brightness distribution increases the relative fluctuation amplitude, the concentrations are underestimated. Interestingly, quenching is equivalent to the same degree of increase in quantum yield. For example, quenching one component of an equimolar mixture by $\eta = 0.5$ or enhanced quantum yield by $\eta = 2$ increases the amplitude by 10%. However, stronger effects are frequently encountered in binding assays, especially in the case of oligomerization. In addition, the fractions of particles ($i = 1$–3) with different mobility f_i, as obtained from fitting Eq. (3.2), are biased in favor of the brighter component. This can be adjusted by (Thompson, 1991)

$$\begin{cases} \hat{f}_i = \dfrac{c_i}{\sum_i c_i} = \dfrac{f_i/\eta_i^2}{\sum_i f_i/\eta_i^2}, \\ \hat{f}_{i,\times} = \dfrac{c_i}{\sum_i c_i} = \dfrac{f_i/\eta_{i,\mathrm{g}}\eta_{i,\mathrm{r}}}{\sum_i f_i/\eta_{i,\mathrm{g}}\eta_{i,\mathrm{r}}}, \end{cases} \tag{3.17}$$

where circumflexed f are used to indicate the real, corrected molar fractions.

Statistical association of fluorophores at a defined number of equal binding sites per complex leads to a binomial distribution of similarly diffusing particles with different but discrete numbers of bound fluorophores. The effect on correlation functions is governed by an additional factor (Weidemann et al., 2003, 2002)

$$\begin{cases} G_j^{\mathrm{D}}(0) = \dfrac{1}{cV_{\mathrm{eff},j}}\left(1 - \dfrac{1}{n} + \dfrac{p_\mathrm{G}\eta_{\mathrm{G},j}^2 + p_\mathrm{R}\eta_{\mathrm{R},j}^2}{n\left(p_\mathrm{G}\eta_{\mathrm{G},j} + p_\mathrm{R}\eta_{\mathrm{R},j}\right)^2}\right) \\ G_\times(0) = \dfrac{d}{cV_{\mathrm{eff},\times}}\left(1 - \dfrac{1}{n} + \dfrac{\left(p_\mathrm{G}\eta_{\mathrm{G,g}}\eta_{\mathrm{G,r}} + p_\mathrm{R}\eta_{\mathrm{R,g}}\eta_{\mathrm{R,r}}\right)}{n\left(p_\mathrm{G}\eta_{\mathrm{G,g}} + p_\mathrm{R}\eta_{\mathrm{R,g}}\right)\left(p_\mathrm{G}\eta_{\mathrm{G,r}} + p_\mathrm{R}\eta_{\mathrm{R,r}}\right)}\right) \end{cases} \tag{3.18}$$

depending on the number of binding sites per complex, and the probability p_G and p_R that such a binding site carries the respective type of green (G) or red (R) fluorophore. Of course, the correlation amplitude also depends on the molecular brightness of the individual fluorophores in the respective detection channels η, which are now assigned to the type of fluorophore, not to the individual diffusing complexes. Assuming negligible cross talk from the red into the green channel ($\eta_{\mathrm{R,g}} = 0$) and a defined cross talk from the green into the red channel ($\eta_{\mathrm{G,r}} = \kappa\eta_\mathrm{g}$), Eq. (3.18) turns into

$$\begin{aligned} G_\mathrm{g}^{\mathrm{D}}(0) &= \frac{1}{cV_{\mathrm{eff,g}}}\left(1 - \frac{1}{n} + \frac{1}{np_\mathrm{G}}\right), \\ G_\mathrm{r}^{\mathrm{D}}(0) &= \frac{1}{cV_{\mathrm{eff,r}}}\left(1 - \frac{1}{n} + \frac{\left(p_\mathrm{G}\kappa^2\eta_\mathrm{g}^2 + p_\mathrm{R}\eta_\mathrm{r}^2\right)}{n\left(p_\mathrm{G}\kappa\eta_\mathrm{g} + p_\mathrm{R}\eta_\mathrm{r}\right)^2}\right), \\ G_\times(0) &= \frac{d}{cV_{\mathrm{eff},\times}}\left(1 - \frac{1}{n} + \frac{\left(p_\mathrm{G}\eta_\mathrm{g}\eta_\mathrm{r}\right)}{n\left(p_\mathrm{G}\eta_\mathrm{g}\right)\left(p_\mathrm{G}\kappa\eta_\mathrm{g} + p_\mathrm{R}\eta_\mathrm{r}\right)}\right). \end{aligned} \tag{3.19}$$

To understand how the binomial distribution at multiple binding sites changes the correlation functions, it may be elusive to discuss Eq. (3.19) for $\kappa = 0$, then even the molecular brightness values fully cancel

$$\begin{cases} G_j^{\mathrm{D}}(0) = \dfrac{1}{cV_{\mathrm{eff},j}}\left(1 - \dfrac{1}{n} + \dfrac{1}{np_i}\right), \\ G_\times(0) = \dfrac{d}{cV_{\mathrm{eff},\times}}\left(1 - \dfrac{1}{n}\right). \end{cases} \quad [3.20]$$

As mentioned above, statistical labeling increases the autocorrelation amplitudes in comparison with a uniformly labeled standard of the same concentration c. The factor depends on the probability p_i, with which each of the n binding sites at the diffusing entity (labeled polymer, oligomer) carries a fluorophore of type i giving signal for channel j (e.g., $i = \mathrm{G} \rightarrow j = \mathrm{g}$). For $p_i = 1$, the factor is unity and returns Eq. (3.2), and for $n = 1$, the amplitude scales with $1/p_i$ and reflects the reciprocal linear relationship with the total concentration of labeled particles. For $p_i < 1$ and $n > 1$, the increase is governed by the binomial distribution; for example, two binding sites occupied by fluorophores with a probability of 50% will increase the amplitude by 3/2. Note that this formalism fully includes the distributed labels but assumes identical molecular brightness at the binding sites for each type of fluorophore.

For cross-correlation, the dependence is even simpler and one may wonder why the CCF amplitudes solely depend on the number of binding sites n. This is a direct result of our assumption that cross talk is negligible which makes the third term in Eq. (3.20) cancel. For $n = 1$, the correction factor is zero because in the absence of cross talk a single binding site does not allow double-labeled particles. For $n > 1$, the CCF amplitude successively increases toward the limit where the factor is unity and Eq. (3.2) is restored. This dependence suggests an alternative way to establish a 100% cross-correlating standard to characterize chromatic mismatch. One simply has to label polymer containing a sufficient large number of binding sites with both colors. For example, irrespective of the degree of labeling, a dual-color labeled protein with 20 accessible binding sites (e.g., primary amines), will show 95% cross-correlation with respect to a 100% dual-color labeled dimer. In turn, Eqs. (3.18)–(3.20) provide access to the number of accessible binding sites of a protein or the degree of oligomerization of a formed complex and thus to relevant biochemical parameters (Kim, Heinze, Bacia, Waxham, & Schwille, 2005; Weidemann et al., 2002).

5. LIGAND BINDING AT THE CELL SURFACE

FCCS showed a promising potential in the field of signaling, where ligand binding at the cell surface trigger cascades of successive noncovalent interactions (Weidemann & Schwille, 2009). From a biophysical perspective, many signaling pathways vary a common theme. First, ligands bind to their cognate cell surface receptors, thereby inducing the recruitment of additional receptor subunits in the membrane. Receptor oligomerization is usually linked to the activation state of cytoplasmic kinase domains, which in turn establish binding sites for a second wave of downstream factors. At the end of the signaling cascade, cytoplasmic signal transducers are activated and transported to the nucleus where they act as transcription factors. In recent years, some of these interactions at the cell surface (Larson, Gosse, Holowka, Baird, & Webb, 2005; Liu et al., 2007; Neugart et al., 2009; Savatier, Jalaguier, Ferguson, Cavailles, & Royer, 2010; Stromqvist et al., 2011; Weidemann et al., 2011; Worch et al., 2010) as well as in the cytoplasm or the nucleus (Baudendistel et al., 2005; Kim et al., 2005; Kim, Heinze, Waxham, & Schwille, 2004; Maeder et al., 2007; Slaughter et al., 2007) were experimentally characterized by FCCS.

Figure 3.6 illustrates a typical example for an FCCS experiment addressing ligand–receptor interactions in the plasma membrane (Weidemann et al., 2011). The cytokine Interleukin-4 (IL-4) binds the single-pass transmembrane receptor IL-4 receptor α with subnanomolar affinity. We transiently expressed an eGFP-tagged and a nonfluorescent cytoplasmic deletion variant of the receptor for improved surface membrane partitioning. Genetic engineering of suitable constructs can be crucial for successful FCCS application in the plasma membrane since large fractions of receptor trapped within vesicles or endogenous membrane systems (Golgi, endoplasmic reticulum) can render FCS measurements impossible. Administration of the 15 n*M* of labeled IL-4-A647 then produced a high degree of surface binding and colocalization, which is a good argument for receptor binding (Fig. 3.6A). However, another nontagged receptor at the cell surface could also be recognized by the ligand. Since codiffusion can be measured by FCCS, the technique is useful to discriminate between these two options (Fig. 3.6B).

Before starting with cellular measurements, one first has to characterize the optical setup. Figure 3.6C shows amplitude-normalized ACFs of standard solutions of A488 and A647, as well as the labeled ligand IL-4-A647.

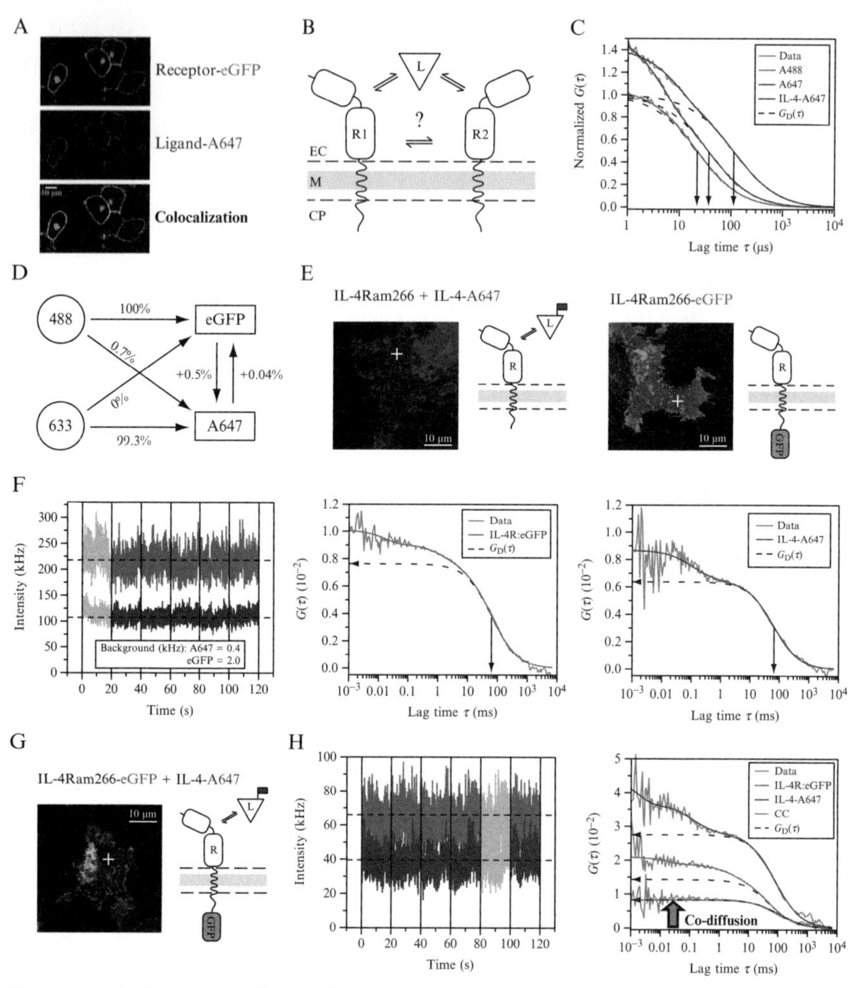

Figure 3.6 Experimental outline to quantify ligand–receptor interactions. (A) Colocalization of a Alexa647 mono-labeled Interleukin-4 (kindly donated by Walter Sebald) and a GFP-tagged IL-4 receptor construct (Weidemann et al., 2011). (B) Problems which can be addressed: (i) which of the receptors is bound by the ligand, (ii) which of the receptors is recruited into a ternary complex, and (iii) what are the associated affinity constants? (C) Characterization of the optical setup with diffusion standards A488 and A647 in free solution, as well as the free ligand. (D) Cross-excitation and cross talk scheme. (E) Confocal images of single-stained cells and (F) corresponding intensity traces measured at indicated positions (crosses). Runs not used for correlation are gray shaded. Arrows point out the slow, receptor related diffusion time and the associated amplitude (dashed line with arrow head). (G) Dual-color FCCS experiment. The positive CCF amplitude indicates codiffusion. (See Color Insert.)

Arrows indicate the diffusion times of A488 (22.5 μs), A647 (37.5 μs), and the ligand (118 μs). Both dyes have been measured at excitation powers producing about ~8 kHz per particle. Nevertheless, A647 shows a much more significant triplet fraction (~38%) compared to A488 (~11%). Using the diffusion coefficients $D_{A488}=400$ and $D_{A647}=330$ μm^2/s (Dertinger et al., 2007; Kapusta, 2010), Eqs. (3.2)–(3.4) returned a beam waist $2w_0=380$ and 446 nm and the corresponding detection volumes $V_{eff}=0.19\pm0.01$ and 0.31 ± 0.01 femtoliters ($n=3$ measurements) for the 488 and 633 nm channel, respectively. Thus, the linear dimensions of the focal volumes were scaled by 1.17, and thus by a significantly lower factor than the ratio of the excitation wavelengths (633/488) suggest. One has to bear in mind that the effective observation volume is a convolution of excitation and detection efficiency, which may explain this discrepancy. Having V_{eff} calibrated, one can calculate the diffusion coefficient of IL-4-A647 $D=105\pm6$ μm^2/s ($n=6$) according to Eq. (3.4), which is in good agreement with literature (Weidemann et al., 2011). More importantly, IL-4-A647 exhibited rather even intensity traces and therefore no signs of significant self-aggregation (not shown). The standard dyes also allow to characterize the color matrix. Simultaneous recording in both detection channels for each laser line and combined excitation resulted in the numbers as shown in Fig. 3.6D. The filter configuration as described in Fig. 3.1A in conjunction with the dye pair leaves 0.5% cross talk from the eGFP into the A647 channel and 0.04% cross talk vice versa. As shown in Figs. 3.4 and 3.5, such cross talk values have small effect and were therefore neglected. Note that it is not necessary to characterize cross-excitation in order to perform a cross talk correction according to Eqs. (3.15) and (3.16).

Although cross talk can be addressed with standard dyes, it is, nevertheless, advisable to perform cellular single-color experiments (Fig. 3.6E and F) and compare with dual-color experiments (Fig. 3.6G and H). Single-color experiments help to assign diffusion times and to properly quantify the noncorrelated background for a certain excitation power, since signal from the other channel can be used for accurate positioning. In addition, a nontagged receptor is useful to probe for recruitment of additional receptor subunits into a ternary complex (Weidemann et al., 2011). The confocal images display the bottom membrane, proximal to the glass surface as taken with APD detectors (Fig. 3.6E and G). In both channels, the stained membrane shows homogeneous regions intermitted by membrane ruffles of higher intensity. For stable FCS measurements, the focal volume was placed within the homogeneous regions, preferably 2 μm from the cell border.

We adjusted the excitation power in each channel such that the intensity trace showed no visible decay due to photobleaching (Fig. 3.6F and H). Surprisingly, IL-4-A647 was less photostable and was measured at CPP = 1.1 kHz while eGFP could be measured at 1.8 kHz. In the eGFP-channel, we obtained 2 kHz background and in the A647-channel 0.4 kHz, both values were small in comparison to the average fluorescence intensities. We took six runs for 20 s; some of the runs were discarded due to instabilities (gray shaded). Evaluating the particle numbers of the slow, membrane-related decay yielded 1700 bound IL-4-A647 molecules and 2500 eGFP receptors/μm^{-2} for each particular cell and therefore revealed a rather high-expression level. Note that such information cannot be derived from confocal imaging alone. The ACFs in each channel clearly show a biphasic decay (Fig. 3.6F and H). For the eGFP-tagged receptor, the fit contained a triplet fraction (17 μs), a fast component (1.5 ms), and a slow component related to membrane diffusion (63 ms). The IL-4-A647 ACF showed a triplet fraction (1 μs), a fast component (106 μs) and also the slow component related to membrane diffusion (62 ms). While the fast component in the A647 channel can be attributed to free ligand (see above), the ms-component of eGFP was less conclusive.

A closer look at the CCF (Fig. 3.6H) shows that the ms-component for eGFP is not cross-correlating. Therefore, it is not related to receptor diffusion and could either represent unusually long blinking times or residual eGFP-fragments freely diffusing in the cytoplasm, not accessible for the ligand (Weidemann et al., 2011). In contrast, the membrane-related component showed significant cross-correlation due to codiffusion. Taking w_0^2 as the two-dimensional observation area, we obtained similar diffusion coefficients in both color channels $D = 0.1 \pm 0.05$ $\mu m^2/s$ ($n = 5$) and, in agreement to Eq. (3.10), a slightly smaller diffusion coefficient in the cross-correlation channel $D = 0.08 \pm 0.008$ $\mu m^2/s$ ($n = 5$).

Evaluating the ACF amplitudes, we encountered a complication because the origin of the fast ms-fraction of eGFP is not yet clear. In the literature, it is common to treat eGFP blinking with the triplet formalism (Eq. 3.2). Accounting for triplet leads to lower amplitudes and therefore increased particle numbers (dashed line in Fig. 3.6H). Applying this approach to data for this particular cell, the receptor density is about fivefold larger as compared to bound ligand (2000 receptors vs. 400 ligands/μm^{-2}). Correspondingly, the ratio $CC_g = 60\%$ is significantly higher than $CC_r = 17\%$. Following the second scenario, the fast ms-fraction was treated as fluorescent particles in the cytoplasm, and the slow, membrane-related fraction was derived according to $N_2 = f_2 c V_{\text{eff}}$ (Eq. 3.2). Then, the green receptor density turns out lower but still more abundant (1100 receptors vs. 400 ligands/ μm^{-2}), and the cross-correlation ratio is $CC_g = 30\%$ versus $CC_r = 17\%$. Additional experiments, like comparison to a proper FCCS control, FCS in the cytoplasm, or varying focal volume FCS (Wawrezinieck, Rigneault, Marguet, & Lenne, 2005), could be used to gain further insight into the origin of the fast ms-fluctuations of IL-4Ram266-eGFP and to resolve the chromatic mismatch issue. In either case, the data clearly proof codiffusion between ligand and receptor although a considerable fraction of receptors at the cell surface were unoccupied.

6. CONCLUSION

In this chapter, we have reviewed the application of dual-color FCCS to characterize bimolecular interactions in free solution or living cells using a conventional confocal microscope equipped with continuous dual-color laser excitation and single-photon counting detectors. While the principle of the measurement is intuitive, to extract binding-related thermodynamic parameters requires a number of careful evaluation steps. To fit the data, the correlation functions have to be correctly modeled, such that correlation times are properly assigned to the underlying molecular processes. Amplitudes have to be corrected for noncorrelating background and cross talk. The chromatic mismatch issue has to be treated by separate measurements using a 100% cross-correlating control made of the same pair of fluorophores. After all these steps, one can link the ratio CC_j to different binding models for which we have explicitly evaluated the most simple case of statistical labeling and oligomerization. Finally, our example of ligand–receptor interaction illustrates a typical data set and discusses the issues encountered in more complex environments like the plasma membrane. In summary, we believe that dual-color FCCS bears great potential to quantitatively assess interactions between molecules in living cells and therefore is an asset for cell biology.

ACKNOWLEDGMENTS

We thank Walter Sebald (Biocenter, University of Würzburg, Germany) for kindly donating the mono-labeled IL-4-A647 and Paul Müller (TU Dresden, Germany) for critical reading of the manuscript. We want to mention that this review discusses the procedures encountered in FCCS rather than providing a comprehensive literature review. Being aware of our own bias, we apologize for not having mentioned many important, and in particular technical contributions to the field. This work was supported by a CRTD seed grant "Development of reporter constructs for the live imaging of Jak/Stat signaling."

REFERENCES

Bacia, K., Kim, S. A., & Schwille, P. (2006). Fluorescence cross-correlation spectroscopy in living cells. *Nature Methods, 3*, 83–89.

Bacia, K., Petrášek, Z., & Schwille, P. (2012). Correcting for spectral cross-talk in dual-color fluorescence cross-correlation spectroscopy. *ChemPhysChem, 13*, 1221–1231.

Bacia, K., & Schwille, P. (2007). Practical guidelines for dual-color fluorescence cross-correlation spectroscopy. *Nature Protocols, 2*, 2842–2856.

Baudendistel, N., Müller, G., Waldeck, W., Angel, P., & Langowski, J. (2005). Two-hybrid fluorescence cross-correlation spectroscopy detects protein-protein interactions in vivo. *ChemPhysChem, 6*, 984–990.

Brock, R., Hink, M. A., & Jovin, T. M. (1998). Fluorescence correlation microscopy of cells in the presence of autofluorescence. *Biophysical Journal, 75*, 2547–2557.

Dertinger, T., Pacheco, V., von der Hocht, I., Hartmann, R., Gregor, I., & Enderlein, J. (2007). Two-focus fluorescence correlation spectroscopy: A new tool for accurate and absolute diffusion measurements. *ChemPhysChem, 8*, 433–443.

Eigen, M., & Rigler, R. (1994). Sorting single molecules: Application to diagnostics and evolutionary biotechnology. *Proceedings of the National Academy of Sciences of the United States of America, 91*, 5740–5747.

Elson, E. L., & Magde, D. (1974). Fluorescence correlation spectroscopy. I. Conceptual basis and theory. *Biopolymers, 13*, 1–27.

Foo, Y. H., Naredi-Rainer, N., Lamb, D. C., Ahmed, S., & Wohland, T. (2012). Factors affecting the quantification of bimolecular interactions by fluorescence cross-correlation spectroscopy. *Biophysical Journal, 102*, 1174–1183.

Heinze, K. G., Rarbach, M., Jahnz, M., & Schwille, P. (2002). Two-photon fluorescence coincidence analysis: Rapid measurements of enzyme kinetics. *Biophysical Journal, 83*, 1671–1681.

Hwang, L. C., & Wohland, T. (2005). Single wavelength excitation fluorescence cross-correlation spectroscopy with spectrally similar fluorophores: Resolution for binding studies. *The Journal of Chemical Physics, 122*, 114708.

Hwang, L. C., & Wohland, T. (2007). Recent advances in fluorescence cross-correlation spectroscopy. *Cell Biochemistry and Biophysics, 49*, 1–13.

Kapusta, P. (2010). Absolute diffusion coefficients: Compilation of reference data for FCS calibration. Application Note, PicoQuant GmbH.

Kim, S. A., Heinze, K. G., Bacia, K., Waxham, M. N., & Schwille, P. (2005). Two-photon cross-correlation analysis of intracellular reactions with variable stoichiometry. *Biophysical Journal, 88*, 4319–4336.

Kim, S. A., Heinze, K. G., & Schwille, P. (2007). Fluorescence correlation spectroscopy in living cells. *Nature Methods, 4*, 963–973.

Kim, S. A., Heinze, K. G., Waxham, M. N., & Schwille, P. (2004). Intracellular calmodulin availability accessed with two-photon cross-correlation. *Proceedings of the National Academy of Sciences of the United States of America, 101*, 105–110.

Kohl, T., Haustein, E., & Schwille, P. (2005). Determining protease activity in vivo by fluorescence cross-correlation analysis. *Biophysical Journal, 89*, 2770–2782.

Kohl, T., Heinze, K. G., Kuhlemann, R., Koltermann, A., & Schwille, P. (2002). A protease assay for two-photon crosscorrelation and FRET analysis based solely on fluorescent proteins. *Proceedings of the National Academy of Sciences of the United States of America, 99*, 12161–12166.

Koltermann, A., Kettling, U., Bieschke, J., Winkler, T., & Eigen, M. (1998). Rapid assay processing by integration of dual-color fluorescence cross-correlation spectroscopy: High throughput screening for enzyme activity. *Proceedings of the National Academy of Sciences of the United States of America, 95*, 1421–1426.

Larson, D. R., Gosse, J. A., Holowka, D. A., Baird, B. A., & Webb, W. W. (2005). Temporally resolved interactions between antigen-stimulated IgE receptors and Lyn kinase on living cells. *The Journal of Cell Biology*, *171*, 527–536.

Liu, P., Sudhaharan, T., Koh, R. M., Hwang, L. C., Ahmed, S., Maruyama, I. N., et al. (2007). Investigation of the dimerization of proteins from the epidermal growth factor receptor family by single wavelength fluorescence cross-correlation spectroscopy. *Biophysical Journal*, *93*, 684–698.

Maeder, C. I., Hink, M. A., Kinkhabwala, A., Mayr, R., Bastiaens, P. I., & Knop, M. (2007). Spatial regulation of Fus3 MAP kinase activity through a reaction-diffusion mechanism in yeast pheromone signalling. *Nature Cell Biology*, *9*, 1319–1326.

Magde, D., Elson, E. L., & Webb, W. W. (1974). Fluorescence correlation spectroscopy. II. An experimental realization. *Biopolymers*, *13*, 29–61.

Neugart, F., Zappe, A., Buk, D. M., Ziegler, I., Steinert, S., Schumacher, M., et al. (2009). Detection of ligand-induced CNTF receptor dimers in living cells by fluorescence cross correlation spectroscopy. *Biochimica et Biophysica Acta*, *1788*, 1890–1900.

Ohrt, T., Staroske, W., Mütze, J., Crell, K., Landthaler, M., & Schwille, P. (2011). Fluorescence cross-correlation spectroscopy reveals mechanistic insights into the effect of 2'-O-methyl modified siRNAs in living cells. *Biophysical Journal*, *100*, 2981–2990.

Riçka, J., & Binkert, T. (1989). Direct measurement of a distinct correlation function by fluorescence cross correlation. *Physical Review A*, *39*, 2646–2652.

Rigler, R., Foldes-Papp, Z., Meyer-Almes, F. J., Sammet, C., Volcker, M., & Schnetz, A. (1998). Fluorescence cross-correlation: A new concept for polymerase chain reaction. *Journal of Biotechnology*, *63*, 97–109.

Rigler, R., Mets, Ü., Widengren, J., & Kask, P. (1993). Fluorescence correlation spectroscopy with high count rate and low background: Analysis of translational diffusion. *European Biophysics Journal*, *22*, 169–175.

Rippe, K. (2000). Simultaneous binding of two DNA duplexes to the NtrC-enhancer complex studied by two-color fluorescence cross-correlation spectroscopy. *Biochemistry*, *39*, 2131–2139.

Ruan, Q., & Tetin, S. Y. (2008). Applications of dual-color fluorescence cross-correlation spectroscopy in antibody binding studies. *Analytical Biochemistry*, *374*, 182–195.

Savatier, J., Jalaguier, S., Ferguson, M. L., Cavailles, V., & Royer, C. A. (2010). Estrogen receptor interactions and dynamics monitored in live cells by fluorescence cross-correlation spectroscopy. *Biochemistry*, *49*, 772–781.

Schwille, P., Haupts, U., Maiti, S., & Webb, W. W. (1999). Molecular dynamics in living cells observed by fluorescence correlation spectroscopy with one- and two-photon excitation. *Biophysical Journal*, *77*, 2251–2265.

Schwille, P., Meyer-Almes, F. J., & Rigler, R. (1997). Dual-color fluorescence cross-correlation spectroscopy for multicomponent diffusional analysis in solution. *Biophysical Journal*, *72*, 1878–1886.

Slaughter, B. D., Schwartz, J. W., & Li, R. (2007). Mapping dynamic protein interactions in MAP kinase signaling using live-cell fluorescence fluctuation spectroscopy and imaging. *Proceedings of the National Academy of Sciences of the United States of America*, *104*, 20320–20325.

Stromqvist, J., Johansson, S., Xu, L., Ohsugi, Y., Andersson, K., Muto, H., et al. (2011). A modified FCCS procedure applied to Ly49A-MHC class I cis-interaction studies in cell membranes. *Biophysical Journal*, *101*, 1257–1269.

Thompson, N. L. (1991). Fluorescence correlation spectroscopy. In: J. R. Lankowicz (Ed.), *Topics in fluorescence spectroscopy*, Vol. 1, (pp. 337–378). New York: Plenum Press.

Wachsmuth, M., Weidemann, T., Müller, G., Hoffmann-Rohrer, U. W., Knoch, T. A., Waldeck, W., et al. (2003). Analyzing intracellular binding and diffusion with continuous fluorescence photobleaching. *Biophysical Journal*, *84*, 3353–3363.

Wawrezinieck, L., Rigneault, H., Marguet, D., & Lenne, P. F. (2005). Fluorescence correlation spectroscopy diffusion laws to probe the submicron cell membrane organization. *Biophysical Journal, 89*, 4029–4042.

Weidemann, T., & Schwille, P. (2009). Fluorescence correlation spectroscopy in living cells. In P. Hinterdorfer & A. Van Oijen (Eds.), *Handbook of Single Molecule Biophysics*. Heidelberg: Springer.

Weidemann, T., Wachsmuth, M., Knoch, T. A., Müller, G., Waldeck, W., & Langowski, J. (2003). Counting nucleosomes in living cells with a combination of fluorescence correlation spectroscopy and confocal imaging. *Journal of Molecular Biology, 334*, 229–240.

Weidemann, T., Wachsmuth, M., Tewes, M., Rippe, K., & Langowski, J. (2002). Analysis of ligand binding by two-colour fluorescence cross-correlation spectroscopy. *Single Molecules, 3*, 49–61.

Weidemann, T., Worch, R., Kurgonaite, K., Hintersteiner, M., Bökel, C., & Schwille, P. (2011). Single cell analysis of ligand binding and complex formation of interleukin-4 receptor subunits. *Biophysical Journal, 101*, 2360–2369.

Widengren, J., Mets, Ü., & Rigler, R. (1995). Fluorescence correlation spectroscopy of triplet states in solution: A theoretical and experimental study. *The Journal of Physical Chemistry, 99*, 13368–13379.

Worch, R., Bökel, C., Höfinger, S., Schwille, P., & Weidemann, T. (2010). Focus on composition and interaction potential of single-pass transmembrane domains. *Proteomics, 10*, 4196–4208.

CHAPTER FOUR

Brightness Analysis

Patrick Macdonald*, Jolene Johnson†, Elizabeth Smith†, Yan Chen†, Joachim D. Mueller*,†,1

*Department of Biomedical Engineering, University of Minnesota, Minneapolis, Minnesota, USA
†School of Physics and Astronomy, University of Minnesota, Minneapolis, Minnesota, USA
[1]Corresponding author: e-mail address: mueller@physics.umn.edu

Contents

Abstract

Brightness analysis provides a powerful tool for the study of protein interactions both in solution and in living cells. We provide a brief survey of some widely used techniques for extracting brightness from fluorescent fluctuation spectroscopy experiments. While all the techniques are equivalent under ideal conditions, we touch upon their relative strengths and discuss in detail a specific scenario wherein the photon-counting histogram (PCH) separates the brightness of rare, bright particles from a dominant background. In a practical vein for ensuring quantitative and unbiased brightness data, we address a number of potential issues stemming from both theoretical assumptions and experimental realities. Two additional issues arising from geometry are examined in greater detail. An oil-immersion objective skews the geometry of the excitation volume as a function of

Methods in Enzymology, Volume 518
ISSN 0076-6879
http://dx.doi.org/10.1016/B978-0-12-388422-0.00004-2

penetration depth. The bias can be characterized and corrected or avoided through the use of a water-immersion objective. Brightness measurements in thin sample geometries, frequently encountered in cells, may be biased. We use *z*-scan FFS to characterize sample geometry and correct any resulting bias in the brightness.

1. INTRODUCTION

Proteins provide the cell with structure and execute a broad range of biological functions. These processes are carried out by transient molecular assemblies that are built and organized through protein–protein interactions. Fluorescence fluctuation spectroscopy (FFS) is a collection of sensitive spectroscopic techniques that can characterize many of these interactions. This research area began 40 years ago when Magde, Elson, and Webb introduced fluorescence correlation spectroscopy (FCS) and determined diffusion coefficients and chemical kinetics from the autocorrelation function (Elson & Magde, 1974; Magde, Elson, & Webb, 1972). Since then, FCS has been used to study the kinetics of a multitude of samples including biological molecules *in vitro* as well as inside living cells. While FCS is widely used in the community, a branch of fluorescence fluctuation research is now dedicated to the development of alternative analysis techniques to extract additional information from fluctuation experiments. This chapter specifically examines methods that determine the molecular brightness, λ, a parameter which characterizes the average fluorescent intensity of a single molecule.

We start by providing background that motivates the development of brightness analysis and follow up with a brief discussion of the most widely used brightness analysis methods. The proper use of any brightness analysis technique is based on careful consideration of a variety of experimental factors. Some critical aspects, such as accounting for background fluorescence and the need for a stationary fluorescence signal over the duration of the experiment, will be discussed in more detail. Quantitative interpretation of brightness experiments relies on a calibration procedure and requires brightness standards, which will be addressed. Nonideal detector effects, like deadtime and afterpulsing, are an inextricable part of the photon collection process (Hillesheim & Müller, 2003). We review their influence on the measured brightness and the associated corrections. Finally, we examine some additional brightness artifacts that can confound the interpretation of FFS experiments. These artifacts occur because the geometry of the sample or the geometry of the excitation light is not properly considered (de Grauw, Vroom, van der

Voort, & Gerritsen, 1999; Dong, Koenig, & So, 2003; Macdonald, Chen, Wang, Chen, & Mueller, 2010). Our discussion of brightness will focus mainly on two-photon excitation, which we found to be especially suitable for cell-based FFS experiments. We hope to demonstrate that brightness, when applied correctly, is a quantitative and extremely powerful technique for the determination of protein stoichiometry.

2. BACKGROUND

We begin with a brief overview of the history and motivation behind the development ofbrightness analysis. Early work in the field focused on FCS and in particular on the diffusion coefficient D, which is inversely related to the residence time τ_D of the molecule in the observation volume. Association of two molecules increases the hydrodynamic size, which leads to slower diffusion and therefore an increase in the residence time. This change in the residence time is readily observed for a small labeled dye binding to a large molecule. However, if two proteins of comparable size associate, the change in the residence time is small. Thus, resolving mixtures of monomers and dimers by the shape of the autocorrelation function is impractical.

In addition to the residence time, FCS experiments also determine the time zero value of the autocorrelation function, $G(0)$. The time zero value is also called the fluctuation amplitude and is proportional to the normalized variance $\langle \Delta N^2 \rangle / \bar{N}^2$. When the number fluctuations obey Poisson statistics, the fluctuation amplitude is inversely proportional to the average number of molecules $\bar{N}$ in the detection volume, $(0) = \gamma_2 / \bar{N}$, where γ_2 is a beam shape factor. Dimerization of a protein changes the number of molecules in the detection volume from $\bar{N}$ to $\bar{N}/2$, which corresponds to a doubling of the fluctuation amplitude. Thus, if the total protein concentration is fixed, the fluctuation amplitude directly measures protein–protein association. Following earlier work by Qian and Elson and Palmer and Thompson, we used the fluctuation amplitude to probe ligand–protein binding equilibria (Chen, Mueller, Tetin, Tyner, & Gratton, 2000). Unfortunately, the fluctuation amplitude is difficult to interpret in the presence of multiple brightness species (Thompson, 1991). For example, two species A and B with average number of molecules $\bar{N}_A$ and $\bar{N}_B$ and molecular brightness λ_A and λ_B result in the following $G(0)$ value:

$$G(0) = \gamma_2 \frac{\lambda_A^2 \bar{N}_A + \lambda_B^2 \bar{N}_B}{(\lambda_A \bar{N}_A + \lambda_B \bar{N}_B)^2}. \quad [4.1]$$

Not only does brightness contribute nonlinearly to the fluctuation amplitude, but also the number concentrations of $\bar{N}_A$ and $\bar{N}_B$ need to be known to interpret the fluctuation amplitude. For example, a 50/50 mixture of monomers and dimers at two different concentrations yields different fluctuation amplitudes. Thus, $G(0)$ does not provide a direct measure of the oligomerization between proteins. This experience taught us that resolving mixtures requires more information than provided by the autocorrelation function. In addition, we started to appreciate brightness as a crucial FFS parameter for quantitative analysis of protein oligomerization experiments.

Several research groups have worked on extracting additional information from FFS experiments to study molecular aggregation. A particular promising approach is based on higher-order photon count moments, which are determined either by direct calculation from the data or by extrapolating the fluctuation amplitude of higher-order autocorrelation functions (Palmer & Thompson, 1987; Qian & Elson, 1990a, 1990b). However, we found that accurate and unbiased extrapolation of the fluctuation amplitude of higher-order correlation functions was difficult to achieve with the instrumentation available to us at the time. As an alternative, we explored the distribution of photon counts, which mathematically contains the same information as the photon count moments. We developed the theory, observed good agreement with experimental data, and referred to the technique as photon-counting histogram (PCH) analysis (Chen, Mueller, So, & Gratton, 1999). PCH determines the brightness of a species and its number of molecules in the detection volume. Because PCH offers access to higher moments which are embedded in the shape of the photon count distribution, a mixture of brightness species can be resolved directly as long as the signal-to-noise ratio (SNR) is sufficiently high.

Our initial applications of PCH demonstrated that the molecular brightness of fluorescent dyes is often sensitive to the local environment. On the one hand, this sensitivity can be exploited to monitor conformation changes of a protein (Perroud, Bokoch, & Zare, 2005); on the other hand, this effect significantly complicates the quantitative interpretation of brightness experiments. We were extremely fortunate to be developing quantitative brightness analysis at a time when the enhanced green fluorescent protein (EGFP) was gaining widespread use. To our surprise, the brightness of EGFP was a stable parameter; it remained unchanged if tagged to another protein as well as measured in the nucleus, cytoplasm, or *in vitro* (Chen, Mueller, Ruan, & Gratton, 2002). Because EGFP's chromophore is embedded inside a protein

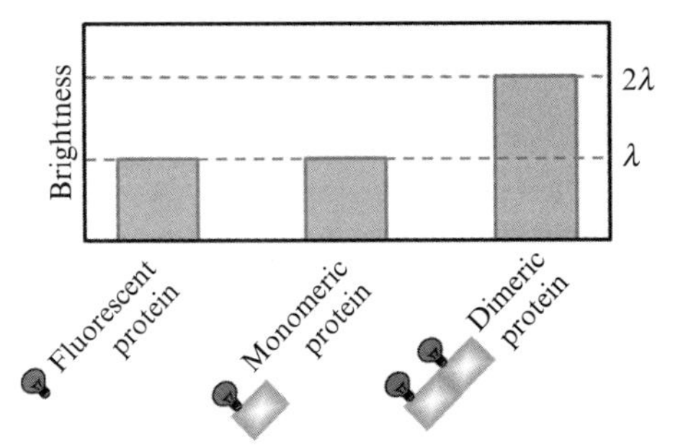

Figure 4.1 *Brightness and stoichiometry.* A fluorescent protein with a brightness λ is attached to the protein of interest. The brightness of the monomeric protein is equivalent to that of the fluorescent protein alone. A dimer has a brightness of 2λ because it carries two fluorescent proteins. (For color version of this figure, the reader is referred to the online version of this chapter.)

barrel, its molecular brightness is largely insensitive to its local environment, provided the pH stays constant. The molecular brightness of EGFP is therefore a robust parameter that can be used to quantify the stoichiometry or aggregation state of a complex. By measuring the molecular brightness of a monomer as calibration, we can determine its oligomerization (Chen, Wei, & Mueller, 2003). For example, protein homodimerization leads to a doubling of the brightness (Fig. 4.1).

As we mentioned earlier, direct resolution of brightness species requires very good SNR. It is often difficult to achieve the necessary experimental conditions. This is especially true for cell measurements which are conducted at relatively high concentrations and low brightness compared to typical *in vitro* experiments. Under such conditions, an FFS measurement of a mixture of brightness species will return a single apparent brightness that is given by a nonlinear combination of brightness λ_i and concentration $\bar{N}_i$ for each species,

$$\lambda = \frac{\sum_i \lambda_i^2 \bar{N}_i}{\sum_i \lambda_i \bar{N}_i}. \qquad [4.2]$$

This brightness of a mixture, unlike the fluctuation amplitude, characterizes the averaged oligomerization state of the protein sample. For example, a 50/50 mixture of monomers and dimers at two different concentrations yields the same apparent brightness. This general property of brightness allows us to measure protein binding curves inside living cells, which we refer to as brightness titration (Chen, Johnson, Macdonald, Wu, & Mueller, 2010; Chen et al., 2003). After collecting data from many cells

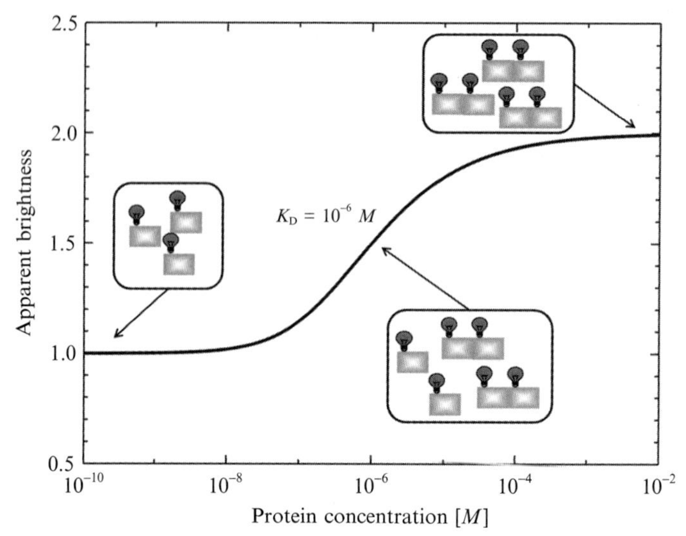

Figure 4.2 *Brightness titration.* A brightness titration curve is modeled for a monomer–dimer transition as a function of concentration. Assuming a dissociation constant of 1 μ*M*, we plot a concentration range that starts when the population is effectively monomeric and continues up to an almost completely dimeric population. At intermediate concentrations, a mixture of monomers and dimers is present. The *y*-axis displays the apparent brightness in monomeric units, thus beginning at 1 and climbing to 2. (For color version of this figure, the reader is referred to the online version of this chapter.)

expressing the protein at different expression levels, the brightness is graphed versus the protein concentration. This brightness titration plot characterizes the binding curve of a cellular protein in its native environment. An example of a brightness titration curve of a monomer/dimer transition is shown in Fig. 4.2. Thus, brightness is an experimental indicator for detecting and quantifying protein interactions.

3. OVERVIEW OF ANALYSIS TECHNIQUES

There are many analysis techniques capable of determining the brightness of a biological complex. In FCS, one can extrapolate the $G(0)$ value from the autocorrelation fit to determine the number of molecules in the sample as we mentioned above (Thompson, 1991). Care needs to be exercised to account for afterpulsing, triplet state, or other fast kinetic processes to ensure unbiased estimation of the fluctuation amplitude. The brightness is determined by multiplying the fluctuation amplitude with the average fluorescence intensity $\langle F \rangle$ divided by γ_2,

$$\lambda = \frac{G(0)\langle F \rangle}{\gamma_2}. \quad [4.3]$$

Moment analysis is another way to extract brightness information from fluctuation data. The first moment (average fluorescence intensity) and the second moment (variance in fluorescence intensity) can be used to determine the average number of molecules within the sample volume (Qian & Elson, 1990a, 1990b). Autocorrelation and moment analysis of brightness are of course related because $G(0)$ is determined by the first two moments of the fluorescence intensity,

$$\lambda = \frac{\langle \Delta F^2 \rangle - \langle F \rangle}{\gamma_2 \langle F \rangle}. \quad [4.4]$$

An alternative but equivalent technique to moment analysis is fluorescence cumulant analysis (FCA) (Mueller, 2004). Cumulants are closely related to moments but have properties that are particularly useful for fluorescence spectroscopy. Cumulants of the fluorescence intensity scale with the number of molecules in the sample volume, and the cumulant for a mixture of species is given by the sum of the cumulants of each species. Higher-order cumulants are simple functions of the brightness and number of molecules. The simple nature of the theory also gives rise to an exact treatment of experimental uncertainties of cumulants, a feature that has been previously missing in moment analysis. Access to higher-order cumulants coupled with a rigorous error analysis provides the necessary framework to experimentally resolve brightness mixtures.

The ability to resolve a heterogeneous mixture of biomolecules is of significant interest for protein interaction studies. FCS can separate multiple species based on the translational diffusion coefficient, but it becomes difficult to separate a species from a mixture when the molecules have similar diffusion times. Moment and FCA can resolve multiple species by calculating the higher-order moments beyond the average and variance. In order to resolve species (i.e., identify the two parameters, number and brightness) using moment/cumulant analysis, two moments must be determined for each species present. Thus, resolving two species from a mixture requires that the first four cumulants be known. Unfortunately, the third and fourth cumulants are often not directly accessible because of insufficient SNR. Other techniques, such as time-integrated fluorescence cumulant analysis (TIFCA), were developed to improve the SNR by resampling the data and thereby optimizing access to higher-order moments.

The PCH and fluorescent-intensity distribution analysis (FIDA) are equivalent methods that were independently developed as a way to resolve multi-species interactions (Chen et al., 1999; Kask, Palo, Ullmann, & Gall, 1999). Both methods analyze the distribution of photon counts but with slightly different mathematical approaches. For simplicity, we will hereafter refer only to PCH. PCH can distinguish species based on their concentration and brightness, even if they have similar diffusion properties (Mueller, Chen, & Gratton, 2000). This makes PCH complementary to the autocorrelation function, which separates species based on size and temporal behavior. PCH relies on a data sampling frequency high enough that the position of the fluorophore within the detection volume is, to good approximation, stationary. FIMDA was the first technique to extend photon count distribution analysis to many sampling times (Palo, Mets, Jäger, Kask, & Gall, 2000). An equivalent extension of PCH has been introduced as well (Perroud, Huang, & Zare, 2005). These methods correct the influence of sampling time on brightness by modifying the distribution based on the second moment of the photon counts. However, this correction is an approximation and may lead to biases in the resolution of a mixture of species, which relies on higher-order information.

TIFCA builds on FCA by extending it to arbitrary sampling times, much like FIMDA builds on FIDA. Both methods implement rebinning procedures so that diffusion coefficients can be determined (Mueller, 2004; Wu & Mueller, 2005). TIFCA takes the influence of sampling time on each cumulant into account. Thus, unlike FIMDA, TIFCA is not an approximation, but an exact description of photon count statistics at arbitrary sampling times. TIFCA analysis is performed for a number of different sampling times generated by rebinning the FFS data. The rebinning of the data increases the SNR of cumulants. This is especially important for higher-order cumulants, which play an important role in distinguishing species. Finally, a number of imaging-based techniques that determine brightness exist. These will not be discussed further because they typically rely on the moments or distribution of the photon counts and thus employ similar concepts as described above.

4. BRIGHTNESS ANALYSIS

All of the above techniques are suitable for brightness analysis. In fact, for samples that are optimal for FFS (bright and stable dye, low concentration, negligible background signal), they return the same brightness value.

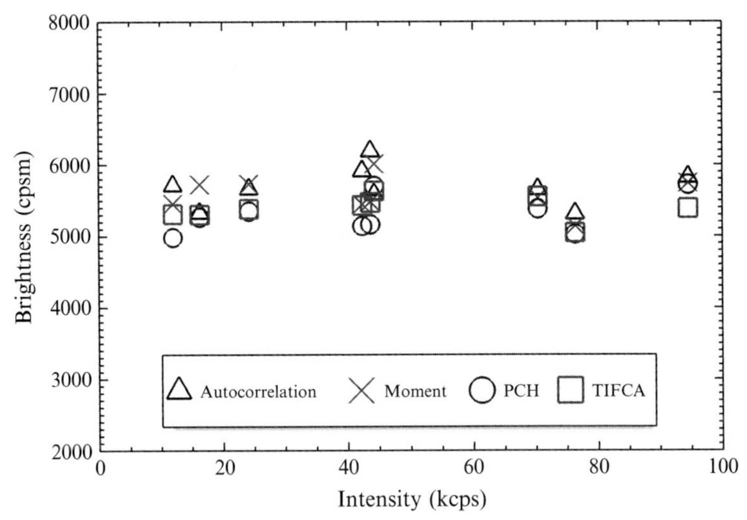

Figure 4.3 *Comparison of different analysis techniques.* A dilution experiment was performed on Alexa 488 dye in a mixture of 50% glycerol and 50% water. The brightness was determined using autocorrelation, moment, PCH, and TIFCA analyses. All analysis techniques return the same brightness within uncertainty over this intensity range. (For color version of this figure, the reader is referred to the online version of this chapter.)

To demonstrate this, we performed a dilution study on a fluorescent dye and extracted the brightness using each analysis method. Alexa 488 dye was diluted into a mixture of glycerol and water, and the total intensity was kept low—less than 100,000 counts per second (cps)—to limit detector artifacts that will be discussed later in this chapter. All analysis methods return the same brightness within uncertainty over the measured intensity range (Fig. 4.3). Thus while there are situations where choosing the analysis method is not a concern, we suggest that it is helpful to have a battery of techniques at ones disposal. Analysis methods differ in their sensitivity to specific artifacts; thus, comparison of techniques is often a useful way to investigate unexpected results. Furthermore, experimental situations arise where some techniques have advantages over others; we illustrate this through an example from our own work.

4.1. Brightness and background

It is important to be aware of the background signal in any brightness measurement, especially with a low brightness or a low concentration sample. The background signal is nearly impossible to eliminate and comes from multiple sources. Detectors have a certain level of dark counts. There is an additional level of room light counts since it is generally difficult to shield all other sources of light around an experimental setup. Finally, there is the background autofluorescence from the sample itself. Aqueous buffers have low but nonzero background counts. Cells are notoriously autofluorescent although this can be greatly reduced by selecting cell lines, cellular location and excitation wavelength. At low concentrations of sample, or low brightness, the total background can represent a significant fraction of the signal. When background is less than 10% of the signal, we find that it can usually be neglected. At higher percentages of background, the recovered brightness will be artificially reduced (absent a strangely bright autofluorescent species) and must be corrected when possible.

One particular situation where it is important to account for background is the measurement of rare but bright particles, such as virus-like particles (VLPs). These samples have concentrations that can be quite low. Concentrations of tens of picomolar or less are common. Although each particle produces large fluorescent-intensity fluctuations due to the presence of many fluorescent labels, the average signal is still dominated by background on account of the low event rate. Figure 4.4A shows the raw photon count data for a VLP sample. During the majority of the measurement time, no VLPs are observed and only background counts are collected. Depending on the sample and instrument setup, the background can account for 50–95% or more of the total signal. When background is responsible for such a large percentage of the collected photon counts, there are issues in properly, interpreting the fluctuation amplitude and brightness. The largely uncorrelated background signal overwhelms the sample signal and diminishes the recovered brightness. One way to account for the background would be to do an additional measurement to estimate the background count rate. The background intensity can then essentially be subtracted off, leaving behind the correlated sample fluctuations. Unfortunately, this is not always possible for biological samples, and even when it can be done, a large background contribution leads to rather large correction factors. The slightest uncertainty in estimating the background leads to huge uncertainties in the corrected brightness. Quantitative analysis demonstrates that the brightness uncertainty exceeds 2500% for a sample with 95% background signal (Johnson, Chen, & Mueller, 2010). Therefore, meaningful autocorrelation analysis of such samples is not feasible.

PCH overcomes this difficulty because it acts as a brightness filter. Rare, bright particles generate a low-amplitude distribution shifted toward high count rates. A strong background signal should consist of a high amplitude,

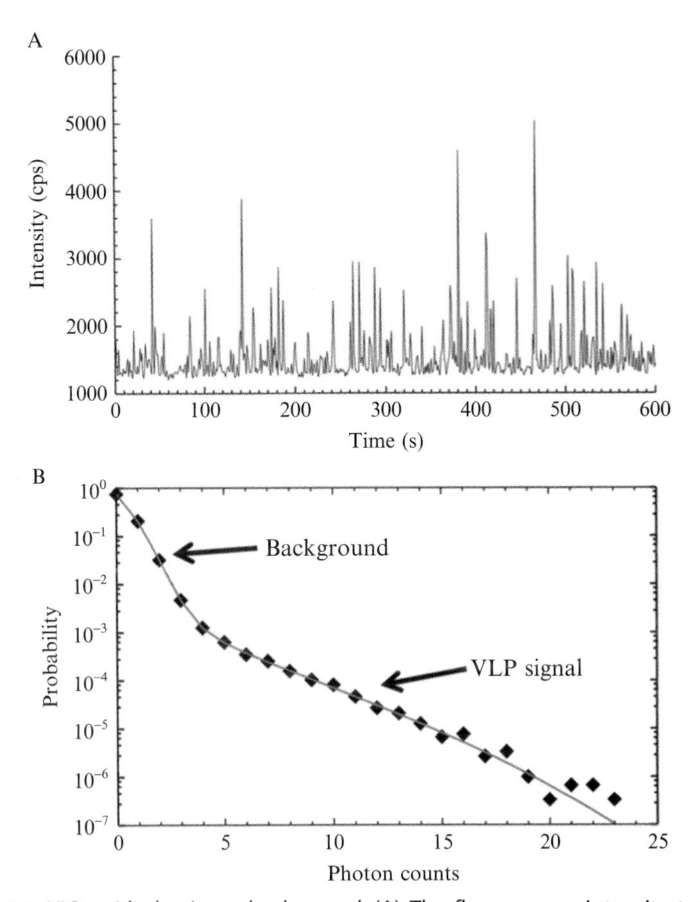

Figure 4.4 *VLPs with dominant background.* (A) The fluorescence intensity trace of a solution containing HIV-1 virus-like particles. Events are relatively rare and the signal is dominated by background. (B) The photon-counting histogram of the data (diamonds) is plotted together with a fit (line) to a background species and a VLP species. The steep section of the curve (<4 counts per bin) comes from the background, while the shallow section of the curve represents the VLP signal. PCH analysis permits the background and the VLP signal to be fit and distinguished in a single step. (For color version of this figure, the reader is referred to the online version of this chapter.)

but fairly narrow distribution located at low count rates since the background is dim. PCH analysis separates the background signal and sample signal into two separate species without any additional background calibration. A graph of the PCH provides a good visual representation of this signal separation (Fig. 4.4B). The steep part of the curve at low counts (<4 counts) corresponds to the background. VLPs are rarely detected, but when they are detected, they produce high photon count rates. The section of the PCH curve with the shallow slope at high counts captures the VLP species. When the PCH is fit to a two-species model, the two subdistributions are accurately identified, thus separating the background and sample signal without any *a priori* knowledge of the source or brightness of the background.

The inherent brightness filter that PCH provides has allowed us to extend brightness analysis to a broader range of samples. For instance, the combination of hydrodynamic sample flow and PCH analysis extends the range of measurable sample concentrations down to 10 fM for bright particles (Johnson et al., 2010). Additionally, we are able to measure VLP samples directly in the highly autofluorescent cell medium without the need for time-intensive purification steps. We have exploited the ability of PCH to determine the copy number of the structural protein (Gag) found in HIV-1 VLPs (Chen, Wu, Musier-Forsyth, Mansky, & Mueller, 2009) and gained far greater insights than can be achieved with autocorrelation analysis alone.

4.2. Segment analysis

FFS theory is based on the assumption of a stationary fluorescence signal. While this assumption is generally easy to satisfy for *in vitro* experiments, FFS measurements in cells frequently encounter drifts in the fluorescence intensity. For example, organelles or large vesicles moving into the excitation volume can create an excluded volume effect that transiently decreases fluorescence, and large autofluorescent complexes can temporarily increase the background signal. These drifts lead to a nonstationary signal that may introduce biases in the calculated brightness. Luckily, these events are typically rare. Thus, by dividing the sequence of photon counts into smaller segments, we map these events to particular segment numbers. Next, we calculate the brightness of each segment independently. If no bias occurred, we expect that the brightness across the segments is uniform. However, if we see significant deviations in some of the segments, this is an indication that the measurement suffers from artifacts introduced by a nonstationary signal. A typical data acquisition time is approximately a minute. With a reasonable SNR, brightness can be calculated from a few seconds of data. Analyzing the data by segments is a good way to test the reliability of the data. Even a simple test of analyzing the first half of the data versus the second can be revealing, and for experimental situations where intensity drifts are ongoing, analyzing the data by small segments—over which the intensity is stationary—and averaging the results can provide a much more stable brightness than is possible when analyzing the data set as a whole (Chen et al., 2002). This approach is also useful for experiments that exhibit mild photobleaching (Caccia, Camozzi, Collini, Zaccolo, & Chirico, 2005).

5. EXPERIMENTAL CONSIDERATIONS

In this section, we address some practical issues and concerns common to all brightness experiments. First, we wish to emphasize the dependence of brightness on the excitation point spread function (PSF). Brightness is a spatiotemporal average of the photons emitted by particles in the excitation volume. Thus, analysis techniques must incorporate the PSF into the theory in order to extract the brightness. PCH analysis starts by selecting a PSF. Our software, which is written in interactive data language (IDL) and is available upon request, offers three different PSF models as well as optional PSF corrections. In autocorrelation analysis, the PSF dependence enters through the value of the shape factor γ_2.

A second issue that sometimes goes unnoticed is the dependence of recovered brightness on the ratio between the data acquisition sampling time T and the time τ_D it takes particles to diffuse through the excitation volume. PCH theory assumes $T \ll \tau_D$, that is, the particles are "frozen" during T. When the fast sampling condition is not met, the undersampling effect will bias brightness data. As an upper limit, sampling time should be adjusted to be at least two times faster than diffusion time. In this regime, the undersampling effect is still small and easily corrected (Palo et al., 2000; Wu & Mueller, 2005). The above issues deal with the analysis and design of brightness experiments. Similarly, there are important practical considerations for the execution of brightness measurements, two of which we now address in more detail.

5.1. Calibrations

A complete calibration is necessary to ensure both the optimal performance of the instrument and proper interpretation of the results no matter which analysis method is used. Each day we perform two calibrations. First, we measure the brightness of a standard dye to ensure the long-term stability of the instrument. Second, we calibrate the monomer brightness of the selected fluorescent protein to establish a baseline brightness in cells. The dye calibration tracks the instrument stability over time and alerts the user to any changes in instrument alignment or overall performance. Since a dye of known concentration is used, we can determine how the average number of molecules measured is related to molar concentration. This provides valuable information for concentration-dependent studies and an additional check on the size of the excitation volume. The fluorescent label calibration involves measuring a number of cells expressing the monomeric fluorescent protein alone. The standard deviation of this experiment identifies the uncertainty in determining the brightness. The monomer calibration measurements should cover cells with a broad range of protein expression to ensure that the brightness is independent of concentration. In addition, it is important to perform a dimer calibration using a construct of two linked fluorescent proteins. As discussed in a previous review (Chen et al., 2010), the doubling of brightness for the dimer is dependent on complete maturation of the protein label, the absence of long-lived dark states, and good cell health. The dimer calibration is therefore necessary to identify robust brightness conditions from which meaningful quantitative information can be extracted. Regardless of the specific type of experiment, proper calibration of the instrument and cellular conditions is necessary for accurate interpretation of the measured brightness values.

5.2. Deadtime and afterpulsing

Photodetectors used in photon-counting experiments are never ideal. Here we will focus specifically on avalanche photodiodes (APDs) because they are the detector most commonly used in fluorescence fluctuation experiments. All brightness experiments that employ APDs must contend with two primary artifacts, deadtime and afterpulsing. Deadtime is a fixed period after the collection of a photon during which the detector cannot register any other events (Fig. 4.5A). This temporary "blindness" leads to a decrease in the number of photons detected and is of particular concern in high count rate experiments. Deadtime has the effect of narrowing the photon count distribution. At high count rates, more events are lost since there is a greater probability of additional photons arriving while the detector is blind to new events. Typical APDs have deadtimes of 50 ns, although newer models have deadtimes as low as 20 ns. For an average intensity of 850 kilocounts/second and a detector with 50 ns deadtime, the uncorrected brightness will be 15% lower than the true brightness. Thus in high count rate experiments, the effect of deadtime on brightness is not negligible and will result in lower brightness if there is no correction applied.

Afterpulses are spurious events that occur with a small probability after the detection of an actual photon (Fig. 4.5B). This leads to an increase in the number of apparent photons detected. Afterpulses will have the effect of broadening the photon count distribution because the afterpulses will artificially

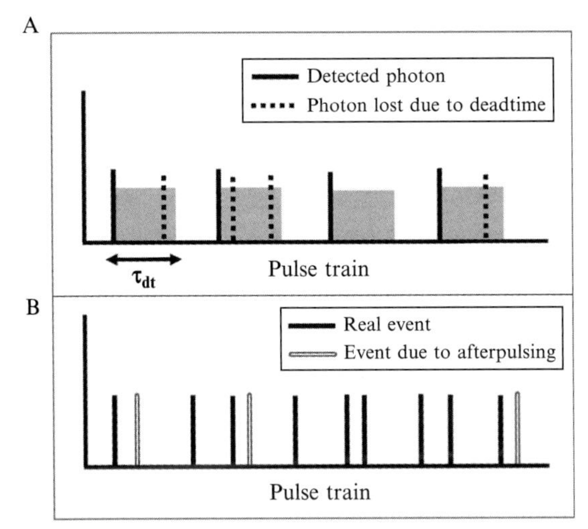

Figure 4.5 *Non ideal detector effects: deadtime and afterpulsing.* (A) When a photon reaches the detector, a deadtime of duration τ_{dt} is initiated and any further photons reaching the detector during that time are lost. (B) Afterpulses are spurious pulses that are generated by the detection of a real event with a probability p_{ap}.

increase the counts in the higher channels, which leads to an artificially high brightness. For newer APD detectors, the afterpulse probability can be sufficiently low that its effect on many experiments is negligible. For instance, APDs used in our lab typically have an afterpulse probability of 0.2–0.5%.

While detectors often include factory specifications for the deadtime and afterpulsing, for the most accurate experiments, it is best to independently measure these parameters. After installing the detector in the experimental setup, we use a very highly concentrated fluorescent dye solution to imitate a constant fluorescence intensity light source (Hillesheim & Müller, 2003). It is useful to employ Mandel's Q-parameter (Mandel, 1979) to characterize the width of the photon count distribution. For a constant intensity light source, the photon count distribution will be Poissonian which results in Q equal 0. However, the nonideal detector effects modify the Q value of a constant intensity light source,

$$Q = -2\tau_{dt} I + 2p_{ap}, \tag{4.5}$$

where I is the intensity, τ_{dt} is the deadtime, and p_{ap} is the afterpulsing probability (Finn, Greenless, Hodapp, & Lewis, 1988). A graph of Q (obtained using moment analysis) versus collected photon intensity is produced by changing the intensity of the excitation light using a neutral density filter. This graph is then fit to the above equation, from which the deadtime

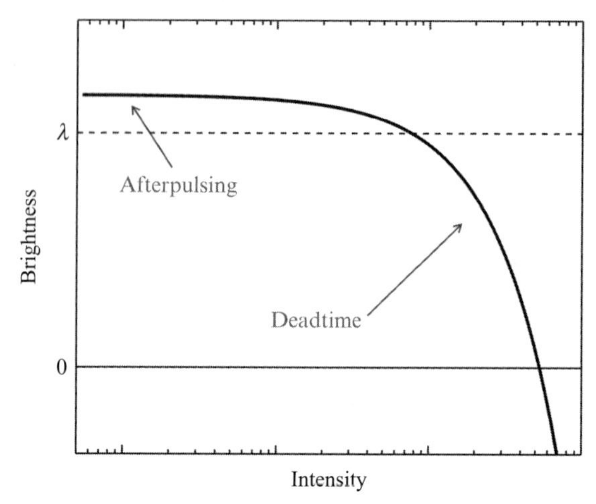

Figure 4.6 *Afterpulsing- and deadtime-biased brightness.* The conceptual graph depicts brightness versus sample intensity plotted on a logarithmic scale. The unbiased brightness from an ideal detector is independent of intensity (dashed line). The bold line displays the combined effect of deadtime and afterpulsing on the experimentally determined brightness. Afterpulsing uniformly increases the brightness. Deadtime reduces the brightness, particularly at high intensities, where it can drive the calculated brightness below zero.

and afterpulsing can be determined. Note that this equation is the first-order approximation. The linear relationship between Q and I breaks down at sufficiently high intensities.

Methods to correct photon count distributions and moments have been developed and tested extensively (Hillesheim & Müller, 2003; Wu & Mueller, 2005). If these corrections are not applied, the brightness of a fluorescent sample varies with the concentration of the fluorophore. A schematic depiction of the brightness versus intensity is shown in Fig. 4.6. At low intensities, the deadtime effect is negligible, but afterpulsing leads to an artificially elevated brightness. As the intensity rises, the brightness decreases due to the growing importance of deadtime on the photon count statistics. At a sufficiently high intensity, the brightness will become negative. Applying the methods described in the above references corrects this artifact and returns a constant brightness across a broad range of count rates.

6. BRIGHTNESS AND GEOMETRY

The previous sections provide an overview of brightness analysis techniques and some general considerations for obtaining reliable brightness data. In this section, we discuss two common situations where geometry—either of the excitation light or the sample—leads to a biased or misinterpreted brightness. The first and more obvious issue arises from the index of refraction mismatch and the spherical aberrations typically found when performing FFS experiments with an oil-immersion objective. The second deals with thin-layer sample geometries, like cell cytoplasm. When the focal volume of the excitation light is taller than the sample, the brightness of the sample appears larger than the actual value. Both effects are particularly troublesome for brightness studies which attempt to determine protein stoichiometry.

6.1. Focal depth dependence of oil-immersion objectives

Oil-immersion objectives are designed to function in a narrow layer of focus immediately at the sample side of a coverslip. Index of refraction mismatch and spherical aberrations become a rapid concern when trying to focus deeper into solution, living cells, or tissue. This leads to a distortion of the PSF and a change in the recovered FFS parameters. Other studies have demonstrated that the PSF is stretched dramatically in the axial direction and significantly in the lateral direction (de Grauw et al., 1999) which has a very noticeable effect on brightness. We recall that brightness λ is defined as

$$\frac{\langle F \rangle}{\bar{N}} \propto \lambda = \frac{\langle F \rangle G(0)}{\gamma_2}, \qquad [4.6]$$

where $\langle F \rangle$ is average intensity, $\bar{N}$ is the average number of molecules in the excitation volume and is inversely proportional to $G(0)$.

The enlarged PSF decreases the efficiency of the two-photon excitation and leads to decreasing intensity $\langle F \rangle$ as a function of focal depth. Figure 4.7A shows intensity measurements at increasing focal depths for Texas Red dye in aqueous solution, using a Zeiss 63× Plan Apochromat oil-immersion objective (N.A. = 1.4). The intensity effect is pronounced, with a 15% decrease within 5 μm and a 70% loss of intensity at a depth of 60 μm. Additionally, the larger PSF covers more area and excites a larger number of molecules. This is reflected in autocorrelation analysis through a decrease in the correlation amplitude $G(0)$ (Fig. 4.7B). Similarly, it takes longer for a particle to cross the inflated excitation volume, resulting in increasing diffusion times as shown in Fig. 4.7C. Since average brightness is reduced by both the decrease in intensity and the increase in number of molecules, brightness values are extremely sensitive to focal depth. Texas Red brightness values fall by 20% within 5 μm. Focal depth differences of ±1 μm cause brightness deviations that fall within the 10% noise typically associated with brightness experiments in living cells. However, for samples

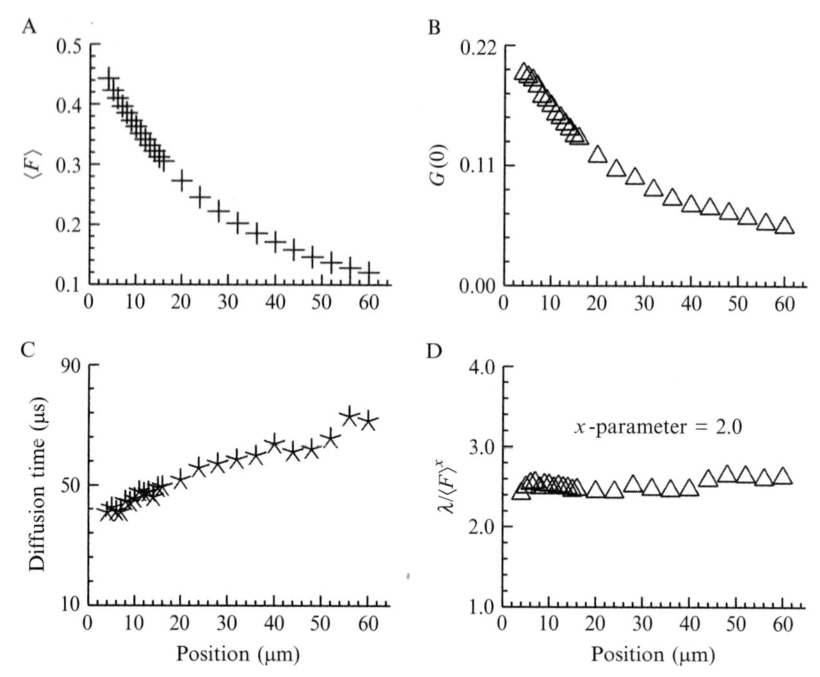

Figure 4.7 *Focal depth dependence of oil-immersion objective.* A dye solution is measured with two-photon excitation FFS using an oil-immersion objective. (A) The intensity falls off strongly as the laser is focused more deeply into the sample. This is the result of an index of refraction mismatch and spherical aberrations that change the shape of the excitation PSF. (B) The autocorrelation amplitude is also plotted as a function of penetration depth into solution. (C) The PSF stretches in both the lateral and the axial directions, and fluorescent particles require a longer time to traverse the PSF. (D) Brightness values also decrease sharply as a function of depth, but a calibration experiment can be used to find a relationship between brightness and intensity for a given FFS instrument. For our setup, the brightness falls as the square of the fluorescence intensity.

beyond a few micrometers in thickness, the effects of the changing PSF cannot be ignored.

We note an observed relationship between the average intensity and the correlation amplitude. Figure 4.7D shows $\lambda/\langle F\rangle^x$ plotted as a function of focal depth, where x is an experimentally determined parameter dependent on objective type, magnification, and numerical aperture. For the Zeiss 63 × oil objective, we find $x=2.0\pm0.1$. In the case of a 100 × oil objective (not shown), $x=1.4\pm0.15$. Performing the above calibration on an objective provides a quick method for correcting the depth-dependent brightness bias in subsequent experiments.

6.2. Focal depth dependence of water-immersion objectives

We perform the same experiments as above but now use a Zeiss 63 × C-Apochromat water-immersion objective (N.A. = 1.2). Note that the water objective comes equipped with a correction collar that is tuned to account for the thickness of the coverglass between the objective and the sample. This collar must be set to the proper position for the objective to function as designed. Intensity (Fig. 4.8A), correlation amplitude (Fig. 4.8B), and diffusion time (Fig. 4.8C) are essentially constant over the measured range of 60 μm. The brightness measured with this water-immersion objective is independent of depth as illustrated in Fig. 4.8D. Thus, water-immersion objectives that have been corrected for spherical aberrations are advantageous for brightness experiments where the focal depth is changed. The rest of the experiments in this section are performed with such an objective to take advantage of FFS measurements that are independent of focal depth.

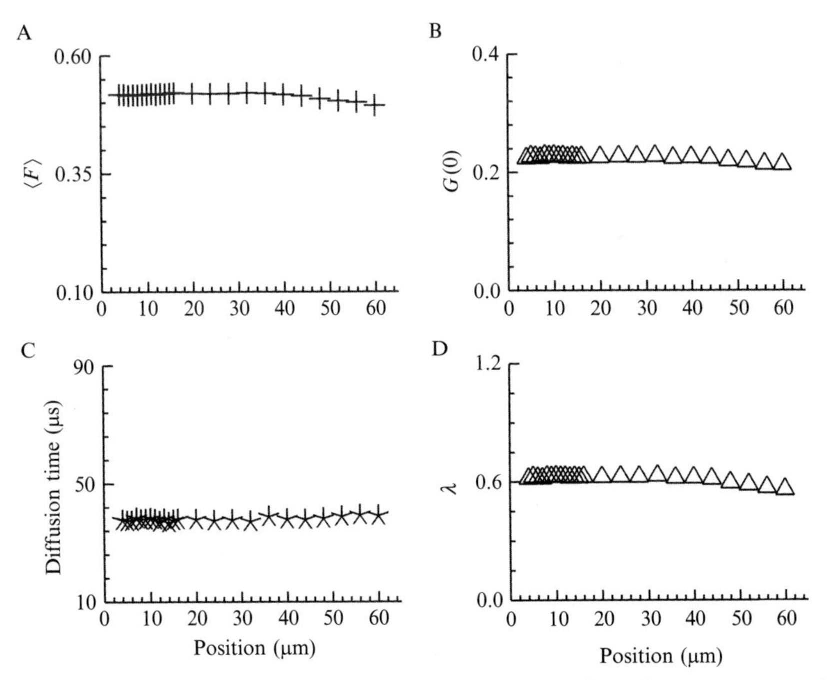

Figure 4.8 *Focal depth dependence of water immersion*. The dye solution experiment is repeated using a water-immersion objective. (A) Fluorescence intensity is shown to be independent of focal depth. (B) Autocorrelation amplitude $G(0)$ and (C) diffusion time are also constant over focal depth. (D) Brightness is essentially constant throughout the 60-μm range measured and is plotted directly as a function of focal depth.

6.3. Thin-layer geometry

We now examine the effects of finite sample geometry on FFS brightness analysis, specifically thin-layer samples like cell cytoplasm. The γ_2 beam shape factor (Thompson, 1991), which we previously mentioned, is used to calculate brightness. The general formula is

$$\gamma_j = \frac{\int PSF^j(r)\mathrm{d}\Omega}{\int PSF(r)\mathrm{d}\Omega} = \frac{V_{\mathrm{PSF}^j}}{V_{\mathrm{PSF}}}. \qquad [4.7]$$

Conventional FFS theories assume a uniform, infinite sample geometry and thus the associated gamma factor $\gamma_{2,\infty}$. In the case of solution samples or cell nuclei—which are large in comparison to the laser excitation volume—this is a perfectly valid assumption. However, the cytoplasmic regions of cells are often smaller than the typical axial or z-dimension of the fluorescence excitation volume. In such a case, some of the excitation light never hits the sample so that $\gamma_{2,\infty}$ no longer accurately describes the shape of the excited sample. We therefore require a gamma factor that can account for sample geometry.

The result of the finite sample volume is that the fluorescent molecules cannot access all parts of the excitation light. Because most applications focus the laser at the midpoint of the sample, the molecules are limited to the central, higher intensity region of the PSF which leads to a skewed spatiotemporal average. Due to these geometric constraints, there is a higher probability of finding molecules in the central focus of the PSF which increases the calculated brightness. This bias can be quite pronounced, which is of concern when trying to use brightness analysis to identify stoichiometry. We take a series of measurements of EGFP expressed in COS-1 cells. EGFP has been demonstrated to exist as a monomer in both the nucleus and the cytoplasm (Chen et al., 2003). Therefore, in terms of normalized brightness $b = \lambda/\lambda_{\mathrm{monomer}}$, the experiment should yield $b = 1$ in all parts of the cell. We begin by focusing in the middle of the nucleus and then performing additional measurements at locations moving outward laterally toward the edge of the cell. The nuclear measurement does indeed return a normalized brightness $b = 1$, but as the measurements move outside the nucleus, the brightness values increase, reaching a value of $b \sim 2$ at the thinnest parts of the cytoplasm. This gives the appearance that EGFP is forming a dimer at the edges of the cell although the increase in brightness is entirely due to the thin-layer brightness bias.

6.4. Sample geometry-dependent FFS

In order to correct for a brightness bias due to confined geometry, it is necessary to introduce a sample geometry function. In this case, we approximate a thin section of cell as a rectangular slab and define the geometry function as follows:

$$S(z;h) = \begin{cases} 1, & h_0 < z < h_0 + h \\ 0, & \text{otherwise}, \end{cases} \qquad [4.8]$$

where z is the axial focus position of the PSF, h is the height of the slab, and h_0 is the z-position of the bottom of the cell or coverglass surface. We can then redefine the accessible PSF volume as

$$V_{\mathrm{PSF}} = \int \mathrm{PSF}\ S(z)\mathrm{d}\Omega. \qquad [4.9]$$

Incorporating this expression into FFS theory, we can derive several shape-dependent definitions of which the two most relevant are average fluorescence intensity and shape factor (Macdonald et al., 2010). The average intensity is defined as

$$\langle F(z;h)\rangle = \lambda\langle c\rangle V_{\mathrm{PSF}}(z;h), \qquad [4.10]$$

where λ is the molecular brightness and c is the average concentration of the fluorescent particles, and the shape factor as

$$\gamma_2(z;h) = \frac{V_{\mathrm{PSF}^2}(z;h)}{V_{\mathrm{PSF}}(z;h)}. \qquad [4.11]$$

With these definitions, it is possible to accurately describe the FFS parameters recovered from a thin-layer sample. However, we are left with the difficulty of determining where the excitation volume is located with respect to the sample along the z-axis. We introduce a z-scan approach, where the beam is scanned uniformly along the z-axis to provide intensity information about the geometry of the fluorescent sample. Figure 4.9A shows a cartoon of a plated cell with a z-scan that begins beneath the coverglass and proceeds up through the top of the cell. For a uniform cytoplasmic layer, the intensity signal (Fig. 4.9C) reaches its maximum when the excitation volume is located at the center of the layer as seen in Fig. 4.9B. Similarly, when the beam has passed entirely through the cell, the intensity signal returns to background. The intensity z-profile (Fig. 4.9C) can be fit to $F(z;h)$ using the slab geometry model to recover the thickness h of the cytoplasmic layer.

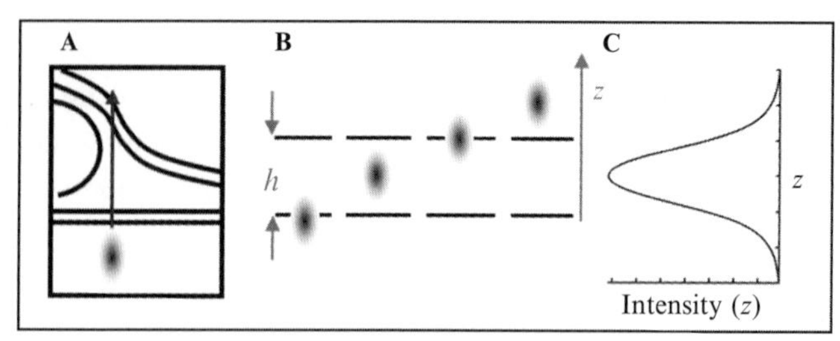

Figure 4.9 *z-Scan approach.* (A) A cartoon displays the side view of a plated cell. The laser is focused beneath the coverslip surface and is subsequently scanned through the cell to gain information about sample geometry. (B) The beam's location relative to the fluorescent sample is frozen at various points during the scan and corresponds to the measured intensity profile. (C) The intensity profile is fit to a model to determine the height of the sample. (For color version of this figure, the reader is referred to the online version of this chapter.)

As a demonstration of the technique, we use the z-scan approach on an EGFP cell. We first position the laser in the x–y plane at the center of the nucleus and perform an intensity z-scan. Following that, we focus the beam at the center of the layer using maximum intensity and take a stationary FFS measurement. A typical z-scan takes only 5–10 s and is performed using a piezo stage and a ramp signal. z-Scan measurements are thus not difficult to perform, nor do they add an appreciable time to the measurement. We repeat these paired measurements at various points moving toward the outer edge of the cell, obtaining brightness and cell thickness at each location. The inset of Fig. 4.10 shows an overlay of the intensity z-profiles ranging from light gray in the nucleus to black at the edge of the cell. The same curves normalized to average peak intensity are shown in the primary plot of Fig. 4.10. The z-profile is given by a convolution of the PSF and the slab geometry. As the slab decreases in thickness, the z-profiles become narrower. While not easy to discern, the final two z-profiles essentially lie directly on top of one another. For sample layers thinner than $\sim$0.5 μm, our present experimental setup can no longer distinguish thickness and so, from the perspective of our technique, these layers may be considered to be two-dimensional. As such, the convolution of the PSF and slab returns an intensity z-profile that is a directly proportional to the radially integrated PSF of our instrument.

6.5. Modified Gaussian–Lorentzian PSF model

FFS theory requires a mathematical description of the PSF of the laser. There are sophisticated and direct methods to measure the PSF of a fluorescence instrument using EM-wave propagation through the microscope objective

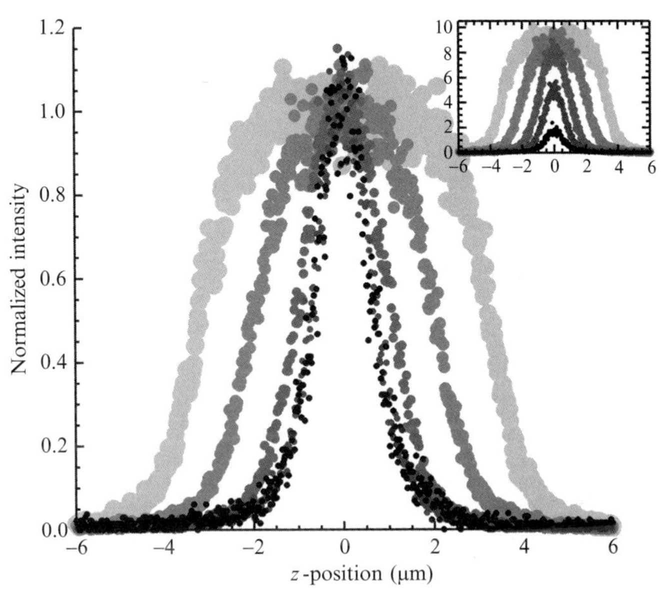

Figure 4.10 *Intensity z-scans.* (Inset) In a cell expressing EGFP, intensity *z*-scan measurements were taken through the center of the nucleus (light gray) and then at four additional locations moving laterally outward toward the thin outer edge of the cell (black). The inset shows the raw intensity *z*-scans. Note that the amplitude decreases as smaller volumes of sample are excited in thin sections. (Primary figure) We normalized the intensities to better observe the profile shape. The convolution of the PSF with a thick section of cytoplasm results in a broad intensity *z*-scan (light gray). The scans from the thinnest two sections lie directly on top of one another as we have reached an effective 2D geometry with regards to the sensitivity of the setup ($\sim$0.5 μm).

(Gregor & Enderlein, 2005; Hess & Webb, 2002). However, because it requires extensive modeling and is numerically expensive, most research groups choose a simple analytical form to describe the PSF. The three-dimensional Gaussian (3DG) and the Gaussian–Lorentzian (GL) are the most widely used models. We endeavored to fit both models to the measured *z*-scan profiles; however, neither model proved to be a good fit. The residuals of the fits showed strong correlations, specifically in the tails of the curve (Macdonald et al., 2010). The 3DG model decays too rapidly, while the GL model decays too slowly to accurately fit the data from our instrument. The inaccuracies from these PSF models are negligible in many FFS applications because the bulk of the fluorescence signal comes from the central region of the PSF which can be well approximated by either of the two models. It is only at the beginning and end of the *z*-scan when the tails of the PSF are the sole source of excitation that discrepancies come to light. Therefore, to perform *z*-scan FFS, it is necessary to develop a model that provides a good fit.

The observations that the GL model decays too slowly in the z-axis led us to develop a modified Gaussian–Lorentzian model (mGL) that can be tuned to match a given experimental setup (Macdonald et al., 2010). We define the mGL model as follows:

$$\mathrm{PSF}_{\mathrm{mGL}}(r,z) = \left(\frac{w_0}{w(z)}\right)^{(1+y)n} \exp\left(-2n\frac{r^2}{w^2(z)}\right), \quad [4.12]$$

where $w^2(z) = w_0^2\left(1 + \frac{z^2}{z_0^2}\right)$ and the radial and axial beamwaists of the PSF are w_0 and z_0, respectively. We introduced an experimentally determined y-parameter which allows us to vary the decay rate of the PSF along the z-axis. Note that for $y=1$, the mGL model reduces to the standard GL model. This mGL model can be calibrated to a given fluorescence instrument by measuring a selection of cells and fitting the z-scan profiles with y as a free parameter. It is necessary to demonstrate that the modified mGL model fits both thin and thick layers. This is best done by taking a series of intensity z-profiles in a single cell, as pictured in Fig. 4.10, and globally fitting all the curves.

6.6. Thin-layer brightness correction

We now address the original issue of the thin-layer brightness bias. We discussed that measurements of monomeric EGFP in very thin sections of cytoplasm return an inflated normalized brightness when using conventional FFS. For our specific setup, we observe an increase by a factor of 1.8. We find it helpful to discuss the brightness bias in terms of the gamma ratio $\gamma_2(z;h)/\gamma_{2,\infty}$ which reflects the observed brightness behavior since the gamma ratio equals 1 for thick samples and it increases as h decreases. We perform z-scan FFS measurements in 10 cells expressing EGFP and plot $\gamma_2(z;h)/\gamma_{2,\infty}$ (crosses) as a function of h in Fig. 4.11. Overlaid are best fits to the gamma ratio data for the 3DG, GL, and mGL models. It is evident that neither the 3DG nor the GL models fit, while the mGL model describes the experimental data very well. For our experimental setup, the mGL model predicts a brightness bias of 1.86 for very thin samples, in close agreement with the data.

Having demonstrated that the z-scan FFS model accurately predicts the brightness bias for thin sections of cytoplasm, it is straightforward to correct that bias. We divide each conventional FFS brightness measurement by the gamma ratio for the corresponding thickness. Under such a z-scan correction, EGFP brightness in cells displays $b=1$ uniformly throughout all sections of the cell.

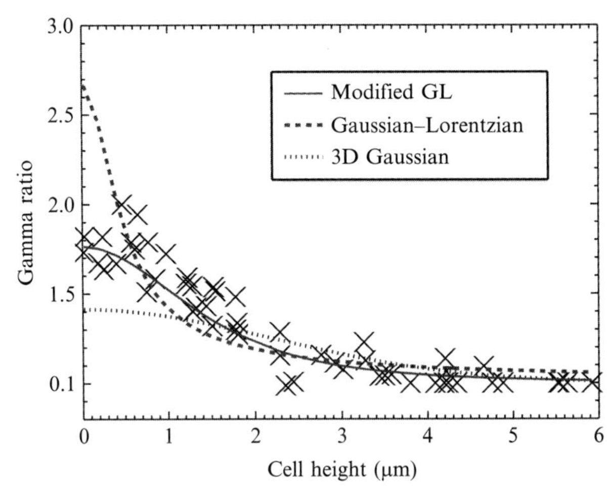

Figure 4.11 *Geometry-dependent shape factor.* The gamma ratio is plotted for different positions in several cells containing EGFP. Thick sample sections, like the nucleus, effectively mimic an infinite volume and thus yield a gamma ratio of 1. Thinner samples lead to an increase in the gamma ratio. The 3D Gaussian and Gaussian–Lorentzian PSF models do not fit the experimental data. The mGL PSF model matches the behavior of experimental data and accurately predicts a brightness increase of almost a factor of 2 at the 2D limit. *Reprinted from Macdonald et al. (2010), Copyright (2010), with permission from Elsevier.* (For color version of this figure, the reader is referred to the online version of this chapter.)

We define a critical thickness as the point at which the brightness bias reaches 20% of the true value and consider that beyond this point, the bias cannot be ignored. For our experimental setup, the critical thickness is 2 μm. COS-1 and U2OS cells have cytoplasmic thicknesses that range from 2.5 μm to under 1 μm so that in these cells only measurements performed next to the nuclear envelope will be above the critical thickness. In CV-1 cells, the thickest cytoplasmic regions are already at or below the critical thickness. The thin-layer bias will also appear in image-based brightness techniques. Since the *z*-scan FFS approach requires only a short intensity scan, it would not be difficult—although more computationally intensive—to generalize intensity profile fitting to a pixel-by-pixel map.

We discussed *z*-scan FFS for only the most basic geometry, a simple slab. The technique has the power to measure more complex geometries, for example, proteins that localize both in the cytoplasm and in the plasma membrane. Brightness analysis in thin-layer geometries is subject to this potentially subtle bias that one should be alert to when measuring in cells or other similar thin samples. *z*-Scan FFS is a quick and simple method for the recovery of unbiased brightness data.

7. SUMMARY

In this chapter, we considered a variety of FFS techniques in the context of quantitative brightness analysis. The text introduced the motivation behind the development of brightness analysis and discussed practical aspects, such as segment analysis. We stressed that in order to correctly interpret brightness data a number of experimental conditions must be considered. Among these factors are repeatable calibration procedures, proper detector characterization, and correct interpretation of the sample and excitation light geometry. While this chapter only focused on single color brightness analysis, the information provided directly extends to dual-color experiments. Overall, we demonstrate that brightness is a powerful and robust indicator of protein stoichiometry provided that theoretical and experimental factors are properly taken into account.

ACKNOWLEDGMENTS

This work was supported by grants from the National Institutes of Health (GM64589 & GM091743) and the National Science Foundation (PHY-0346782).

REFERENCES

Caccia, M., Camozzi, E., Collini, M., Zaccolo, M., & Chirico, G. (2005). Photon moment analysis in cells in the presence of photo-bleaching. *Applied Spectroscopy*, *59*, 227–236.

Chen, Y., Johnson, J., Macdonald, P., Wu, B., & Mueller, J. D. (2010). Observing protein interactions and their stoichiometry in living cells by brightness analysis of fluorescence fluctuation experiments. *Methods in Enzymology*, *472*, 345–363.

Chen, Y., Mueller, J. D., Ruan, Q., & Gratton, E. (2002). Molecular brightness characterization of EGFP in vivo by fluorescence fluctuation spectroscopy. *Biophysical Journal*, *82*, 133–144.

Chen, Y., Mueller, J. D., So, P. T., & Gratton, E. (1999). The photon counting histogram in fluorescence fluctuation spectroscopy. *Biophysical Journal*, *77*, 553–567.

Chen, Y., Mueller, J. D., Tetin, S. Y., Tyner, J. D., & Gratton, E. (2000). Probing ligand protein binding equilibria with fluorescence fluctuation spectroscopy. *Biophysical Journal*, *79*, 1074–1084.

Chen, Y., Wei, L.-N., & Mueller, J. D. (2003). Probing protein oligomerization in living cells with fluorescence fluctuation spectroscopy. *Proceedings of the National Academy of Sciences of the United States of America*, *100*, 15492–15497.

Chen, Y., Wu, B., Musier-Forsyth, K., Mansky, L. M., & Mueller, J. D. (2009). Fluorescence fluctuation spectroscopy on viral-like particles reveals variable gag stoichiometry. *Biophysical Journal*, *96*, 1961–1969.

de Grauw, C. J., Vroom, J. M., van der Voort, H. T. M., & Gerritsen, H. C. (1999). Imaging properties in two-photon excitation microscopy and effects of refractive-index mismatch in thick specimens. *Applied Optics*, *38*, 5995–6003.

Dong, C. Y., Koenig, K., & So, P. (2003). Characterizing point spread functions of two-photon fluorescence microscopy in turbid medium. *Journal of Biomedical Optics*, *8*, 450–459.

Elson, E. L., & Magde, D. (1974). Fluorescence correlation spectroscopy. I. Conceptual basis and theory. *Biopolymers*, *13*, 1–27.

Finn, M. A., Greenless, G. W., Hodapp, T. W., & Lewis, D. A. (1988). Real-time elimination of dead-time and afterpulsing in counting systems. *The Review of Scientific Instruments*, *59*, 2457–2459.

Gregor, I., & Enderlein, J. (2005). Focusing astigmatic Gaussian beams through optical systems with a high numerical aperture. *Optics Letters*, *30*, 2527–2529.

Hess, S. T., & Webb, W. W. (2002). Focal volume optics and experimental artifacts in confocal fluorescence correlation spectroscopy. *Biophysical Journal*, *83*, 2300–2317.

Hillesheim, L. N., & Müller, J. D. (2003). The photon counting histogram in fluorescence fluctuation spectroscopy with non-ideal photodetectors. *Biophysical Journal*, *85*, 1948–1958.

Johnson, J., Chen, Y., & Mueller, J. D. (2010). Characterization of brightness and stoichiometry of bright particles by flow-fluorescence fluctuation spectroscopy. *Biophysical Journal*, *99*, 3084–3092.

Kask, P., Palo, K., Ullmann, D., & Gall, K. (1999). Fluorescence-intensity distribution analysis and its application in biomolecular detection technology. *Proceedings of the National Academy of Sciences of the United States of America*, *96*, 13756–13761.

Macdonald, P. J., Chen, Y., Wang, X., Chen, Y., & Mueller, J. D. (2010). Brightness analysis by z-scan fluorescence fluctuation spectroscopy for the study of protein interactions within living cells. *Biophysical Journal*, *99*, 979–988.

Magde, D., Elson, E., & Webb, W. W. (1972). Thermodynamic fluctuations in a reacting system—Measurement by fluorescence correlation spectroscopy. *Physical Review Letters*, *29*, 705–708.

Mandel, L. (1979). Sub-Poissonian photon statistics in resonance fluorescence. *Optics Letters*, *4*, 205–207.

Mueller, J. D. (2004). Cumulant analysis in fluorescence fluctuation spectroscopy. *Biophysical Journal*, *86*, 3981–3992.

Mueller, J. D., Chen, Y., & Gratton, E. (2000). Resolving heterogeneity on the single molecular level with the photon-counting histogram. *Biophysical Journal*, *78*, 474–486.

Palmer, A. G., 3rd, & Thompson, N. L. (1987). Molecular aggregation characterized by high order autocorrelation in fluorescence correlation spectroscopy. *Biophysical Journal*, *52*, 257–270.

Palo, K., Mets, U., Jäger, S., Kask, P., & Gall, K. (2000). Fluorescence intensity multiple distributions analysis: Concurrent determination of diffusion times and molecular brightness. *Biophysical Journal*, *79*, 2858–2866.

Perroud, T. D., Bokoch, M. P., & Zare, R. N. (2005). Cytochrome c conformations resolved by the photon counting histogram: Watching the alkaline transition with single-molecule sensitivity. *Proceedings of the National Academy of Sciences of the United States of America*, *102*, 17570–17575.

Perroud, T. D., Huang, B., & Zare, R. N. (2005). Effect of bin time on the photon counting histogram for one-photon excitation. *Chemphyschem*, *6*, 905–912.

Qian, H., & Elson, E. L. (1990a). On the analysis of high order moments of fluorescence fluctuations. *Biophysical Journal*, *57*, 375–380.

Qian, H., & Elson, E. L. (1990b). Distribution of molecular aggregation by analysis of fluctuation moments. *Proceedings of the National Academy of Sciences of the United States of America*, *87*, 5479–5483.

Thompson, N. L. (1991). Fluorescence correlation spectroscopy. In J. R. Lakowicz (Ed.), *Topics in fluorescence spectroscopy* (pp. 337–378). New York: Plenum.

Wu, B., & Mueller, J. D. (2005). Time-integrated fluorescence cumulant analysis in fluorescence fluctuation spectroscopy. *Biophysical Journal*, *89*, 2721–2735.

CHAPTER FIVE

Time-Integrated Fluorescence Cumulant Analysis and Its Application in Living Cells

Bin Wu*,†, Robert H. Singer*,†, Joachim D. Mueller‡,§,1

*Department of Anatomy and Structural Biology, Albert Einstein College of Medicine, Bronx, New York, USA

†Gruss-Lipper Biophotonic Center, Albert Einstein College of Medicine, Bronx, New York, USA

‡School of Physics and Astronomy, University of Minnesota, Minneapolis, Minnesota, USA

§Department of Biomedical Engineering, University of Minnesota, Minneapolis, Minnesota, USA

[1]Corresponding author: e-mail address: mueller@physics.umn.edu

Contents

Abstract

Time-integrated fluorescence cumulant analysis (TIFCA) is a data analysis technique for fluorescence fluctuation spectroscopy (FFS) that extracts information from the cumulants of the integrated fluorescence intensity. It is the first exact theory that describes the effect of sampling time on FFS experiment. Rebinning of data to longer sampling times helps to increase the signal/noise ratio of the experimental cumulants of the photon counts. The sampling time dependence of the cumulants encodes both brightness and diffusion information of the sample. TIFCA analysis extracts this information by fitting the cumulants to model functions. Generalization of TIFCA to multicolor FFS experiment is straightforward. Here, we present an overview of the theory, its implementation, as well as the benefits and requirements of TIFCA. The questions of why, when, and how to use TIFCA will be discussed. We give several examples of practical applications of TIFCA, particularly focused on measuring molecular interaction in living cells.

Methods in Enzymology, Volume 518
ISSN 0076-6879
http://dx.doi.org/10.1016/B978-0-12-388422-0.00005-4

1. INTRODUCTION

Fluorescence fluctuation spectroscopy (FFS) is a promising tool for measuring the concentration, mobility, and interactions with great spatiotemporal resolution directly in living cell (Digman & Gratton, 2011; Slaughter & Li, 2010). FFS exploits fluorescence intensity fluctuations of fluorophores passing through a small observation volume created by a confocal or two-photon microscope. Each passage of a fluorescent molecule through the small volume leads to a short burst of detected photons. Collectively, these diffusing molecules give rise to a stochastic fluorescence signal. Various statistical analysis tools are used to extract physical and chemical properties of the fluorescently labeled molecules from the stochastic fluorescence signals. For example, the amount of time it takes for the molecule to diffuse through the observation volume depends on its diffusion constant. Fluorescence correlation spectroscopy (FCS; Berland, So, & Gratton, 1995; Magde, Elson, & Webb, 1972; Rigler, Mets, Widengren, & Kask, 1993; Schwille, Kummer, Heikal, Moerner, & Webb, 2000; Tetin et al., 2006; Webb, 2001) is widely used to measure the diffusion time from the autocorrelation function of the fluorescence signal. The amplitude of the fluorescent burst depends on the number of fluorophores carried by the molecule. The brightness, defined as the average number of photons per second emitted by the molecule, captures the fluctuation amplitude information. The photon counting histogram (PCH) analysis (Chen, Müller, So, & Gratton, 1999) and fluorescence intensity distribution analysis (FIDA; Kask, Palo, Ullmann, & Gall, 1999) measure the molecular brightness by fitting the experimental PCH to a theoretical distribution. Both brightness and diffusion information have been used to characterize fluorescent samples, as described in previous chapters. Here, we focus on time-integrated fluorescence cumulant analysis (TIFCA), a method that unifies both brightness and diffusion into an exact and simple analytical model.

The capability of FFS to accurately measure experimental parameters depends, just like any other techniques, on the signal/noise ratio (SNR) of the data (Müller, Chen, & Gratton, 2000; Saffarian & Elson, 2003). Unfortunately, the achieved SNR in the cellular environment is sufficiently low that resolving heterogeneous biological samples is typically not feasible (Müller et al., 2000), which severely limits the potential of FFS application in cells. However, many factors that affect SNR are either already optimized or beyond our direct experimental control. Here, we focus on the sampling time and consider its effect on the SNR. Conventional FFS analysis assumes a short sampling time compared to the characteristic timescale in order to capture the dynamics of fluctuation. However, this leads to low SNR because the number of photons detected per molecule is small. A longer sampling time results in an improved signal, but the existing theory breaks down because of particle diffusion during the prolonged sampling time. Fluorescence cumulant analysis (Müller, 2004), in contrast to histogram analysis, allows an exact treatment for any sampling time. Cumulants are a set of measures that provide an alternative to the moments of a distribution and have properties particularly suitable for studying random variables (Kendall & Stuart, 1977a; Saleh, 1978). For example, cumulants are additive for independent random variables and each cumulant of a different order contains independent information. TIFCA is based on factorial cumulants of the photon counts that is modeled exactly for arbitrary sampling times (Wu & Müller, 2005).

TIFCA offers advantages compared to conventional FFS analysis tools. *First*, statistical analysis shows that extending the sampling time by rebinning increase the SNR of cumulants (Wu & Müller, 2005). This result is especially important for higher order cumulants, which are notoriously difficult to measure experimentally, but are essential for resolving species. While typical cell experiments only provide two statistically significant cumulants, by choosing a longer sampling time, we are often able to determine the next higher order of cumulants. The additional information provided by the higher order cumulants is crucial for the resolution of mixtures. *Second*, TIFCA collectively analyzes the cumulants for a range of sampling times by rebinning the original data. This approach preserves the temporal information of the original data and at the same time increases the SNR. The sampling time-dependent analysis of TIFCA integrates brightness and diffusion time into the same theory, which effectively combines the strength of both FCS and PCH. Fluorescence intensity multiple distribution analysis (FIMDA) was introduced to extend PCH/FIDA to long sampling times (Palo, Mets, Jäger, Kask, & Gall, 2000; Perroud, Huang, & Zare, 2005) by introducing a sampling time-dependent brightness and number of molecules. However, this approach is an approximation that effectively corrects the first two cumulants of the probability distribution, but the higher order cumulants are not exact. Because resolution of species relies on higher order cumulants, the approximation introduced by FIMDA is of concern and may introduce biases in the analysis of FFS experiments. TIFCA is free of such potential biases. *Third*, TIFCA introduces a simple relationship between cumulants and the FFS parameters brightness, number of molecules, and diffusion time. Since each order of cumulant contains independent information, the number of statistical significant cumulants directly specifies the number of independent parameters that can be determined from the data. We derived a theory to calculate the statistical error of cumulants based on the moments-of-moments technique. With this theoretical error analysis, one can predict the number of statistically significant cumulants for specific experimental condition. This capability is very useful for feasibility studies and experimental design. *Fourth*, TIFCA is particularly suitable for multicolor experiments (Wu, Chen, & Müller, 2006). It is straightforward to generalize the TIFCA theory to an arbitrary number of colors, each measured in a separate detection channel. *Fifth*, the number of data points fitted by TIFCA is largely independent of the intensity. In contrast, the number of data points in PCH analysis scales with intensity, which is an especially important consideration for multicolor experiments, since the number of data points scales with the maximum photon counts raised to the power of the number of detection channels. To illustrate this point consider a two-color experiment (equals two-detection channels) with a maximum photon count of 100 for a given short sampling time. Signal/noise considerations typically restrict statistically significant cumulants up to the fourth order, which leads to 14 distinct cumulants that need to be fitted. Rebinning of data to longer sampling times does not affect the number of cumulants. PCH, on the other hand, has to fit 10^4 data points. In contrast to TIFCA, rebinning increases the maximum photon counts of PCH. For example, rebinning by a factor of 100 leads to a maximum photon count of $\sim 10{,}000$, which translates into a PCH function with $\sim 10^8$ data points. So fitting of data by PCH is computationally far more expensive than for TIFCA.

So what can TIFCA do and when to use TIFCA? This is summarized in Box 5.1.

2. THEORY AND IMPLEMENTATION OF TIFCA

Consider a fluorescent species with brightness λ in a single-color experiment. The nth factorial cumulant of photon counts $\kappa_{[n]}$ is given by Wu and Müller (2005):

$$\kappa_{[n]}(T) = \gamma_n N \lambda^n B_n(T; \tau_d, r), \qquad [5.1]$$

where N is the average number of molecules in the observation volume and τ_d is the diffusion time of the molecule. The parameters γ_n and r describe the

BOX 5.1 What TIFCA does and when to use TIFCA

a. Theoretical analysis. TIFCA provides a simple expression for the factorial cumulants of photon counts. Each cumulant can be determined for an arbitrary sampling time and arbitrary number of detection channels.
b. Improves signal/noise of FFS experiment. By systematically changing the sampling time, the signal/noise of cumulants can be optimized.
c. Analyze experimental data. TIFCA also presents a practical algorithm to fit experimental data. Brightness, diffusion constant, and the number of molecules are obtained simultaneously from the fit. Since the sampling time effect has been explicitly taken into account, there is no undersampling bias (Müller, 2004). Our programs written in IDL or Fortran are freely available for download.
d. TIFCA is particularly suited for multicolor experiment. The method offers a succinct way of data reduction for multicolor experiment. It is easily generalized to any number of colors.

point spread function (PSF) of the observation volume and are determined through a calibration experiment (Palmer & Thompson, 1989; Wu & Müller, 2005). The parameter T is the sampling time and the function $B_n(T;\tau_d,r)$ is called the binning function (Wu & Müller, 2005), which summarizes the dependence of cumulants on the sampling time. Mathematically, the nth binning function involves an integration of the nth order correlation function. The integration cannot be solved analytically for arbitrary PSFs. In practice, the binning function is calculated numerically for the 3D Gaussian PSF and saved in a data table. Specific values of the binning function are extracted from the data table by interpolation. For short sampling time $T \ll \tau_d$, the binning function is approximated by $B_n(T;\tau_d,r) \approx T^n$. In this scenario, the cumulant is reduced to a very simple analytical function $\kappa_{[n]}(T) = \gamma_n N(\lambda T)^n$. In Fig. 5.1, we plot B_n/T^n up to sixth order as a function of sampling time for $\tau_d = 1$. When the sampling time is short, B_n/T^n goes to one as expected. In general, B_n/T^n decays as a function of T, with the higher order binning function decaying faster. The cumulants of a mixture of noninteracting fluorescent species are given by the sum of the cumulants of each individual species according to the additive property of cumulants for independent random variables.

It is straightforward to generalize the theory of TIFCA to multivariate cumulants of arbitrary number of channels, that is, bivariate cumulants describe dual-color FFS data. For simplicity, we limit our discussion to

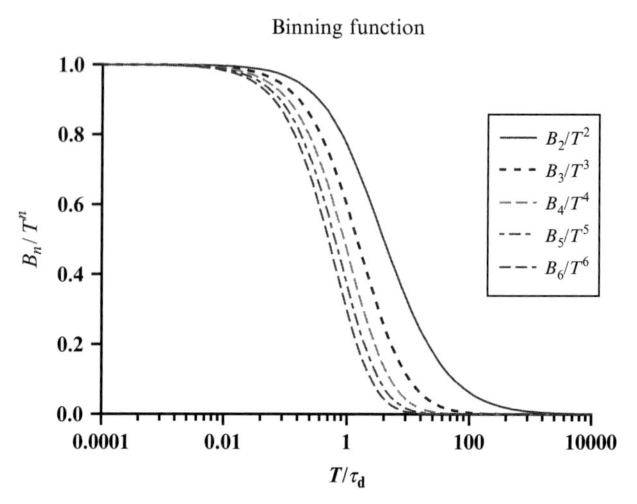

Figure 5.1 The theoretical binning function $B_n(T;\tau_d,r)$ up to sixth order. The function B_n/T^n is plotted as a function of the binning time T. The binning functions are calculated for a diffusion time $\tau_d=1$ and a PSF squared beam waist ratio $r=25$. (For color version of this figure, the reader is referred to the online version of this chapter.)

cumulants for dual-color experiments. The fluorescence is split with a dichroic mirror into two paths with different detectors, which produce two streams of photon counts. As a convention, we use R to refer to the red channel and G to refer to the green channel. Each molecule is characterized by the brightness values in each channel (λ_R,λ_G) and the [m,n]th order bivariate factorial cumulant $\kappa_{[m,n]}$ given by Wu et al. (2006):

$$\kappa_{[m,n]} = N\gamma_{m+n}\lambda_R^m\lambda_G^n B_{m+n}(T;\tau_d,r), \qquad [5.2]$$

where the number of molecule N, the γ-factors, and the binning function are defined the same as in the case of single-color TIFCA.

One important aspect of TIFCA is the error analysis (Kendall & Stuart, 1977b; Wu et al., 2006; Wu & Müller, 2005). The variance of a factorial cumulant is a measure of its statistical accuracy and used as the weight in the nonlinear least-square fit to the theoretical model to evaluate the goodness-of-fit of the data. The variance is also a good indicator of how many statistical significant cumulants are present in the data. One can determine the variance of cumulants by dividing the data into small segments; calculate the cumulants of each segment and the experimental variance of the average. In addition, we can use a technique called moments of moments to calculate the variance directly as a function of cumulants (Kendall & Stuart, 1977b).

The theory we presented so far assumes that the photodetectors are ideal. Real detectors are never ideal. Particularly, dead time and after pulsing cause significant changes in the photon counting statistics and have to be taken into account to obtain the correct description of experimental data (Hillesheim, Chen, & Müller, 2006; Hillesheim & Müller, 2003). An after pulse is a fake pulse following the detection of a real photon count. Deadtime describes a period of time after the registration of a photon in which the detector is unable to generate photon signals. A detailed description of these nonideal effects on fluorescent fluctuation experiments, especially PCH analysis, has been worked out (Hillesheim et al., 2006; Hillesheim & Müller, 2003). To calculate nonideal detector influenced cumulants, we use Taylor expansion to express the deadtime/after pulsing influenced cumulant in terms of ideal cumulants (Wu et al., 2006). Practically, correct brightness and concentration values are recovered over a concentration range of three orders of magnitudes when nonideal detector effects are taken into account.

So far, we discussed the theoretical underpinning of TIFCA. Next, we describe how to implement TIFCA practically. Since the cumulant function is not analytical, nonlinear least-squares data fitting has to be done to determine TIFCA parameters from the data. In Box 5.2, we summarize the process of data analysis. We have written data analysis software in IDL and in Fortran, which we distribute freely (http://singerlab.org/supplements).

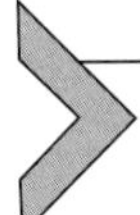

3. APPLICATION OF TIFCA

3.1. TIFCA improves the signal/noise of FFS experiments

The cumulant $\hat{\kappa}_{[n]}$ is calculated from the photon counting data by software. The relative error $\delta\hat{\kappa}_{[n]}$ of the factorial cumulant $\hat{\kappa}_{[n]}$ is defined as $\delta\hat{\kappa}_{[n]} = \sqrt{\mathrm{Var}\left[\hat{\kappa}_{[n]}\right]}/\hat{\kappa}_{[n]}$ and is a measure of the noise-to-signal ratio. A relative error larger than one indicates that the cumulant is not statistical significant. By rebinning the neighboring photons, TIFCA is able to decrease the relative error of the cumulant. This is due to the increased number of photons collected per sampling time for a single molecule. On the other hand, with each rebinning step, the number of data points decreases, which increases the relative error. Which of the two factors dominates depends on the reduced binning time T/τ_{d}. We have shown that for a short sampling time, the relative error $\delta\hat{\kappa}_{[n]}$ scales as $\sqrt{T^{1-n}}$ and for a long

BOX 5.2 How to do TIFCA

a. Calculate experimental factorial cumulants

Since commercial FCS systems do not directly provide the experimental factorial cumulants, they have to be calculated after data acquisition. We used the software MathStatica to derive formulas of factorial cumulants up to the 20th order and the variance of the factorial cumulants up to the 10th order by the technique of moments-of-moments (Kendall & Stuart, 1977b; Wu et al., 2006; Wu & Müller, 2005). The unbiased estimator of a factorial cumulant is just an algebraic function of raw moments and the total number of data points. Typically, the raw photon counting data are acquired at short sampling time. We rebin the data to determine the factorial cumulants for different sampling times. The procedure is performed as follows: we feed the recorded sequence of photon counts into software to calculate the experimental factorial cumulants of photon counts of sampling time *T*. To get cumulant for a sampling or binning time of 2*T*, we add neighboring photon counts together to get a new sequence of photon counts with binning time 2*T*. This process is repeated to calculate the cumulants for binning times of specific integer multiples of *T*. By rebinning, we calculate the factorial cumulants over binning times that cover three orders of magnitude.

b. Fitting of the cumulants

We fit the experimentally determined factorial cumulants $\hat{\kappa}_{[n]}$ to theoretical cumulants $\kappa_{[n]}$ determined with a nonlinear least squares fitting program. The reduced χ^2 of the fit is given by

$$\chi^2 = \frac{1}{(K-p)} \sum_T \sum_n^{r_0} \frac{\left(\hat{\kappa}_{[n]}(T) - \kappa_n(T)\right)^2}{\mathrm{Var}\left[\hat{\kappa}_{[n]}(T)\right]}.$$

The value of *K* is the total number of cumulants used in the fit and *p* is the number of free fitting parameters of the model.

c. Calibration

The theoretical expression of cumulants contains parameters that are best determined empirically. The γ-factors depend on the point spread function (PSF) of the instrument. We use a 3D Gaussian PSF to calculate the binning function, which is sufficient to describe the temporal behavior of the cumulants. However, the absolute value of the γ-factors have to be determined empirically, because the experimental PSF deviates to some degree from the 3D Gaussian model. We fix the first two γ-factors to that of the 3D Gaussian, $\gamma_1 = 1$ and $\gamma_2 = 0.3535$. To calibrate high order γ-factors, we perform experiments on a simple fluorescent dye solution that serves as a good representation of a single brightness sample. We fit the first four cumulants simultaneously to determine brightness λ, the diffusion time τ_d, the average number of molecules *N*, γ_3, and γ_4. Alternatively, we can derive a theoretical expression of γ-factors based

BOX 5.2 How to do TIFCA—Cont'd

on a parametric PSF. For example, using the algorithm of PCH calibration (Huang, Perroud, & Zare, 2004), we derive $\gamma_n = \gamma_n^{3DG}(1+F_1)^{n-2}/(1+F_2)^{n-1}$, where the F-parameters are used to calibrate the PSF for PCH function. The advantage of this approach is that it predicts higher order γ-factors that are difficult to determine by an experimental calibration.

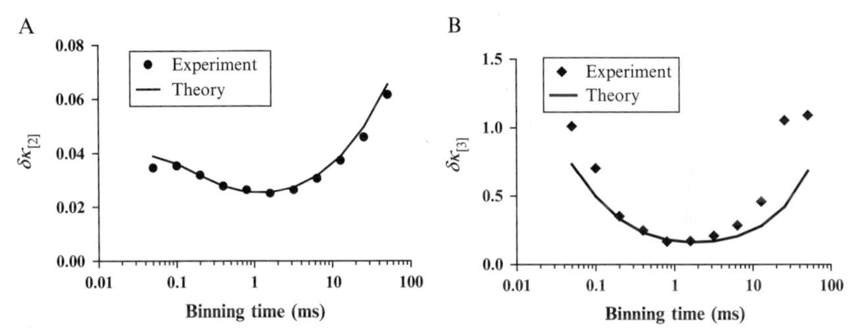

Figure 5.2 The relative error is plotted as a function of binning time for the second (A) and the third (B) factorial cumulant (symbol, experiment; line, theory). A U2OS cell expressing EGFP is measured by FFS in the nucleus for 1 min. The error of the cumulants is calculated with the moments-of-moments technique. Increasing the binning time leads to an initial decrease of the relative error until a minimum is reached, which is followed by a steady increase in the relative error. The decrease of relative error for $\hat{\kappa}_{[3]}$ is particularly important since initially the relative error is larger than 1, which means it is not statistically significant at the original data acquisition frequency.

sampling time as $\sqrt{T^{n-1}}$ (Wu & Müller, 2005). Therefore as long as $n>1$, rebinning reduces the relative error of $\hat{\kappa}_{[n]}$ for short sampling but increase the relative error at long sampling time. This is demonstrated in Fig. 5.2A and B, where we plot the relative error of the second and the third cumulant for EGFP measured in a living cell. The experimentally observed dependence of the relative error exactly mirrors the behavior predicted by theory. The relative error decreases initially with each increase in sampling time, reaches a minimum and then increases at long sampling times. The decrease in relative error is particularly significant for the third cumulant. At the original data sampling time, the relative error is larger than one, indicating that $\hat{\kappa}_{[3]}$ is not statistical significant. Rebinning reduces the relative error and makes $\hat{\kappa}_{[3]}$ significant, and thus available for data fitting.

The concept of choosing a sampling time to minimize the relative error has important implication for FFS experiments, because it maximizes the number of independent parameter that can be determined from the data (Müller, 2004). For singe-color FFS experiment, each species is characterized by three parameters: λ, τ_d, and N. The diffusion time τ_d is determined from the shape of the time-dependent cumulant function $\kappa_{[n]}(T)$. Therefore, two cumulants are needed to identify the brightness and number of molecules of each species. For example, a binary mixture requires the knowledge of four cumulants to identify its components. If the experimental data only contain statistically significant cumulants up to second order, resolving two species is impossible. However, if the third-order cumulant is statistically significant and the brightness of one species is known independently, it is possible to determine the brightness of the second species.

3.2. TIFCA resolves a binary mixture with three cumulants in living cells

Previously, we have demonstrated that it is possible to measure cumulant up to seventh order in an *in vitro* experiment (Müller, 2004). However, the achievable SNR in live-cell measurement is significantly limited. Nevertheless, as we have just shown in the last section, it is possible to measure three statistical significant cumulants with a single-color FFS experiment in living cells. With two cumulants, it is possible to define an apparent brightness (Chen, Wei, & Müller, 2003). The normalized brightness is defined as the ratio between the apparent brightness and the monomer brightness. By measuring the normalized brightness as a function of concentration (brightness titration), the oligomerization and affinity of protein interaction can be directly quantified in living cells. This is a powerful and robust technique and has been successfully applied to measure protein oligomerization (Chen et al., 2003). However, in certain circumstances, the brightness titration is incomplete and the oligomerization cannot be conclusively determined. To illustrate this point, we measured the nuclear receptor retinoid X receptor fused with EGFP (EGFP-RXR) (Chen et al., 2003). The experiment was done without the presence of ligand. In Fig. 5.3A, the normalized brightness is plotted as a function of EGFP-RXR concentration. The brightness increases slightly as a function of EGFP-RXR concentration, indicating that the protein oligomerizes weakly. However, the brightness only increases roughly to 1.5 at the highest concentration. Since the oligomerization number must be an integer number, a fractional number less than two

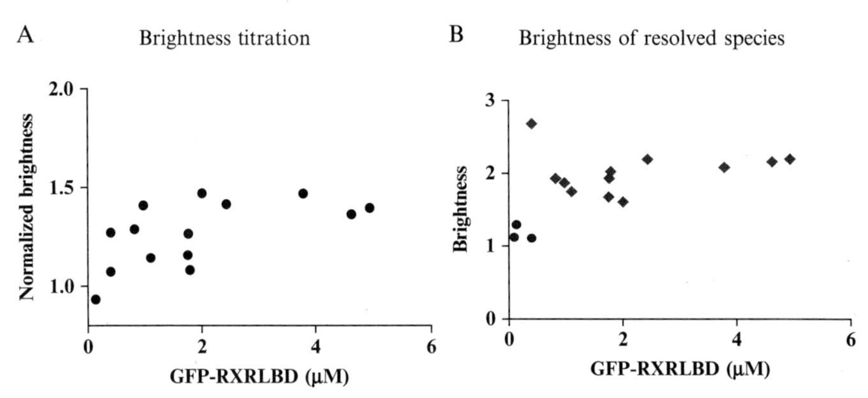

Figure 5.3 Resolving species using single-color TIFCA in cells. EGFP-RXR is transfected in U2OS cells. Each cell was measured for 30 s in the nucleus in the absence of ligand. (A) The normalized brightness of the sample, normalized by the monomer EGFP brightness, was plotted as a function of concentration of EGFP-RXR. The brightness reaches 1.5 at high concentrations, indicating that oligomerization occurs even in the absence of ligand. But the level of oligomerization remains unknown. (B) The same data were fitted with a two-species TIFCA model. Brightness of one species is fixed to EGFP, while all other parameters are allowed to vary freely. The brightness of the second species recovered from the fit is plotted as a function of the total concentration of EGFP-RXR. The data are divided into two groups. In some cells (blue circle), the second species remains to be a monomer. These cells typically have a low concentration of EGFP-RXR. In other cells, the brightness of the second species clusters around 2, suggesting that a fraction of EGFP-RXR in these cells oligomerizes and forms dimers. (For interpretation of the references to color in this figure legend, the reader is referred to the online version of this chapter.)

suggests that the system is a mixture of monomer and oligomers. However, the apparent brightness is unable to reveal the nature of the oligomer.

With the knowledge of the third cumulants, it is possible to get more information from the data. Since one of the species must be monomeric, we fit the data to a two-species model with the brightness of one species fixed to that of monomer EGFP. Three cumulants allows us to determine the brightness of the second species. In Fig. 5.3B, we plot the brightness of the second species as a function of total EGFP-RXR concentration. Note that each point is a measurement of a single cell. To aid the visual interpretation, we divide the brightness into two groups. The first group of brightness (blue circle) is roughly one, which indicates that for these cells the second species is also a monomer. These data correspond to the data points in Fig. 5.3A with apparent brightness close to one. The other group of brightness (red diamond) is scattered around two, which demonstrates that the second species is a dimer. It has been shown that EGFP-RXR forms dimer in the presence of ligand. The current data show that the dimer exists even in the absence of applied ligand.

3.3. Calibrate an mRNA imaging system with TIFCA

Imaging mRNA with single-molecule sensitivity in live cells has become an indispensable tool for the quantitative studying of RNA biology. The MS2/PP7 system has been extensively used due to its unique simplicity and sensitivity (Bertrand et al., 1998; Chao, Patskovsky, Almo, & Singer, 2008; Golding, Paulsson, Zawilski, & Cox, 2005; Larson, Zenklusen, Wu, Chao, & Singer, 2011; Zimyanin et al., 2008). Here, we use the PP7 system as an example (Chao et al., 2008; Larson et al., 2011). In this labeling method, a genetically encoded sequence derived from the bacteriophage PP7 is inserted into the gene of interest. The sequence folds into a unique stem–loop structure that forms the PP7 binding site (PBS) for the PP7 capsid protein (PCP). When cells expressing the gene carrying PBS also express PCP fused to a fluorescent protein (PCP-FP), the mRNA of interest is fluorescently labeled by PCP-FP. To increase the signal of mRNA over the background of free PCP-FP, multiple copies of PBS are utilized. Quantitative fluorescence imaging and spectroscopy require knowledge of the labeling efficiency of mRNA. A uniform labeling of mRNA makes it easy for quantitative interpretation of experimental results. FFS offers a simple method to measure the number of CP-FPs bound to an mRNA by the normalized brightness of an mRNA (Wu, Chao, & Singer, 2012). Furthermore, the mRNA size is significantly larger than free CP-FP and diffuses much slower. Therefore, one can distinguish them by both brightness and diffusion time. TIFCA (Wu & Müller, 2005) is ideal for the analysis since it incorporates both brightness and diffusion time into the same analysis model.

We constructed a plasmid coding for cyan fluorescent protein (CFP), with $24 \times$ PBS inserted after the stop codon in the 3′-untranslated region (Fig. 5.4A). The plasmid was transiently transfected together with nuclear localization signal (NLS)–tandem dimeric version of PP7 coat protein (tdPCP)-EGFP in U2OS cells (Wu et al., 2012). The NLS was used to sequester the nonbound coat protein in the nucleus and we have used a single-chain ttdPCP (Wu et al., 2012). The experiment was done at the two-photon laser wavelength 1010 nm so that CFP will not be excited. We analyzed the data with TIFCA. A one-species model was not able to fit the data, which was expected since both mRNA and free tdPCP-EGFP are present. We proceeded to fit the data with a two-species model, which

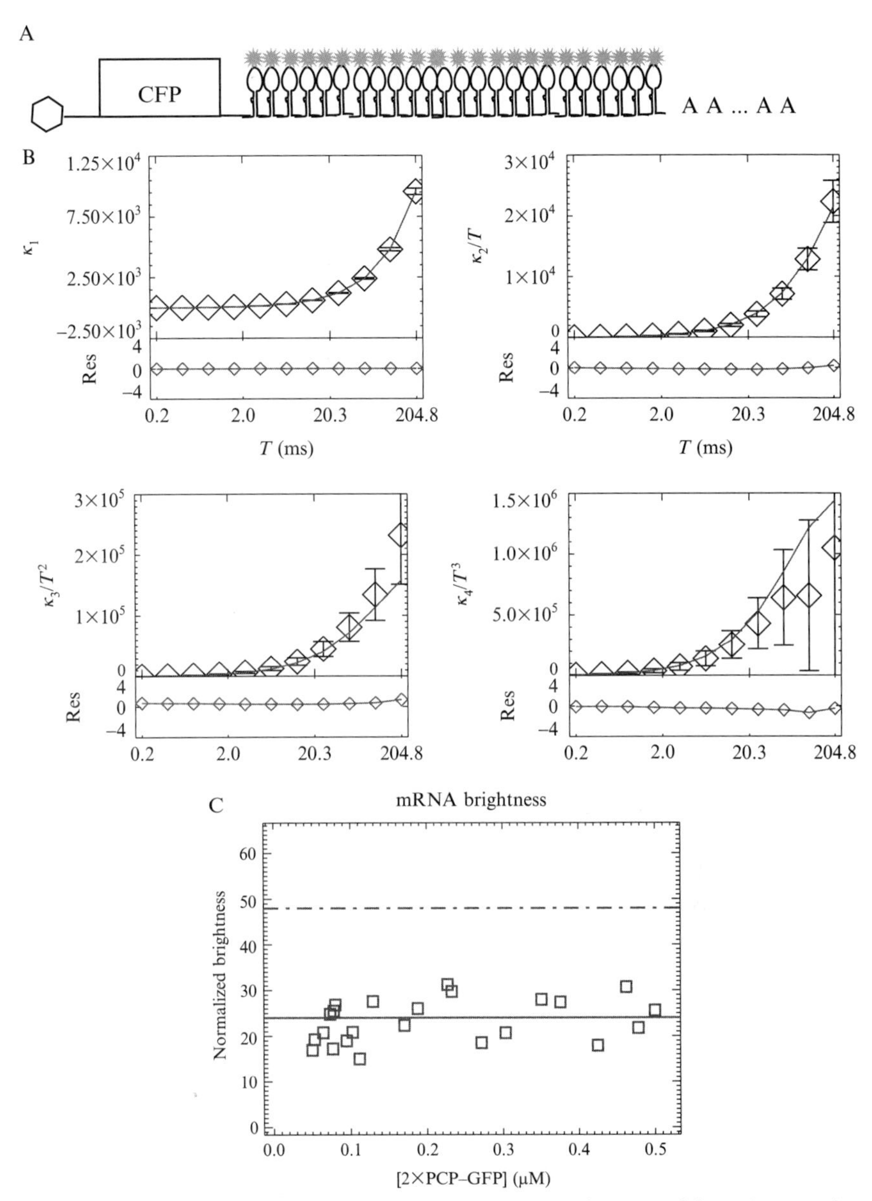

Figure 5.4 Calibrate mRNA brightness. (A) The schematic diagram of the mRNA used in the experiment. The mRNAs encodes cyan fluorescent protein (CFP) in its open reading frame. After the stop codon, 24 × PBS are inserted in the 3′-untranslated region. (B) The plasmid was transiently transfected together with a plasmid which expresses nucleus localization signal nucleus localization signal (NLS)–tandem dimeric version of PP7 coat protein (tdPCP)-EGFP in U2OS. Here, we have used a single-chain tdPCP. The NLS was used to sequester the protein in the nucleus. A cell was measured in the perinuclear

describe the data within experimental uncertainty. An example of the fit is presented in Fig. 5.4B. The brightness of mRNA is determined from the fit. In Fig. 5.4C, we plot the normalized brightness of the CFP-24 $\times$ PBS mRNA labeled by NLS–tdPCP-EGFP as a function of total EGFP concentration. Each symbol represents a measurement of a single cell. Even though there are different concentrations of mRNA and tdPCP between cells, the brightness and therefore the number of coat proteins binding to the mRNA is relatively constant. The average number of NLS–tdPCP-EGFP on one mRNA is 23 ± 5, that is, within error, equal to the expected maximum occupancy for 24 $\times$ PBS.

3.4. Resolve an EGFP/EYFP binary mixture in living cells

Dual-color TIFCA of two-channel FFS data characterizes each species by four parameters: λ_R, λ_G, τ_d, and N. The diffusion time τ_d is again determined from the shape of the time-dependent cumulant function $\kappa_{[m,n]}(T)$. Therefore, three cumulants are needed to identify the remaining three parameters of each species. As we have just demonstrated, it is possible to get statistically significant cumulants up to the third order. Therefore, a total of nine measurable cumulants are present: $\kappa_{[1,0]}$, $\kappa_{[0,1]}$, $\kappa_{[2,0]}$, $\kappa_{[1,1]}$, $\kappa_{[0,2]}$, $\kappa_{[3,0]}$, $\kappa_{[2,1]}$, $\kappa_{[1,2]}$, and $\kappa_{[0,3]}$, which opens up the possibility to directly resolve two species in a cell with a single measurement (Wu et al., 2006).

To test the efficacy of dual-color TIFCA for resolving a binary mixture of fluorescent proteins, we use EGFP and EYFP as a model system. The spectrum of EGFP and EYFP is plotted in Fig. 5.5A together with the transmission curve of the dichroic mirror. We first calibrate their brightness by measuring cells transfected with only one of the proteins. Dual-color TIFCA fits recover the brightness of each of the two proteins in each channel. The brightness are combined into a two dimensional vector and are plotted in Fig. 5.5B (EGFP, +; EYFP, X). Each point in the plot is a unique combination of red and green brightness, which can be viewed as a signature

region for 3 min at the wavelength 1010 nm. A two-species TIFCA model fits the data and yield the mRNA normalized brightness of 26.1, the diffusion constant of 0.35 $\mu m^2/s$ and a concentration of 13 nM. The four panels in the figure represent the first four factorial cumulants of photon counts (symbols) and the theoretical fit to Eq. (5.2) (lines). (C) A series of cells were measured and analyzed as described in (B). The normalized mRNA brightness values, which measure the number of EGFP on mRNA, plotted as a function of total concentration of NLS–tdPCP-EGFP, are determined by dividing the total fluorescence intensity by the EGFP brightness. The data indicate that the average number of EGFP on mRNA is 23 ± 5, implying that 24 $\times$ PBS are fully occupied. (For color version of this figure, the reader is referred to the online version of this chapter.)

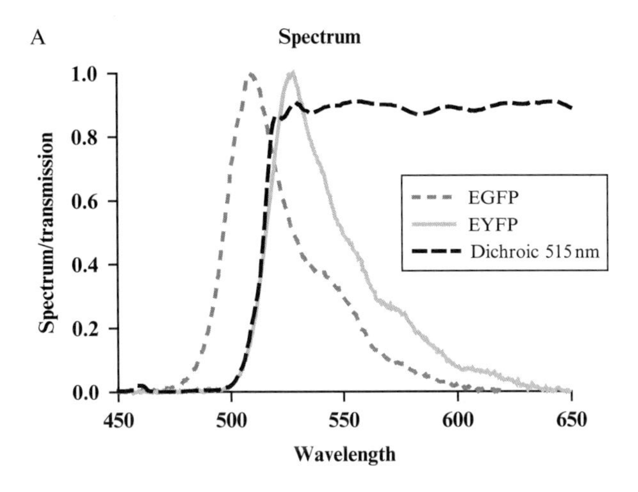

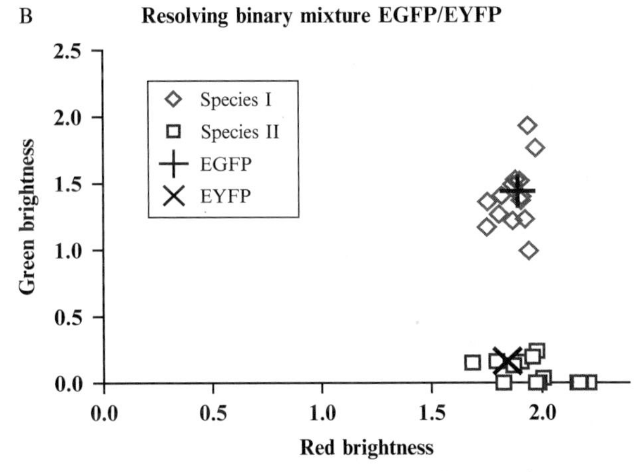

Figure 5.5 Resolving EGFP/EYFP mixtures in living cells. (A) The emission spectra of EGFP (green dotted line) and EYFP (solid line) are plotted together with the transmission curve (dashed line) of the dichroic mirror used to separate the fluorescence into two channels. (B) EGFP and EYFP are transiently expressed in COS cells. Cells were measured for 5 min in the nucleus. The data were fitted to a two-species dual-color TIFCA model. The brightness of each species (λ_R,λ_G) is represented as a point in the brightness signature plot. The brightness of species I is close to that of the EGFP (+), and the brightness of species II is close to that of EYFP (X). The EGFP and EYFP brightness signature are obtained in advance by a calibration experiment. (For interpretation of the references to color in this figure legend, the reader is referred to the online version of this chapter.)

of a molecular species. Next, cells are cotransfected with EGFP and EYFP and are measured for 5 min with a sampling time of 50 μs. The single-species fit to these data returns reduced χ^2 values that range from 10 to 100 depending on the cotransfection ratio of the two proteins, which indicate that we are dealing with mixtures. Figure 5.5B shows the molecular brightness recovered from a two-species fit. It is clear from the brightness signature that species I (diamond) is EGFP and species II (square) is EYFP. The fit also recovers the concentration of the proteins, which are not shown.

3.5. Measure protein interaction with heterospecies partition analysis

In the previous section, we showed that it is possible to obtain nine statistically significant bivariate cumulants in live-cell experiments and a binary protein mixture can be reliably resolved with a single measurement. Thus, dual-color brightness analysis has tremendous potential for studying protein heterointeractions. However, heterointeractions between two proteins, D and A, result generally in a mixture of at least three species (D, A, DA, DA_2, etc.). Unfortunately, a general analysis method for extracting the brightness of three or more species from a heterogeneous sample is not available. Recently, we introduced heterospecies partition analysis to tackle a general interaction scheme between two proteins $D + nA \leftrightarrow DA_n$. We label protein D with EGFP and A with mCherry. With proper choice of filters, it is possible to eliminate the green channel fluorescence of mCherry completely (Fig. 5.6A). HSP combines all heterointeracting molecules into a single heterospecies $H = \{D, DA, DA_2, \ldots\}$ (Fig. 5.6B). Using the two first-order and three second-order cumulants, and with the proper choice of filter set, the red channel brightness of heterospecies λ_H is analytically related to the stoichiometry n and the degree of binding (Wu, Chen, & Müller, 2010).

We apply HSP to study the interaction between RXR and a coactivator transcription intermediate factor-2 (TIF2) (Wu et al., 2010). Cells expressing mCherry-RXR and EGFP-TIF2 are measured in the nucleus of CV-1 cells. The brightness of the heterospecies is determined by dual-color TIFCA fit to the two-species HSP model. The red channel brightness of the heterospecies is shown in Fig. 5.6C as a function of mCherry-RXR concentration. Without ligand the interaction between RXR and coactivator is weak with less than one NR bound per coactivator molecule on average. The strength of interaction increases if the ligand is present. At the lowest concentration, the brightness of the heterospecies is close to that of EGFP, indicating that EGFP-TIF2 and mCherry-RXR do not interact with each other. The brightness increases with growing RXR concentration and saturates at a brightness, which corresponds to the hetero-trimer TIF2–RXR_2. Thus, the experiment demonstrates that RXR and TIF2

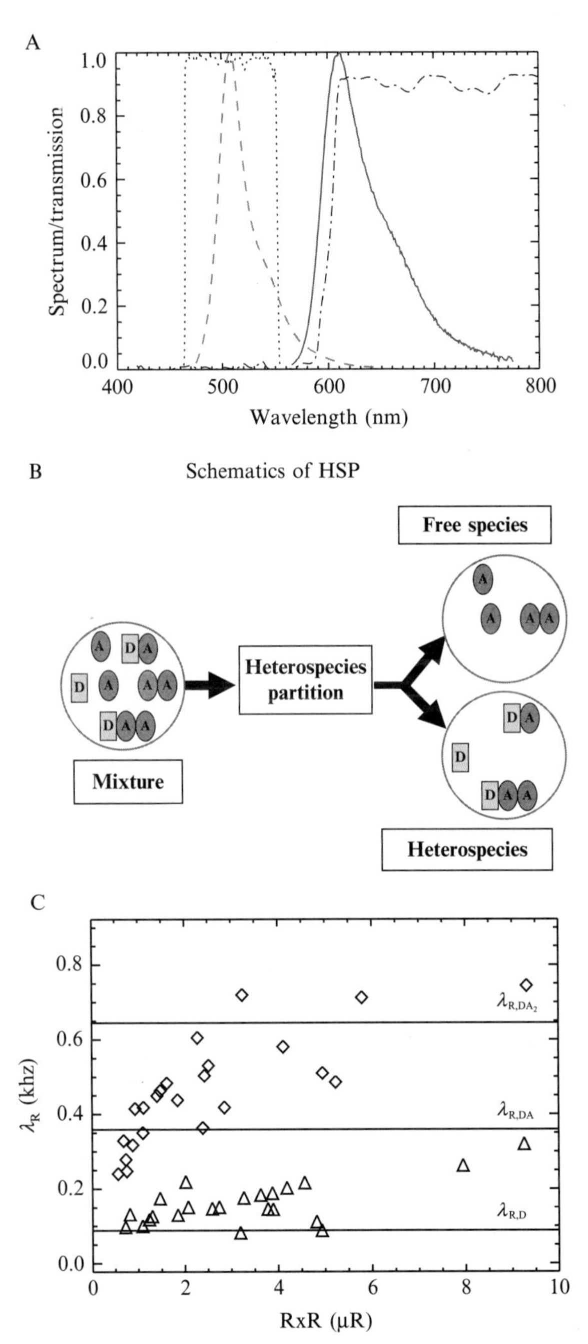

Figure 5.6 (A) The spectra of EGFP (green dash) and mCherry (purple solid) are plotted together with the transmission curve of the dichroic mirror (black dash-dot) and the emission filter in the green channel (blue dot). This filter set completely eliminates

interact in the presence of ligand and two nuclear receptors are recruited by one coactivator. Previous *in vitro* and *in vivo* experiments based on fragments of nuclear receptor and coactivator shows distinct binding stoichiometry (Chen & Müller, 2007; Margeat et al., 2001; Teichert et al., 2009). With full-length proteins, a 2:1 stoichiometry between NRs to coactivator was observed, which confirms the binding model suggested from interaction between NR and its hormone response elements (Xu, Glass, & Rosenfeld, 1999).

4. CONCLUSION

In this chapter, we reviewed the technique of TIFCA for the analysis of FFS experiment. TIFCA differs from other methods by using the factorial cumulants of photon counts to extract information. In contrast to PCH, FIDA or FIMDA, TIFCA is exact for all sampling times. A central concept of the theory is the binning function, which characterizes the influence of sampling time on cumulants. The error analysis of cumulants allows experimentalist to measure or predict whether experimental conditions are sufficient for resolving species and helps in identifying optimal experimental conditions. Nonideal photodetector effects on cumulants are also taken into account in the theory. In practice, parameters are reliably recovered for a concentration range of three orders of magnitude. Statistical significant higher order cumulants established by TIFCA help in resolving binary mixtures and in determining the oligomerization of proteins that is difficult to achieve with other analysis methods. By combining heterospecies

fluorescence contributions of mCherry to the green detection channel. (B) Conceptual picture illustrating the projection of a mixture of brightness species into two different classes. One class contains the molecules of A that are not interacting with D, which is referred to as free species (A, A_2, . . ., A_r). The other class includes all species that contain the molecule D (D, DA, DA_2, . . ., DA_s) and is called heterospecies. The FFS parameters describing the heterospecies characterize the binding between D and A. (C) Full-length EGFP-TIF2 and mCherry-RXR are cotransfected in CV-1 cells. The cells were measured in the nuclei for 1 min. The bivariate factorial cumulants are fitted by dual-color TIFCA to a two-species HSP model. As a result, the brightness of the heterospecies is recovered. The red channel brightness of the heterospecies is plotted as a function of mCherry-RXR concentration. The theoretical brightness of a hetero-dimer and that of a hetero-trimer are shown as solid lines for reference. In the absence of ligand (triangle), there is only weak interaction. In the presence of ligand 9-*cis* retinoic acid (diamond), EGFP-TIF2 interacts strongly with mCherry-RXR and binds to as many as two mCherry-RXR molecules at high concentrations. (For interpretation of the references to color in this figure legend, the reader is referred to the online version of this chapter.)

partition analysis with dual-color TIFCA, we are able to determine the oligomerization and binding curve of a general type of protein hetero-interactions. This paper demonstrates the significant potential of TIFCA as a sensitive and robust technique to characterize molecular interaction in living cells.

ACKNOWLEDGMENTS

We thank Jinhui Li for providing data about EGFP-RXR in the absence of ligand. B. W. is supported by grants from NIH GM84364 and GM86217 to R. H. S. J. D. M. is supported by grants from NIH GM64589 and NSF 0346782.

REFERENCES

Berland, K. M., So, P. T., & Gratton, E. (1995). Two-photon fluorescence correlation spectroscopy: Method and application to the intracellular environment. *Biophysical Journal, 68*(2), 694–701.

Bertrand, E., Chartrand, P., Schaefer, M., Shenoy, S. M., Singer, R. H., & Long, R. M. (1998). Localization of ASH1 mRNA particles in living yeast. *Molecular Cell, 2*(4), 437–445.

Chao, J. A., Patskovsky, Y., Almo, S. C., & Singer, R. H. (2008). Structural basis for the coevolution of a viral RNA-protein complex. *Nature Structural and Molecular Biology, 15*(1), 103–105.

Chen, Y., & Müller, J. D. (2007). Determining the stoichiometry of protein hetero-complexes in living cells with fluorescence fluctuation spectroscopy. *Proceedings of the National Academy of Sciences of the United States of America, 104*(9), 3147–3152.

Chen, Y., Müller, J. D., So, P. T., & Gratton, E. (1999). The photon counting histogram in fluorescence fluctuation spectroscopy. *Biophysical Journal,* 77(1), 553–567.

Chen, Y., Wei, L. N., & Müller, J. D. (2003). Probing protein oligomerization in living cells with fluorescence fluctuation spectroscopy. *Proceedings of the National Academy of Sciences of the United States of America, 100*(26), 15492–15497.

Digman, M. A., & Gratton, E. (2011). Lessons in fluctuation correlation spectroscopy. *Annual Review of Physical Chemistry, 62,* 645–668.

Golding, I., Paulsson, J., Zawilski, S. M., & Cox, E. C. (2005). Real-time kinetics of gene activity in individual bacteria. *Cell, 123*(6), 1025–1036.

Hillesheim, L. N., Chen, Y., & Müller, J. D. (2006). Dual-color photon counting histogram analysis of mRFP1 and EGFP in living cells. *Biophysical Journal, 91*(11), 4273–4284.

Hillesheim, L. N., & Müller, J. D. (2003). The photon counting histogram in fluorescence fluctuation spectroscopy with non-ideal photodetectors. *Biophysical Journal, 85*(3), 1948–1958.

Huang, B., Perroud, T. D., & Zare, R. N. (2004). Photon counting histogram: One-photon excitation. *ChemPhysChem, 5*(10), 1523–1531.

Kask, P., Palo, K., Ullmann, D., & Gall, K. (1999). Fluorescence-intensity distribution analysis and its application in biomolecular detection technology. *Proceedings of the National Academy of Sciences of the United States of America, 96*(24), 13756–13761.

Kendall, M. G., & Stuart, A. (1977a). *Chapter 3: Moments and cumulants. The advanced theory of statistics.* New York: MacMillan Publishing Co., Inc. pp. 57–96.

Kendall, M. G., & Stuart, A. (1977b). *Chapter 12: Cumulants of sampling distributions (1). The advanced theory of statistics.* New York: MacMillan Publishing Co., Inc. pp. 293–328.

Larson, D. R., Zenklusen, D., Wu, B., Chao, J. A., & Singer, R. H. (2011). Real-time observation of transcription initiation and elongation on an endogenous yeast gene. *Science*, *332*(6028), 475–478.

Magde, D., Elson, E., & Webb, W. W. (1972). Thermodynamics fluctuations in a reacting system: Measurement by fluorescence correlation spectroscopy. *Physical Review Letters*, *29*, 705–708.

Margeat, E., Poujol, N., Boulahtouf, A., Chen, Y., Müller, J. D., Gratton, E., et al. (2001). The human estrogen receptor alpha dimer binds a single SRC-1 coactivator molecule with an affinity dictated by agonist structure. *Journal of Molecular Biology*, *306*(3), 433–442.

Müller, J. D. (2004). Cumulant analysis in fluorescence fluctuation spectroscopy. *Biophysical Journal*, *86*(6), 3981–3992.

Müller, J. D., Chen, Y., & Gratton, E. (2000). Resolving heterogeneity on the single molecular level with the photon-counting histogram. *Biophysical Journal*, *78*(1), 474–486.

Palmer, A. G., & Thompson, N. L. (1989). High-order fluorescence fluctuation analysis of model protein clusters. *Proceedings of the National Academy of Sciences of the United States of America*, *86*(16), 6148–6152.

Palo, K., Mets, U., Jäger, S., Kask, P., & Gall, K. (2000). Fluorescence intensity multiple distributions analysis: Concurrent determination of diffusion times and molecular brightness. *Biophysical Journal*, *79*(6), 2858–2866.

Perroud, T. D., Huang, B., & Zare, R. N. (2005). Effect of bin time on the photon counting histogram for one-photon excitation. *ChemPhysChem*, *6*(5), 905–912.

Rigler, R., Mets, U., Widengren, J., & Kask, P. (1993). Fluorescence correlation spectroscopy with high count rate and low background: Analysis of translational diffusion. *European Biophysical Journal*, *22*, 169–175.

Saffarian, S., & Elson, E. L. (2003). Statistical analysis of fluorescence correlation spectroscopy: The standard deviation and bias. *Biophysical Journal*, *84*(3), 2030–2042.

Saleh, B. (1978). *Photoelectron statistics*. New York: Springer-Verlag.

Schwille, P., Kummer, S., Heikal, A. A., Moerner, W. E., & Webb, W. W. (2000). Fluorescence correlation spectroscopy reveals fast optical excitation-driven intramolecular dynamics of yellow fluorescent proteins. *Proceedings of the National Academy of Sciences of the United States of America*, *97*(1), 151–156.

Slaughter, B. D., & Li, R. (2010). Toward quantitative "in vivo biochemistry" with fluorescence fluctuation spectroscopy. *Molecular Biology of the Cell*, *21*(24), 4306–4311.

Teichert, A., Arnold, L. A., Otieno, S., Oda, Y., Augustinaite, I., Geistlinger, T. R., et al. (2009). Quantification of the vitamin D receptor-coregulator interaction. *Biochemistry*, *48*(7), 1454–1461.

Tetin, S. Y., Ruan, Q., Saldana, S. C., Pope, M. R., Chen, Y., Wu, H., et al. (2006). Interactions of two monoclonal antibodies with BNP: High resolution epitope mapping using fluorescence correlation spectroscopy. *Biochemistry*, *45*(47), 14155–14165.

Webb, W. W. (2001). Fluorescence correlation spectroscopy: Inception, biophysical experimentations, and prospectus. *Applied Optics*, *40*, 3969–3983.

Wu, B., Chao, J. A., & Singer, R. H. (2012). Fluorescence fluctuation spectroscopy enables quantitative imaging of single mRNAs in living cells. *Biophysical Journal*, *102*(12), 2936–2944 Epub 2012, June 19.

Wu, B., Chen, Y., & Müller, J. D. (2006). Dual-color time-integrated fluorescence cumulant analysis. *Biophysical Journal*, *91*(7), 2687–2698.

Wu, B., Chen, Y., & Müller, J. D. (2010). Heterospecies partition analysis reveals binding curve and stoichiometry of protein interactions in living cells. *Proceedings of the National Academy of Sciences*, *107*(9), 4117–4122.

Wu, B., & Müller, J. D. (2005). Time-integrated fluorescence cumulant analysis in fluorescence fluctuation spectroscopy. *Biophysical Journal*, *89*, 2721–2735.

Xu, L., Glass, C. K., & Rosenfeld, M. G. (1999). Coactivator and corepressor complexes in nuclear receptor function. *Current Opinion in Genetics and Development*, *9*(2), 140–147.
Zimyanin, V. L., Belaya, K., Pecreaux, J., Gilchrist, M. J., Clark, A., Davis, I., et al. (2008). In vivo imaging of oskar mRNA transport reveals the mechanism of posterior localization. *Cell*, *134*(5), 843–853.

CHAPTER SIX

Raster Image Correlation Spectroscopy and Number and Brightness Analysis

Michelle A. Digman[*,†], **Milka Stakic**[*], **Enrico Gratton**[*,†,1]

[*]Laboratory for Fluorescence Dynamics, Department of Biomedical Engineering, University of California, Irvine, California, USA

[†]Department of Development and Cell Biology, University of California, Irvine, California, USA

[1]Corresponding author: e-mail address: egratton@uci.edu

Contents

Abstract

The raster image correlation spectroscopy (RICS) and number and molecular brightness (N&B) methods are used to measure molecular diffusion in complex biological environments such as the cell interior, detect the formation of molecular aggregates, establish the stoichiometry of the aggregates, spatially map the number of mobile molecules, and quantify the relative fraction of molecules participating in molecular complexes. These methods are based on correlation of fluorescence intensity fluctuations from microscope images that can be measured in a conventional laser-scanning confocal

Methods in Enzymology, Volume 518
ISSN 0076-6879
http://dx.doi.org/10.1016/B978-0-12-388422-0.00006-6

microscope. In this chapter, we discuss the mathematical framework used for data analysis as well as the parameters need for data acquisition. We demonstrate the information obtainable by the N&B method using simulation in which different regions of an image have different numbers of interacting molecules. Then, using an example of two interacting proteins in the cell, we show in a real case how the RICS and N&B analyses work step by step to detect the existence of molecular complexes to quantify their properties and spatially map their interactions. We also discuss common control experiments needed to rule out instrumental artifacts and how to calibrate the microscope in terms of relative molecular brightness.

1. INTRODUCTION

1.1. What can fluctuation analysis reveal about the motion and interactions of single molecules in cells?

Fluctuation analysis methods are relatively common in several areas of physics, chemistry, and biology. Fundamentally, fluctuation spectroscopy is the basis for methods such as dynamic light scattering (DLS) and fluorescence correlation spectroscopy (FCS). A major difference between DLS and FCS is that the former is applied to solutions or homogeneous samples. To generate the DLS signal, many molecules are needed. The fluctuation of the scattered lights is generated by interference among the particles. In general, DLS is used to analyze the distribution of particle sizes. Instead, FCS is sensitive to single molecules and can be applied to complex environment such as the cell interior or tissues but requires the molecules under observation to be fluorescent. Although the FCS method has some disadvantages because fluorescent molecules can bleach due to constant illumination, it can measure multiple-labeled fluorophores which is not possible in DLS. In addition, FCS has the capability to analyze the amplitude of the fluctuations and connect these amplitudes with molecular characteristics such as molecular brightness. As a result of these differences with respect to DLS, FCS has developed as an independent method during the 1970s (Magde, Elson, & Webb, 1972, 1974). A major breakthrough in the FCS approach was due to the realization that FCS measurements can be performed in live cells (Berland, So, & Gratton, 1995; Bacia and Schwille, 2003, 2007; Bacia et al., 2006; Haustein and Schwille, 2007; Kim et al., 2007; Schwille, 1999; Schwille et al., 1997a,b) and that scanning confocal microscopy methods could be exploited to obtain the map of fluctuations in entire cells and in tissues (Digman, Brown, et al., 2005; Digman, Dalal, Horwitz, & Gratton, 2008; Digman & Gratton, 2009a, 2009b; Digman,

Sengupta, et al., 2005; Digman, Wiseman, Choi, Horwitz, & Gratton, 2009; Digman, Wiseman, Horwitz, & Gratton, 2009; Ries, Chiantia, & Schwille, 2009; Ries, Yu, Burkhardt, Brand, & Schwille, 2009).

Although FCS still shares some technological and analysis approach of the DLS method, today, FCS analysis has diverged from DLS. In this contribution, we will briefly describe the conceptual advances in FCS that made possible the development of the imaging fluctuation methods, which are currently driving the research by exploiting fluctuations observed in live cellular environments for the study of molecular interactions.

2. A CONCEPTUAL OVERVIEW OF FLUCTUATION METHODS

The basic concept in FCS is that fluorescence intensity fluctuations in a given volume of observation are caused by single molecules or particles that randomly pass through that volume. Additional fluctuations could also arise from particles in the observation volume which change the fluorescence intensity due to conformational transitions or, in general, change their excitation–emission properties while in the observation volume. Although this is the common explanation found in almost all articles introducing FCS, this view does not convey the information about changes of the spatial location of the particle as a function of time. The consideration of the spatial distribution of molecules brings in the change of paradigm which is needed to correlate fluctuations at one location with the fluctuations in surrounding locations.

3. WHAT IS DIFFERENT IN THE RICS METHOD?

In classical single point FCS analysis, we consider fluctuations occurring only at one volume of illumination. In the spatial correlation approach, we consider (cross) correlations between adjacent (or far) volumes of observation. In this way, the full extent of the probability density of finding a particle at a different location and at a different time is introduced into the description of the correlation of the fluctuations. For example, if the particle diffuses isotropically, measuring the fluctuations at one point is sufficient to characterize the diffusion coefficient of the particle. However, if the particle undergoes anisotropic diffusion, the probability density of finding the particle at a given position and at a given time will depend on the position and the time. To capture the time and spatial evolution of the particle location, we need to introduce more complex correlation functions than those used in conventional FCS. In Fig. 6.1, we illustrate

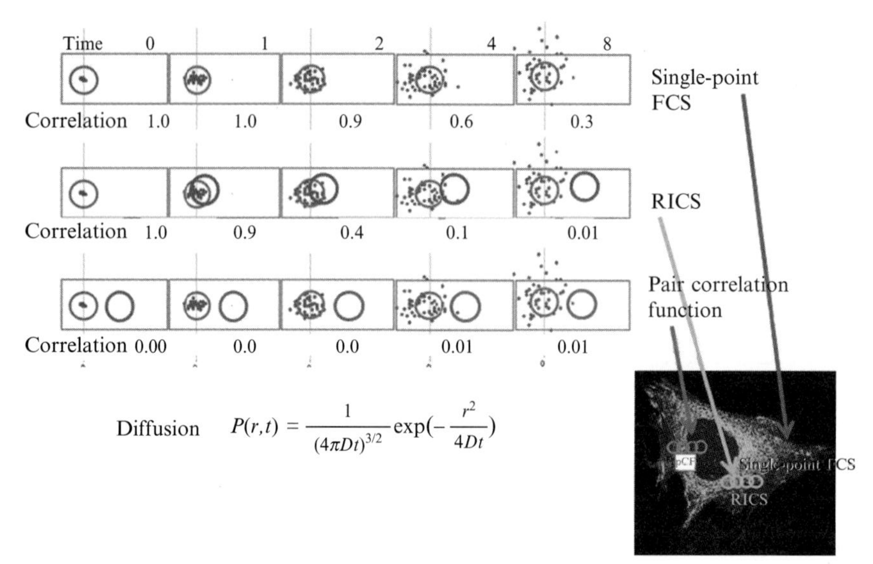

Figure 6.1 Schematic illustration of fluctuation spectroscopy experiments. In single point FCS, the intensity fluctuations at a given observation volume are measured as a function of time due to diffusion. The correlation is the product of the intensity at time 0 (points at left in the figure) to the intensity at a later time. The correlation decreases as a function of time due to diffusion. This time depends on the diffusion coefficient of the molecule and on the size of the observation volume. In the RICS method, the position of the volume of observation is changed with time producing a correlation that is dependent on how fast molecules are moving as well as how fast the position of the volume of observation is changing with time. In the pair correlation approach, two (or more) volumes of observations are used at a given distance. Initially, a molecule is in one of the observation volume. Due to diffusion, a molecule could appear in the other volume at a later time. The time it takes to appear in the other volume depends on how fast the molecule moves, how far the two volumes are, and if there are obstacles to diffusion in between the two volumes. (See Color Insert.)

the principle of the fluctuations techniques discussed in this chapter in the context of the spatial and temporal correlation due to free diffusion. The probability to find a particle at a given distance and time if the particle was at the origin at time 0 is given by the diffusion equation in Fig. 6.1 for diffusion in three dimensions. This probability can be approximated by a Gaussian in which the exponential term represents the broadening of the distribution with a variance that is function of time. The amplitude term decreases with time to maintain the integrated probability to unity.

We set the probability to be unity at time $t=0$ in the volume of observation. In the first row (Fig. 6.1), we schematically show the experiment for the single point FCS experiment. The volume of observation is always the same in this approach. As the time evolves, the probability distribution broadens. At some point in time, the particle has a probability to be found outside the volume of observation (the blue circle in first row of Fig. 6.1). The correlation function implicitly defined in the figure as the product of the number of particles inside the blue circle at time 0 and the number of particles in the blue circle at a later time has the property that the correlation decreases as a function of time since the particle can be found everywhere in the volume. We normalize the correlation function so that the correlation goes to zero at very long time when the probability of finding a particle in a given volume is equal everywhere.

Row 2 of Fig. 6.1 shows the RICS approach. The RICS correlation function is calculated assuming that the position of the volume of observation changes with time according to a specific relationship. This relationship is typically obtained when a raster scan is done as in the common laser-scanning microscope. The position along the horizontal direction changes linearly with time with a characteristic motion given by the linear motion of the scanner. The laser scan speed and pixel size can be manually adjusted. In the vertical direction, the motion is also linear, but the speed of the motion depends on the line time.

Row 3 of Fig. 6.1 shows the pair-correlation function approach. In this method, the probability distribution is sampled simultaneously at two different positions. If the two positions are nonoverlapping, the same particle cannot be found in the two separated locations. Only after some time when the distribution has sufficiently broadened, the same particle can be observed at the second location. The time cross-correlation of the intensity fluctuations at these two locations provides information about the average time to go from one location to the other. Using this approach is possible to explore the probability distribution in a very large volume. The pair-correlation approach gives similar information of the single particle-tracking method but using many particles simultaneously and without the requirement of observing separated particles (Digman & Gratton, 2009b).

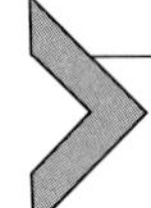

4. RICS AND CROSS-RICS

The RICS and ccRICS correlation functions are shown below

$$G_{\mathrm{RICS}}(\xi,\psi)=\frac{\langle I(x,y)I(x+\xi,y+\psi)\rangle}{\langle I(x,y)\rangle\langle I(x,y)\rangle}-1 \qquad [6.1]$$

$$G_{ccRICS}(\xi,\psi)=\frac{\langle I_1(x,y)I_2(x+\xi,y+\psi)\rangle}{\langle I_1(x,y)\rangle\langle I_2(x,y)\rangle}-1 \qquad [6.2]$$

where I is the intensity at each pixel and the brackets indicate the average over all pixels x, y of an image. The indices 1, 2 indicate channel 1 and channel 2, respectively. As the correlation functions above show, one image is sufficient to determine the RICS correlation function. However, in general, a stack of images is collected with the purpose of separating the mobile from the immobile population of molecules.

The definition of the RICS correlation function is identical to the definition of the image correlation spectroscopy (ICS) given by Petersen et al. (Petersen, 1986; Petersen et al., 1998; Petersen, Höddelius, Wiseman, Seger, & Magnusson, 1993). The difference in the RICS approach is in the way data are collected in the raster scan confocal microscope. There is a relationship between the position of the pixel in the image and the time the pixel is measured.

$$\text{time at pixel } n=y\times\tau_1+x\times\tau_p \qquad [6.3]$$

$$\xi=x\times\tau_p \qquad [6.4]$$

$$\psi=y\times\tau_1 \qquad [6.5]$$

The resulting correlation function can then be expressed in terms of the pixel time τ_p and the line time τ_1.

In relation to Fig. 6.1, it is clear that if the pixel dwell time is very long a molecule could move faster than the scanner when the collection of the line is finished. In this case, fast molecules could not contribute to the correlation function. If this occurs, the diffusion obtained by the RICS analysis will be smaller than the true diffusion coefficient. Also, the amplitude of the correlation function will be smaller because the numerator of Eq. (6.2) will be smaller but the denominator will be the same.

Therefore, it is important that the pixel time to be fast enough and the pixel size to be sufficiently larger to "catch" the molecules before the line return will occur. Figure 6.2 shows the relation between the required pixel dwell time and the diffusion coefficient. For common situations, for example, for a protein diffusing in a cell with a diffusion coefficient of 20 $\mu m^2/s$, a pixel size of 0.05 μm and using a relatively slow pixel dwell time (25 μs), the condition for "catching" the molecules occur at about pixel 2–3. This discussion shows that the RICS analysis can be done using small subframes (e.g., 16 pixels) which correspond to averaging the diffusion coefficient in a

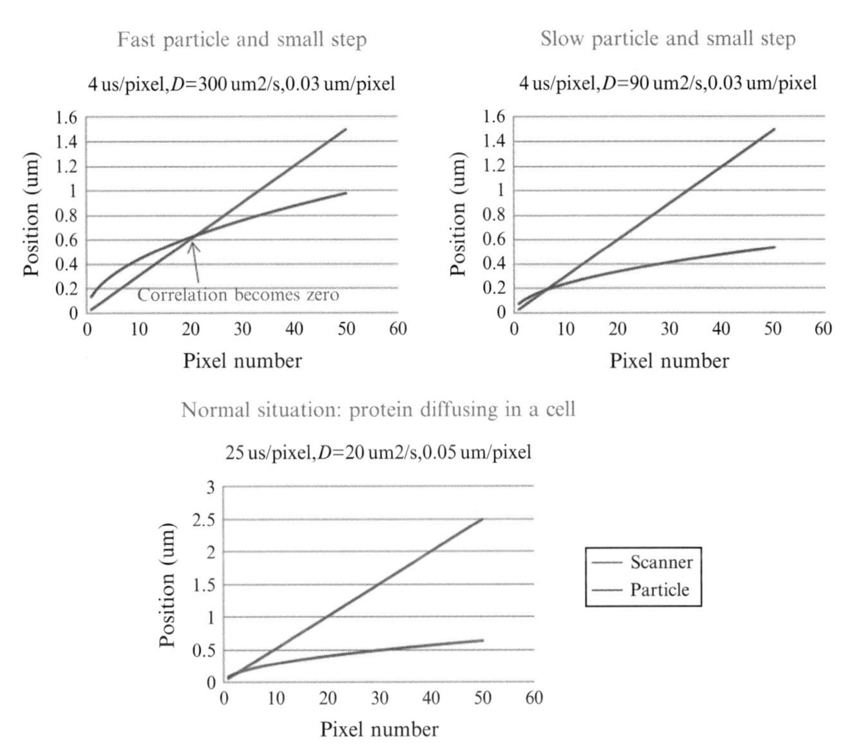

Figure 6.2 Schematic representation of timing relationships in the RICS method. The horizontal axis (pixel) is scanned linearly with time. The position of a diffusing particle changes with the square root of time. At some time (or pixel), the scanner will have traveled a larger distance than the particle. At that time (or pixel distance), the correlation according to Fig. 6.1 goes to zero. This is the minimum number of pixels needed to compute the correlation function. This number of pixels depends on the pixel size, the scan speed, and the diffusion coefficient of the molecule. (For color version of this figure, the reader is referred to the online version of this chapter.)

region of about 0.8 μm. This is the typical "spatial" resolution of the RICS method which is larger than the size of a diffraction limited spot but no so large that it only gives averages over the entire cell.

In the example below, we show how to produce maps of diffusion and (G_0) for a cell expressing EGFP–paxillin. The focal adhesions where paxillin concentrate are visible as areas of relatively large intensity. The image was sequentially analyzed using small regions of analysis (32 pixels) at a time. The apparent diffusion is larger in the regions away from the focal adhesions (Fig. 6.3 Panel D). For this measurement, the spatial resolution of the diffusion parameter map was 1.6 μm.

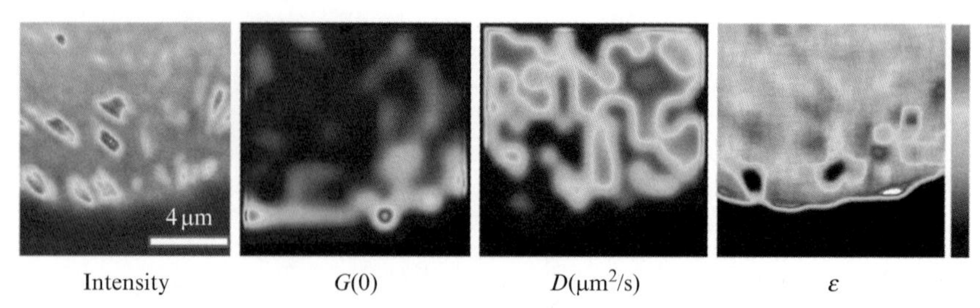

Figure 6.3 Intensity image, (G0) map, Diffusion map and brightness map of paxillin-EGFP in focal adhesions. The same color scheme at right is used for all figures. The corresponding full scale are intensity 0–489 photon/s; $G(0) = 0.0064$, $D = 0$–$46\ \mu m^2/s$, 28,000–52,000 counts/dwell time/molecule. The maps from the fluctuation experiments were obtained using a region of analysis of 32 pixels and sliding this region in steps of 16 pixels across the entire image. (See Color Insert.)

5. PCH AND AMPLITUDE FLUCTUATION ANALYSIS

As we discussed in the previous section, in a fluctuation experiment, there is a time correlation between fluctuations at different locations and at different times. In addition, the amplitude of the fluctuations contains information about the brightness of the particle. To extract this information, we analyze the statistics of the fluctuation amplitude using different mathematical methods. The PCH (photon counting histogram) method was introduced in 1999 (Chen, Müller, So, & Gratton, 1999). This method is based on measuring the photon counting distribution at one location, similar to the single point FCS experimental approach. This method is quite powerful since the photon counting distribution can provide information about the brightness and the number of particles in the volume of observation. However, when analyzing images, the limited statistic available at each pixel does not allow the full photon counting distribution to be analyzed. Instead, for the analysis of images, we use a method that utilizes only the first and second moment of the histogram of the amplitude fluctuations. This method is called the Brightness and Number analysis or N&B (Digman et al., 2008). In this original presentation of the method, the basic of the moment analysis was presented. In a subsequent paper {Digman, 2009 #68}, the concept of cross-correlation based on the brightness of complexes using proteins with two fluorescent labels was introduced. The analysis of the cross-brightness of the complex leads the method to determine the stoichiometry of the complex. The stoichiometry is determined by first "isolating" the particles that are cross-correlated and then analyzing the brightness of the complex in terms of how bright are the cross-correlated species in each color. However, in the 2009 article {Digman, 2009 #68}, the determination of the number of molecules in the complex with respect to the molecules of each species in the cell was not discussed. In this chapter, we present the equations needed to perform the cross-number analysis and we provide an example of a biological system in which a complex is formed and the fractional number of molecules in the complex is calculated.

6. N&B AND CROSS-NB

The data set need for the Number and Brightness analysis consists of a stack of images of the same field of view observed at two different emission band-passes if cross-N&B measurements are required. For each pixel of the stack of images, we define

$$\mathrm{av} = \frac{\sum_k I(x,y)}{K} \quad [6.6]$$

$$\mathrm{var} = \frac{\sum_k (I(x,y) - \mathrm{av})^2}{K} \quad [6.7]$$

$$\mathrm{cross\_var} = \frac{\sum_k (I_1(x,y) - \mathrm{av}_1)(I_2(x,y) - \mathrm{av}_2)}{K} \quad [6.8]$$

Where the sum is over the same pixel in each frame of the stack, K is the number of frames and I is the intensity at one pixel of each frame

$$B_1 = \frac{\mathrm{var}_1}{\mathrm{av}_1}, \quad \text{Brightness of channel 1} \quad [6.9]$$

$$B_2 = \frac{\mathrm{var}_2}{\mathrm{av}_2}, \quad \text{Brightness of channel 2} \quad [6.10]$$

$$B_{cc} = \frac{\mathrm{cross\_var}}{\mathrm{av}_1 \times \mathrm{av}_2}, \quad \text{Cross\_variance} \quad [6.11]$$

$$N_1 = \frac{\mathrm{av}_1^2}{\mathrm{var}_1}, \quad \text{Number of particles in channel 1} \quad [6.12]$$

$$N_2 = \frac{\mathrm{av}_2^2}{\mathrm{var}_2}, \quad \text{Number of particles in channel 2} \quad [6.13]$$

$$N_{cc} = \mathrm{cross\_variance}, \quad \text{Cross number} \quad [6.14]$$

$$\frac{N_{cc}}{N}, \quad \text{Fraction of cross number} \quad [6.15]$$

The equations for the relative fraction of cross-correlated molecules have not been given before in the context of the N&B technique. They follow the same concept used in the single point cross-correlation fluctuation spectroscopy equation to determine the fraction of cross-correlated molecules. A new concept in the cross-correlation N&B analysis is the stoichiometry contour map, which provides the most common brightness composition of molecular complexes found in a sample. To show the utility of the stoichiometry map, let us consider a situation in which we detect two species in two separate channels, for example, green and red. The two species can form a complex in which two molecules of the red species associate with one green molecule. From the point of view of the green channel, the brightness of all components is the same. In the red channels, some of the molecules will appear brighter in the complex. The individual brightness map cannot reveal the nature of the complex. The Bcc histogram (Eq. 6.11) will tell us that there is a positive cross-correlation, but this histogram alone cannot reveal the composition of the complex. To construct the stoichiometry histogram, we scan all pixels that have a positive cross-brightness signal and we detect the individual brightness at the two channels for that pixel. We then construct the histogram of the brightness of the two channels with the condition that they have to be cross-correlated. The brightness of each channel is normalized to the molecular brightness of each species. Using this normalization, we can determine the molecular composition of the complex.

7. SIMULATIONS FOR CROSS-RICS AND CROSS-N&B

In the following, we show simulations of molecules interacting in different number at selected spots of an image (Fig. 6.4). With simulations we can better determine if the algorithms we are using can recover the underlying molecular distributions. For this part, we are using simulated data in which the image is divided in spots with different ratio of cross-correlated molecules according to Table 6.1. This table describes the number of molecules in the various spots of the image according to the simulation and their properties. All molecules diffuse with the same diffusion coefficient of $D = 10\ \mu m^2/s$.

All molecules have the same brightness. Channel 1 has constant number of molecules (100) in each spot but variable number of cross-correlated molecules. Channel 2 has a variable number of molecules, and all of them (100%) correlate with a variable fraction of the molecules in channel 1, depending on the spot. Table 6.1 shows, for each of the eight spots, the expected

Figure 6.4 Definition of the spot number locations for Table 6.1 and Figs. 6.5 and 6.6. (For color version of this figure, the reader is referred to the online version of this chapter.)

Table 6.1 Intensity, Number and Brightness values for the 8 spots of the simulation

Spot	Channel 1	Channel 2	av1	av2	B1	B2	N1	N2	Ncc	Ncc/N1	Ncc/N2
1	100+0	0	100	0	1	0	100	0	0	0	0
2	90+10	10	100	10	1	1	100	10	10	0.1	1
3	80+20	20	100	20	1	1	100	20	200	0.2	1
4	70+30	30	100	30	1	1	100	30	30	0.3	1
5	60+40	40	100	40	1	1	100	40	40	0.4	1
6	50+40	50	100	50	1	1	100	50	50	0.5	1
7	40+60	60	100	60	1	1	100	60	60	0.6	1
8	0	100	0	100	0	1	0	100	0	0	0

intensities, relative brightness, and number of molecules for both channels for this simulation as well as the number of cross-correlated molecules. Figure 6.5 shows the result of the analysis of the above simulations obtained with the SimFCS software (available at www.lfd.uci.edu). The brightness map for the two channels shows that molecules have the same brightness in the two channels, although each spot has a different intensity in channel 2. Accordingly, the number map is different for the two channels and obviously in this example follows the intensity map.

Next in Fig. 6.6, we show the brightness maps of the molecules that are cross-correlated and the map of the fractional number of the cross-correlated

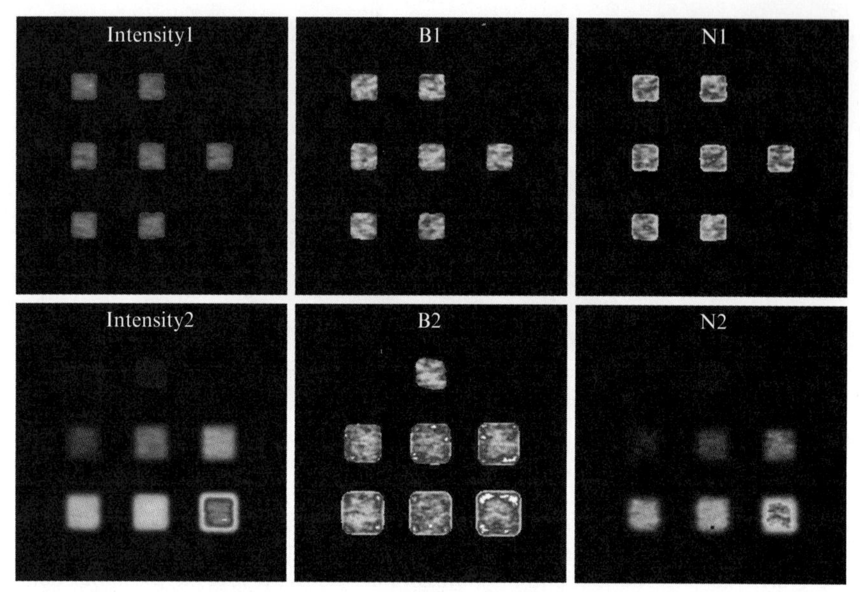

Figure 6.5 According to expected values from Table 6.1, we recover a uniform intensity in all spots for intensity 1, a variable intensity for the spots as seen in channel 2, the same brightness in all spots for channel 1 and channel 2, the same number of molecules for channel 1 in all spots, and a variable number of molecules in the various spots for channel 2. (See Color Insert.)

molecules. The Bcc map (cross-brightness) shows the same cross-brightness in each spot, since all molecules that are cross-correlated have the same brightness. The stoichiometry map shows that there is only one cross-correlated population with a ratio 1:1 of the two molecules. The number of cross-correlated molecules instead is different in the different spots in channel 1 but is the same in each spot of channel 2. The value of the fraction of molecules recovered by this algorithm is identical to the fraction of molecules simulated.

8. Applications of RICS and N&B to Detect Molecular Complexes in Cells

In this section, we show an application of the cross-RICS and cross-N&B using a biological system in which two proteins are known to interact. T3T cells were co-transfected with dynamin-2a and endophilin (see Section 12 at the end of this chapter). What is known for this system is that the two proteins, dynamin-2a labeled with EGFP and endophilin

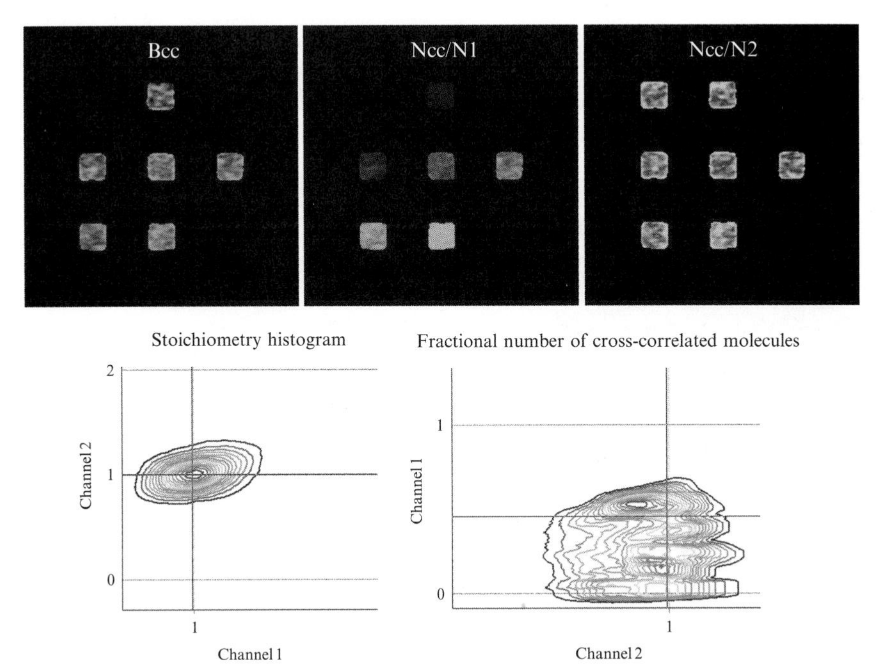

Figure 6.6 The cross-brightness is the same for all molecules since the molecules that are in the complexes have all the same brightness. The ratio of molecules that are in the complex with respect to the total molecules in each channels varies in channel 1 in the different spots, but it is the same (all molecules are cross-correlated) in channel 2. The stoichiometry histogram reports the relative number of molecules of each kind in the complex. In this case, all the complexes have a stoichiometry 1:1. The histogram of the fractional number of cross-correlated molecules varies only along the channel 1 axis, since all the molecules in channel 2 are cross-correlated. (See Color Insert.)

labeled with mCherry should form a complex in the cytoplasm (Ross et al., 2011; Sundborger et al., 2010). Endophilin in the cytoplasm has been found to be a dimer (Ringstad, Nemoto, & De Camilli, 2001; Ross et al., 2011).

We measured the correlation between the two proteins in the cytoplasm. The intensity image shows that the dynamin-2a is concentrated in small vesicles while endophilin is more uniformly distributed in the cytoplasm (Fig. 6.7A and C). First, we perform the RICS analysis to determine if there are mobile molecules that can be detected in the two channels and if they are cross-correlated (Fig. 6.7B and D). The RICS analysis only reports on the mobile molecules. Immobile structures or slowly moving vesicles do not contribute to fast molecular fluctuations. Vesicles carrying both colors are clearly cross-correlated when they move. We eliminate these large particles

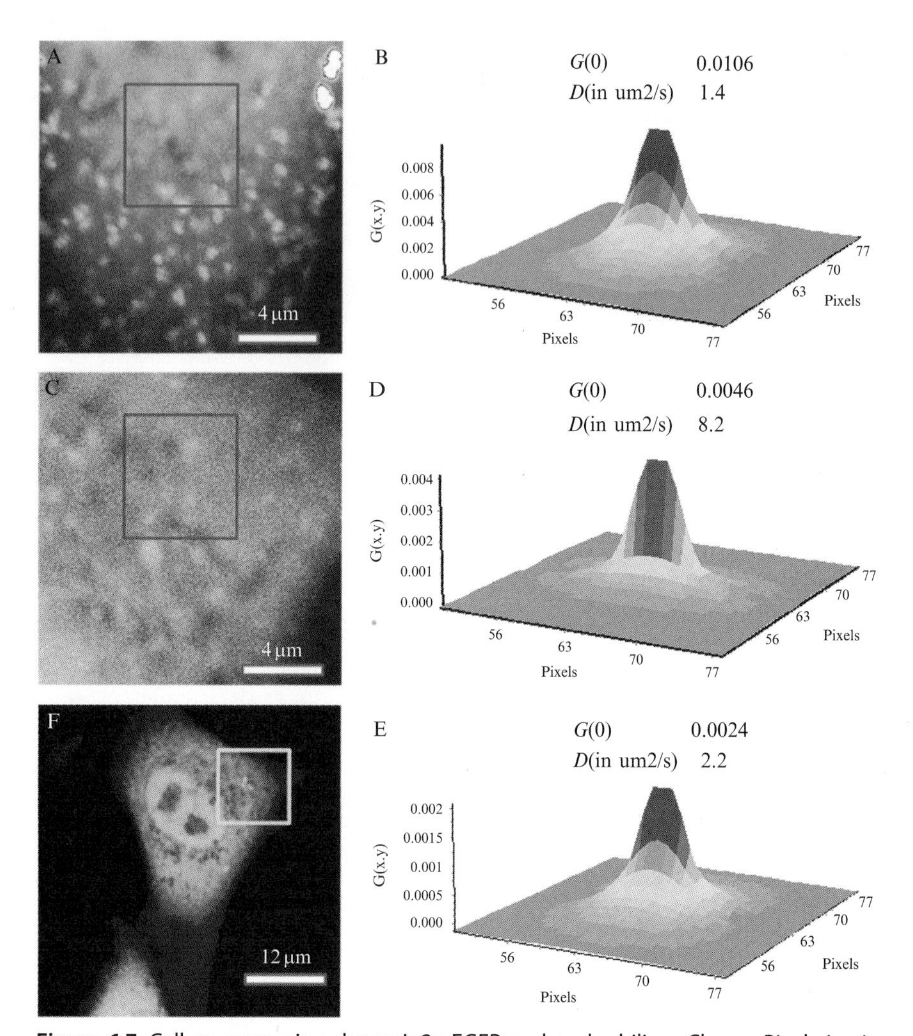

Figure 6.7 Cell co-expressing dynamin2a-EGFP and endophilin-mCherry. Pixel size is 0.065 μm. Optical section in the cytoplasm. Only the part in the red square was used for the RICS analysis. (A) Channel 1 (EGFP) and (C) Channel red (mCherry). (B and C) RICS autocorrelation function and fit using one component diffusion for the green and red channels, respectively. (E) RICS cross-correlation. For the fit, the waist was measured to be 0.23 μm. The same waist was used for both wavelengths. (F) Unzoomed average intensity image of the red channel. The area analyzed in parts A–E is in the yellow square. (See Color Insert.)

in our analysis using the moving average method (Digman et al., 2005b). In both channels, we find mobile molecules. The recovered diffusion coefficients for the individual channels are reported in the RICS analysis of Fig. 6.7. We then analyze this image for cross-correlation between the two proteins. We found that the two proteins cross-correlate, although the coefficient of cross-correlation is relative small (Fig. 6.7E). We then calculate the map of the brightness in the two channels (B1 and B2, Eqs. 6.9 and 6.10) (Fig. 6.8). For dynamin2a, the molecular brightness corresponds to the monomer. This is determined by a calibration procedure using cells expressing only EGFP (Fig. 6.9). For four different cells and several focal planes in the cytoplasm of each cell, we found a relatively narrow distribution of the average brightness for dynamin2a-EGFP, which correspond to the EGFP monomer.

The average brightness of endophilin in the cytoplasm is higher than the brightness expected for a monomer of mCherry (Fig. 6.9) and its value better match that of a dimer in accord with previous results (Ross et al., 2011). We then calculated the brightness of the molecules that cross-correlate using the Bcc (brightness cross-correlation algorithm, Fig. 6.8D). Using the values of

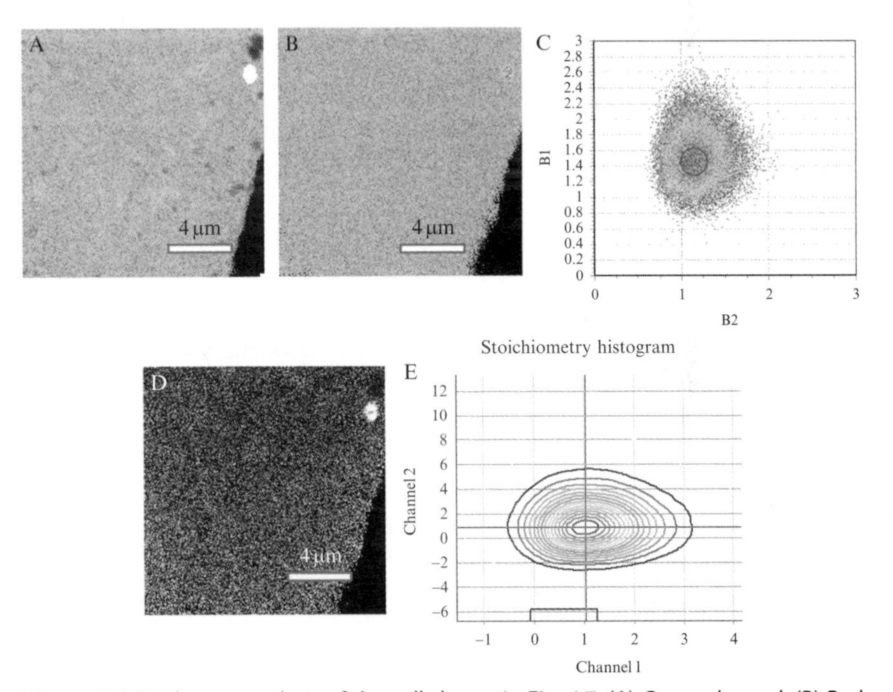

Figure 6.8 Brightness analysis of the cell shown in Fig. 6.7. (A) Green channel. (B) Red channel. (C) 2-dimensional pixel histogram of the values of brightness for the two channels. The *x* and *y* numbers above are used for the calibration of the brightness of mCherry and EGFP, respectively. (D) Cross-correlation brightness map. (E) Contour map of the number of pixels in channel 1 of a given brightness which cross-correlate with channel 2 for all possible brightness. This is the so-called stoichiometry histogram. In this case, the most abundant value is at a ratio 1:1. (See Color Insert.)

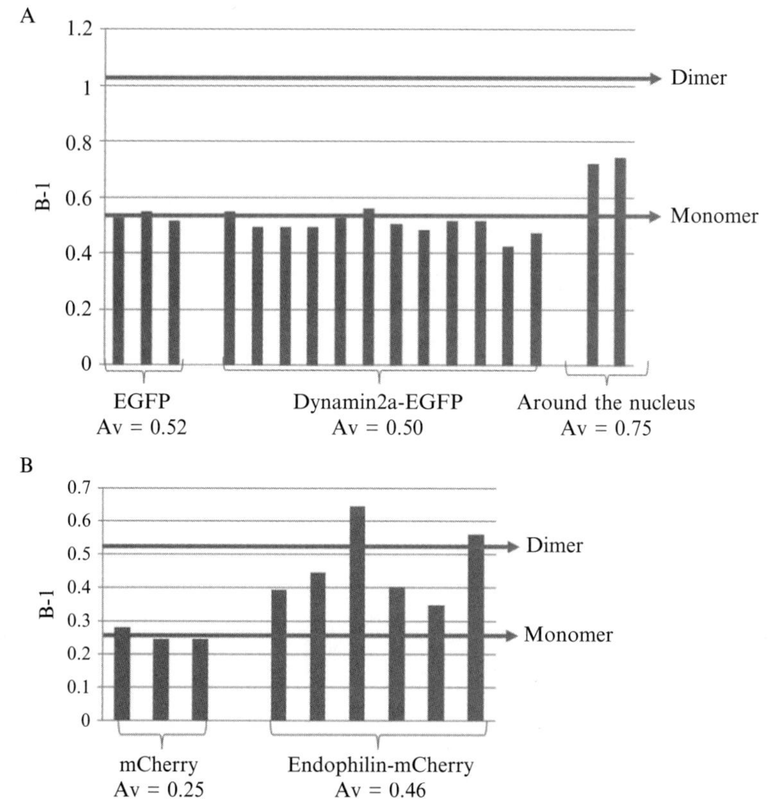

Figure 6.9 Calibration of (A) the EGFP and (B) mCherry channel. Three cells expressing EGFP and mCherry were used to calibrate the brightness scale. The plane of focus was in the cytoplasm of the cells. For the dynamin2a-EGFP, four cells were measured with the plane of focus on the cytoplasm for a total of 12 planes. All cells and planes show the same brightness than for the EGFP only cell. For the same, four cells for a total of six focal planes in the cytoplasm of cells expressing the endophilin-mCherry show larger brightness than the mCherry alone, in the range expected for a dimer of the endophilin protein in accord with previous results. The brightness scale is in units of counts/dwell time/molecule where the dwell time was 12.6 μs. (See Color Insert.)

Bcc and B1-B2, we construct the average stoichiometry histogram which is about one endophilin per dynamin-2a. This observation was not previously reported. The results are interesting for the understanding of the role of dynamin2a and endophilin in the cytoplasm of cells, but we caution that only four cells were analyzed for determining the stoichiometry of this cytoplasmic complex. Since there is an unknown fraction of endogenous material, in these type of experiments, we need to measure many cells to explore a range of concentrations to avoid conclusions based perhaps on a largely under or over expressing cell. At present, it is unclear how a species that appears to be a dimer could form a complex as a monomer. However, we note that the fraction of molecules in the complex is a small fraction of the total molecules.

The Number analysis shows that there are more molecules of dynamin-2a than endophilin by about a factor of 3–4 (Fig. 6.10).

About 36% of the total endophilin molecules are in a complex with dynamin-2a, but less than 12% of the total dynamin-2a molecules are in the complex. In other words, there is an abundant pool of dynamin-2a in the cytoplasm which is not cross-correlating with endophilin. This is in agreement with the ccRICS analysis which shows also about 50% of

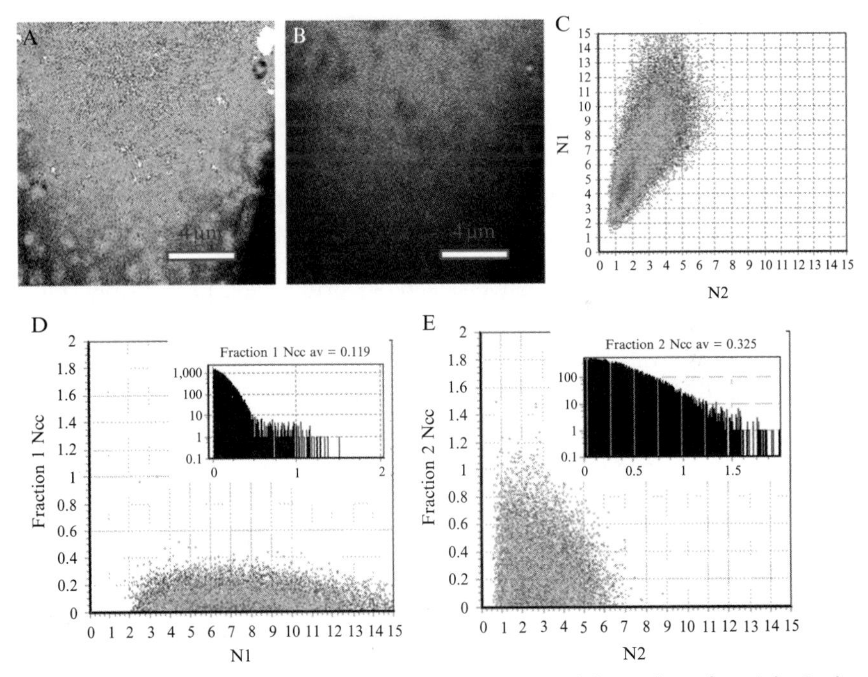

Figure 6.10 Number analysis of the cell shown in Fig. 6.7. (A) Number of particles in the green channel. (B) Number of particles in the red channel. (C) Two-dimensional pixel histogram of the values of the number of particles for the two channels. The x and y numbers above are used for the calibration of the maximum number of particles of mCherry and EGFP, respectively. (D) Fraction of molecules cross-correlated in channel 1. The inset is the histogram. (E) Fraction of molecules cross-correlated in channel 2. The inset is the histogram. (See Color Insert.)

endophilin and 24% dynamin-2a in the complex. The difference between the two estimates could arise from the fact that the ccRICS analysis which calculates the fraction of the mobile complexes with respect to the total mobile molecules while the number analysis calculates the ration between the cross-correlated molecules to the total number of molecules, mobile, or not.

9. CALIBRATION MEASUREMENTS USING EGFP AND mCHERRY AND EFFECTIVE BLEEDTHROUGH

We calibrated the brightness scale using a cell expressing only EGFP (Fig. 6.9). We found that EGFP has a B value of 0.55 and mCherry was 0.25 (Fig. 6.9). Remember that the B value is 1 plus the molecular brightness in units of counts/dwell time/molecule. In the sample with EGFP-only, we also estimated the amount of bleedthrough to the red channel. We found that brightness of the EGFP in the red channel was $B=0.028$, which is about 11% of the brightness of mCherry.

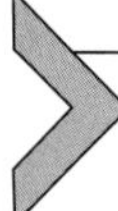

10. DIFFERENCE OF DISTRIBUTION IN THE NUCLEUS VERSUS CYTOPLASM OF DYNAMIN-2A

For this demonstration, we selected a bright cell, presumably overexpressing both proteins. The reason for using overexpressing cells in this example is to avoid interference with the endogenous population. For the purpose of this demonstration of the method, it is instructive to have an excess of the labeled proteins. Of course from the biological point of viewpoint, overexpressing cells could behave very differently than wild-type cells. For the overexpressing cells, we found that dynamin-2a tends to be less concentrated in the nucleus but with a high concentration in the perinuclear region. Instead, endophilin distributes equally in the nucleus and in the cytoplasm. For this example, we would like to establish whether or not there is association between the two proteins in the nucleus where the concentration of dynamin-2a is very different than in the cytosol. Furthermore, our relative fraction analysis should give the information about the relative number of molecules (concentrations) in the various cell compartments and the fraction of molecules in the aggregates.

Figure 6.10 shows the intensity images in the two channels where there is a large difference for the EGFP channel indicating the relative low concentration of the dynamin-2a in the nucleus. The brightness map is relatively uniform, compared to the intensity analysis. The brightness analysis shows that on the average, the increased brightness of the dynamin-2a in the perinuclear region is due to aggregates of about two to four dynamin-2a molecules. This could be due to bleedthrough of the high brightness of the vesicles into the surrounding cytoplasmic compartment rather than truly cytoplasmic proteins. In principle, everything that is immobile should give a value of $B=1$. However, if there is a relatively fast moving pool of vesicles, then the moving average filtering algorithm used to compensate for slow moving components will not be able to completely remove them. Instead, the brightness of the endophilin is uniform in the entire cell and corresponds to the brightness of the dimer (Fig. 6.10). Also for this overexpressing cells, we found that stoichiometry of the complex is 1:1 in the nucleus (Fig. 6.11).

In regard to the number analysis, we first show the map of the number of molecules in the two channels. Clearly, there is a large change in the number of molecules in channel 1 in the perinuclear region but very few in the nucleus. When we analyze for the cross-correlated molecules, we find them everywhere. However, we can see that the fewer molecules of dynamin-2a in the nucleus are in percentage more cross-correlated than the large pool of dynamin2a in the cytoplasm. On the contrary, since there is an excess of endophilin in the nucleus, the measured fraction of interacting molecules for endophilin is smaller in the nucleus (Fig. 6.12).

11. CONCLUSIONS AND FUTURE PROSPECTS

In this method article, we discuss the RICS and N&B methodology used to measure the diffusion of single molecules in cells and their aggregation from a stack of raster scan images. The RICS and the N&B methods were first proposed few years ago. In this contribution, there is an original new addition to the number analysis with the introduction of the concept of number of cross-correlated molecules. The cross-correlation is detected by the simultaneous appearance of fluctuation in the same pixel of an image. The concept is similar to the cross-correlation of fluctuations used in fluctuation correlation spectroscopy. Here, instead of measuring the fluctuation in a single point, we use a stack of raster scan images and we detect correlation of fluctuations in every pixel of an image. By correlating the fluctuations

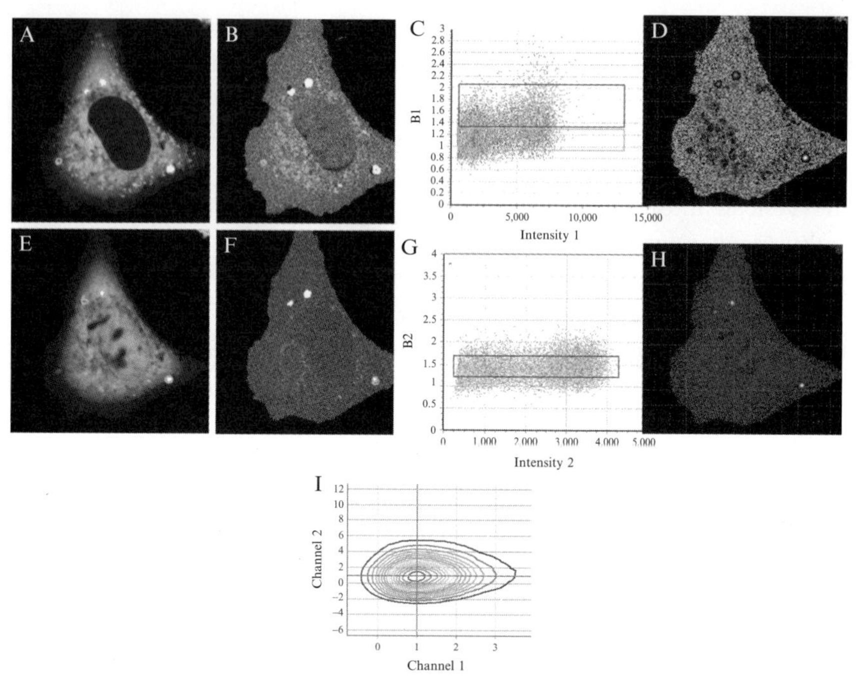

Figure 6.11 Cell over expressing dynamin2a and endophilin. (A) Intensity image of green channel. (B) Brightness map of the green channel. (C) Selection map, green cursor corresponds to the brightness of monomers, and red cursor corresponds to higher aggregates. (D) Selection map coded with the color of the cursors in (C). (E) Intensity image of red channel. (F) Brightness map of the red channel. (G) Selection map, red cursor corresponds aggregates of 2 endophilin molecules. (H) Selection map coded with the color of the cursors in (G). (I) Stoichiometry map showing that the most populated species is the 1:1 molecular complex. (See Color Insert.)

among pixels and among frames in the stack, we obtain information about the diffusion of molecules, the number of fluorescent molecules in a pixel, and their brightness. The cross-correlation of fluctuations between two molecular species emitting in two different channels gives us information about their interactions, the stoichiometry of the molecular aggregate, and the fraction of molecules interacting of each independent species. The method of data analysis makes use of fast Fourier transform algorithm which result is a very fast computation and quasi real-time display. Data are collected using commercial laser-scanning microscope without any change in the hardware and optics.

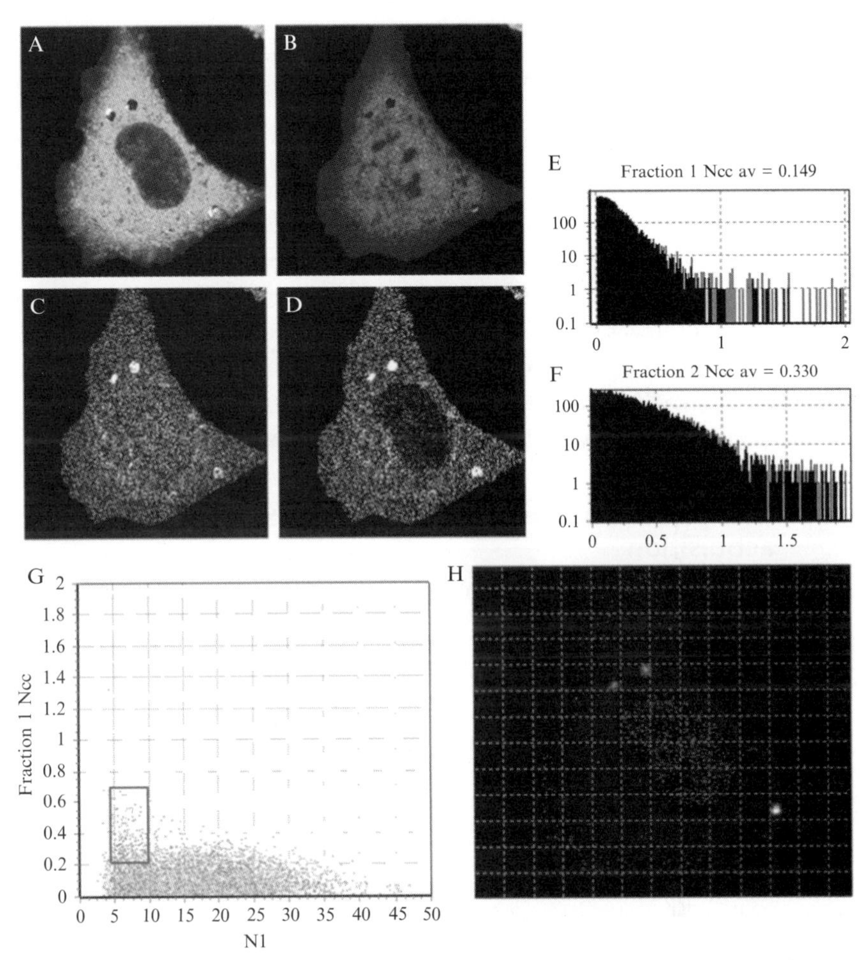

Figure 6.12 Cross-number analysis of the cell showing in Fig. 6.11. (A) and (B) fluorescence intensity in the green and red channels, respectively. (C) and (D) Cross-number maps for the green and red channels, respectively. The scale in the figure is obtained from the histograms in (E) and (F), for the two channels. (G) Selection of the Ncc1 versus N1 histogram for low number N1 (in the nucleus) has the larger fraction of cross-correlated molecules. The region selected by the pixels in the red square is shown in panel H. (See Color Insert.)

With the recent introduction of very fast and sensitive cameras based on the cMOS technology, the fluctuation methods could be used with these parallel detectors, potentially improving by order of magnitude the speed of data acquisition.

12. MATERIALS AND METHODS

12.1. NIH3T3 cell cultures

NIH3T3 cells were grown in High Glucose Dulbecco's Modified Eagle Medium (DMEM) (Invitrogen, Carlsbad, CA) supplemented with 10% FBS and 1× pen/strep at 37 °C in a humidified 5% CO_2 incubator. Twenty-four hours prior to imaging cells were plated at 50–80% confluency in a pre-coated fibronectin [2 mg/ml] 35-mm glass bottom dish (No. 1.5, 0.17 mm thickness) (MatTek Corp, Ashland, MA). Cells were transfected with 0.4–1 μg of DNA and Lipofectamine(tm) 2000 Reagent according to manufacturer's protocol (Life Technologies, Grand Island, NY).

12.2. One-photon microscopy, RICS, ccRICS, N&B, and ccN&B acquisition

All cells were imaged using an Olympus FV1000 inverted laser-scanning confocal microscope using a 60× 1.2NA water UPlanSApo objective (Olympus America Inc.). To insure cell viability during the course of the experiments, a humidified stage top incubation chamber set to 37 °C with an objective warmer was mounted (Tokai Hit Co., Ltd.) and allowed to stabilize for 20 min before each plate was imaged. EGFP and mCherry were excited with an Argon Ion laser set to 488 nm (1% percent) and a 559-nm diode laser (1% percent), respectively. The 405/488/559/635 primary dichroic filter was selected. Band pass filters at 505–540 and 575–675 nm were configured for green and red emission, respectively. Images were collected at 12.5 μs/pixel at 256 × 256 pixel resolution with a scanning areas ranging from 70 × 70 μm to 16 × 16 μm using the photon counting mode of the Olympus FV1000 microscope. For the raster image correlation spectroscopy (RICS) acquisition, the zoom was set to 50 nm/pixel (zoom 16.5) at 256 × 256 image resolution. In all experiments, a stack of 100 frames was collected with a frame time of 1 s. The parafocal illumination of the two colors was determined using 40 nm yellow-green beads. The x–y and z illumination volumes were coincident in the two colors within 20 nm.

12.3. Data analysis

All data analyses were performed using the SimFCS program (available at www.lfd.uci.edu). For the RICS analysis, correction for cell and organelles motions was achieved using a moving average of 10 frames. This method effectively removes every motion occurring in times longer than 10 s. Bleaching was checked for each measurement. Bleaching never exceed 10% of the original intensity. The RICS analysis is independent form bleaching since the correlation is calculated for each frame. Instead for N&B analysis, a change of 10% of the average intensity can result in substantial artifacts. We apply the pixel detrend routine of SimFCS for random counts using a section of 10 s. Since this procedure slightly modifies the brightness calibration, for ach measurement, we recalibrate internally the S factor of the SimFCS program. This factor must be such that a part of the sample that is immobile must have a brightness of 1. In every field, we imaged a part of the support glass and we modified the value of S to achieve a B value of 1 in that region of the image. Using this procedure, we obtained very consistent results.

ACKNOWLEDGMENTS

Funds were provided by NIH Grants P41-RR03155, P41 GM103540, and P50 GM076516.

REFERENCES

Bacia, K., & Schwille, P. (2003). A dynamic view of cellular processes by in vivo fluorescence auto- and cross-correlation spectroscopy. *Methods*, *29*, 74–85.

Bacia, K., Kim, S. A., & Schwille, P. (2006). Fluorescence cross-correlation spectroscopy in living cells. *Nature Methods*, *3*, 83–89.

Bacia, K., & Schwille, P. (2007). Practical guidelines for dual-color fluorescence cross-correlation spectroscopy. *Nature Protocols*, *2*, 2842–2856.

Berland, K. M., So, P. T., & Gratton, E. (1995). Two-photon fluorescence correlation spectroscopy: Method and application to the intracellular environment. *Biophysical Journal*, *68*, 694–701.

Chen, Y., Müller, J. D., So, P. T., & Gratton, E. (1999). The photon counting histogram in fluorescence fluctuation spectroscopy. *Biophysical Journal*, *77*, 553–567.

Digman, M. A., Brown, C. M., Sengupta, P., Wiseman, P. W., Horwitz, A. R., & Gratton, E. (2005). Measuring fast dynamics in solutions and cells with a laser scanning microscope. *Biophysical Journal*, *89*, 1317–1327.

Digman, M. A., Dalal, R., Horwitz, A. F., & Gratton, E. (2008). Mapping the number of molecules and brightness in the laser scanning microscope. *Biophysical Journal*, *94*, 2320–2332.

Digman, M. A., & Gratton, E. (2009a). Analysis of diffusion and binding in cells using the RICS approach. *Microscopy Research and Technique*, *72*, 323–332.

Digman, M. A., & Gratton, E. (2009b). Imaging barriers to diffusion by pair correlation functions. *Biophysical Journal*, *97*, 665–673.

Digman, M. A., Sengupta, P., Wiseman, P. W., Brown, C. M., Horwitz, A. R., & Gratton, E. (2005). Fluctuation correlation spectroscopy with a laser-scanning microscope: Exploiting the hidden time structure. *Biophysical Journal*, *88*, L33–L36.

Digman, M. A., Wiseman, P. W., Choi, C., Horwitz, A. R., & Gratton, E. (2009). Stoichiometry of molecular complexes at adhesions in living cells. *Proceedings of the National Academy of Sciences of the United States of America*, *106*, 2170–2175.

Digman, M. A., Wiseman, P. W., Horwitz, A. R., & Gratton, E. (2009). Detecting protein complexes in living cells from laser scanning confocal image sequences by the cross correlation raster image spectroscopy method. *Biophysical Journal, 96*, 707–716.

Haustein, E., & Schwille, P. (2007). Fluorescence correlation spectroscopy: Novel variations of an established technique. *Annual Review of Biophysics and Biomolecular Structure, 36*, 151–169.

Kim, S. A., Heinze, K. G., & Schwille, P. (2007). Fluorescence correlation spectroscopy in living cells. *Nature Methods, 4*, 963–973.

Magde, D., Elson, E., & Webb, W. W. (1972). Thermodynamic fluctuations in a reacting system—Measurement by fluorescence correlation spectroscopy. *Physical Review Letters, 29*, 705–708.

Magde, D., Elson, E. L., & Webb, W. W. (1974). Fluorescence correlation spectroscopy. II. An experimental realization. *Biopolymers, 13*, 29–61.

Petersen, N. O. (1986). Scanning fluorescence correlation spectroscopy. I. Theory and simulation of aggregation measurements. *Biophysical Journal, 49*(4), 809–815.

Petersen, N. O., Brown, C., Kaminski, A., Rocheleau, J., Srivastave, M., & Wiseman, P. W. (1998). Analysis of membrane protein cluster densities and sizes in situ by image correlation spectroscopy. *Faraday Discussions, 111*, 289–305.

Petersen, N. O., Höddelius, P., Wiseman, P. W., Seger, O., & Magnusson, K. E. (1993). Quantitation of membrane receptor distributions by image correlation spectroscopy: Concept and application. *Biophysical Journal, 165*(3), 1135–1146.

Ries, J., Chiantia, S., & Schwille, P. (2009). Accurate determination of membrane dynamics with line-scan FCS. *Biophysical Journal, 96*, 1999–2008.

Ries, J., Yu, S. R., Burkhardt, M., Brand, M., & Schwille, P. (2009). Modular scanning FCS quantifies receptor-ligand interactions in living multicellular organisms. *Nature Methods, 6*, 643–645.

Ringstad, N., Nemoto, Y., & De Camilli, P. (2001). Differential expression of endophilin 1 and 2 dimers at central nervous system synapses. *The Journal of Biological Chemistry, 276*, 40424–40430.

Ross, J. A., Chen, Y., Müller, J., Barylko, B., Wang, L., Banks, H. B., et al. (2011). Dimeric endophilin A2 stimulates assembly and GTPase activity of dynamin 2. *Biophysical Journal, 100*, 729–737.

Schwille, P., Bieschke, J., & Oehlenschläger, F. (1997a). Kinetic investigations by fluorescence correlation spectroscopy: the analytical and diagnostic potential of diffusion studies. *Biophysical Chemistry, 66*, 211–228.

Schwille, P., Meyer-Almes, F.-J., & Rigler, R. (1997b). Dual-color fluorescence cross-correlation spectroscopy for multicomponent diffusional analysis in solution. *Biophysical Journal, 72*, 1878–1886.

Schwille, P., Korlach, J., & Webb, W. W. (1999). Fluorescence correlation spectroscopy with single molecule sensitivity on cell and model membranes. *Cytometry, 36*, 176–182.

Sundborger, A., Soderblom, C., Vorontsova, O., Evergren, E., Hinshaw, J. E., & Shupliakov, O. (2010). An endophilin-dynamin complex promotes budding of clathrin-coated vesicles during synaptic vesicle recycling. *Journal of Cell Science, 124*, 133–143.

CHAPTER SEVEN

Global Analysis in Fluorescence Correlation Spectroscopy and Fluorescence Lifetime Microscopy

Neil Anthony*, Keith Berland*,[1]
*Department of Physics, Emory University, Atlanta, Georgia, USA
[1]Corresponding author: e-mail address: kberlan@emory.edu

Contents

Abstract

Fluorescence correlation spectroscopy (FCS) and related fluctuation spectroscopy and microscopy methods have become important research tools that enable detailed investigations of the chemical and physical properties of molecules and molecular systems in a variety of complex environments. Information recovery via curve fitting of fluctuation data can present complicating challenges due to limited resolution and/or problems with fitting model verification. We discuss a new approach to data analysis called τFCS that couples multiple modes of signal acquisition, here specifically FCS and fluorescence lifetimes, with global analysis. We demonstrate enhanced resolution using τFCS, including the capability to recover the concentration of both molecular species in a two-component

Methods in Enzymology, Volume 518
ISSN 0076-6879
http://dx.doi.org/10.1016/B978-0-12-388422-0.00007-8

mixture even when the species have identical diffusion coefficients and molecular brightness values, provided their fluorescent lifetimes are distinct. We also demonstrate how τFCS provides useful tools for model discrimination in FCS curve fitting.

1. INTRODUCTION

Fluorescence correlation spectroscopy (FCS) and related fluctuation spectroscopy and microscopy methods have become important research tools that enable detailed investigations of the chemical and physical properties of molecules and molecular systems in a variety of complex environments (Ehrenberg & Rigler, 1976; Elson, 2011; Elson & Magde, 1974; Magde, Webb, & Elson, 1972; Rigler & Elson, 2001; Thompson, 1991; Webb, 2001). In particular, fluorescence fluctuation measurements can provide detailed quantitative information about hydrodynamic mobility, chemical kinetics, molecular concentrations, interaction stoichiometry and binding energetics, and physical dynamics of molecules in solution or within living cells and tissues. The capability to investigate both kinetic and thermodynamic properties of molecular systems without bulk perturbations of the system away from equilibrium enhances the appeal of fluctuation studies (Elson, 2011), and also facilitates the application of fluctuation methods in complex environments, such as within living cells.

All fluctuation methods are based on the statistical sampling of stochastic spatial and/or temporal fluctuation events, and information recovery requires curve fitting procedures to match experimental data to a physical model. Knowledge of the sample composition is thus necessary to determine its physical parameters by fluctuation analysis. In particular, to complete curve fitting procedures one needs to have an accurate physical model for the mechanisms driving the fluctuation dynamics (such as diffusion or chemical reactions) and must also know whether a sample has single or multiple (and how many) molecular components. While these requirement are often successfully met this is not always the case, and curve fitting can present significant challenges in fluctuation analysis. Two particular factors that can complicate analysis of experimental fluctuation data are discussed here. First, the underlying mechanisms for the fluctuation dynamics and/or the sample composition may not be known *a priori* and the fluctuation data itself does not always provide sufficient cues for model selection and discrimination. Second, in many applications of fluctuation methods, the measurements do not have sufficient resolution to accurately determine the concentration and dynamics of multiple molecular components within samples, even though such information would be very useful. For example, diffusion coefficients and or molecular brightness values may not be sufficiently different to resolve heterogeneous sample populations or detect molecular interactions.

The focus of this chapter is to introduce a multimode global analysis strategy that has the capability to address each of these challenges and to enhance the resolution and model discrimination capabilities of fluctuation spectroscopy methods. We will show that simultaneous global analysis (Beechem, 1992; Knutson, Beechem, & Brand, 1983; Verveer, Squire, & Bastiaens, 2000) of fluorescence lifetime and FCS data, here called τFCS, improves the ability to discriminate between fitting models and can greatly enhance the resolution of fluctuation methods when compared with fitting the fluctuation data alone.

2. BACKGROUND

As introduced above, there are applications of fluctuation methods for which curve fitting and data analysis present significant challenges. For example, FCS processes the statistical information through analysis of the temporal autocorrelation function. While extremely useful in a wide variety of applications, FCS alone can typically only resolve multiple sample components if their diffusion coefficients differ appreciably (Meseth, Wohland, Rigler, & Vogel, 1999). This is often not the case, such as when studying binding interactions between two molecules of similar or identical molecular weights, for which the diffusion coefficient of the dimer is typically not sufficiently different from the monomers to resolve via FCS diffusion analysis. This limited resolution was one of the motivating factors for the development of several alternate approaches to identify molecular interactions using analysis of fluctuation amplitudes, among them dual-color cross correlation spectroscopy and molecular brightness based assays such as PCH, FIDA, and cumulants (Berland, So, Chen, Mantulin, & Gratton, 1996; Chen, Müller, Berland, & Gratton, 1999; Chen, Müller, So, & Gratton, 1999; Eigen & Rigler, 1994; Kask et al., 2000; Müller, 2004; Schwille, Meyer-Almes, & Rigler, 1997; Wu, Chen, & Müller, 2006; Wu & Müller, 2005) . The global analysis introduced below is intended to add to the list of options for experimental approaches to resolve multiple sample components and molecular interactions when FCS alone is insufficient. While not specifically covered in this chapter, the strategy introduced here can also be applied to improve resolution in other fluctuation spectroscopy methods.

2.1. Motivation for global analysis

The general approach of discussed here involves global fitting (Knutson et al., 1983) of fluctuation data and data from at least one other simultaneously acquired fluorescence acquisition mode such as fluorescence lifetime, anisotropy, or energy transfer. The utility of multiparameter or multimode data acquisition has previously been identified (Kudryavtsev et al., 2007; Kuhnemuth & Seidel, 2001; Sisamakis, Valeri, Kalinin, Rothwell, & Seidel, 2010; Weidtkamp-Peters et al., 2009). By employing simultaneous measurements of multiple measurement modes that are each sensitive to different physical properties of an experimental system, it becomes possible to more precisely define the overall composition and behavior of the system than is typically possible with a single methods alone. Measurements of fluorescence anisotropy (molecular rotations), lifetimes (excited state dynamics), and Forster resonance energy transfer (interactions) (Lakowicz, 1999) are examples of other measurement modes that are often easily acquired simultaneously with fluctuation data on the microscope. When combined with fluorescence fluctuation analysis, which can provide unique information about molecular concentrations and stoichiometry of interacting molecules, the multimode approach can become especially powerful. One relevant example that uses this approach is referred to as fluorescence lifetime correlation spectroscopy that combines lifetime and FCS measurements together with filter based mathematical processing of the experimental data to resolve multiple components that are not resolvable by FCS alone (Gregor & Enderlein, 2007; Kapusta, Wahl, Benda, Hof, & Enderlein, 2007). A second example was recently introduced that combines fluctuation and anisotropy analysis and illustrates the capability to dramatically enhance the resolution of sample composition (Nguyen, Sarkar, Veetil, Koushik, & Vogel, 2012).

We refer to the specific implementation of the multimode approach introduced here as τFCS. This method employs both FCS and lifetime data and combines that with global analysis curve fitting to analyze the multimode data. Global fitting is described in the research literature (Beechem, 1989; Rahim, Pelet, Kamm, & So, 2012, Skakun et al., 2005; Verveer, Squire, & Bastiaens, 2001; Verveer et al., 2000). Briefly, the concept is to fit multiple data sets simultaneously, while linking the values of physical parameters that are common across all data sets, ensuring that a physical model is identified that can simultaneously fit all the experimental data sets. One fundamental reason for why the global analysis is effective is that the different experimental modes (e.g., FCS, fluorescence lifetime, Anisotropy, etc.) have unique dependences on the fundamental system parameters of concentration, molecular brightness (defined below), diffusion times, etc. For example, as shown in detail below, the amplitude of correlation functions is proportional to the sum of the concentration multiplied by the *square* of the molecular brightness for each component species, whereas for lifetime data the amplitude depends on the product of concentration and molecular brightness to the first power. There are thus significantly greater constraints on the fitting models that can reasonably fit the both types of data, and as such provides a powerful tool for model discrimination in curve fitting. Global fitting also improves the capability to resolve multiple components in heterogeneous samples. We illustrate this principle below using FCS and fluorescence lifetime to demonstrate the capability to accurately resolve the molecular concentrations of two-component mixtures even for the case where the diffusion coefficients and molecular brightness of the two components are identical and thus not resolvable via FCS or molecular brightness based fluctuation methods alone. As we show below, τFCS analysis for such a system also provides more information than lifetime alone as lifetime data could only be fit for fractional intensities rather than molecular concentrations.

3. THEORY

3.1. Fluorescence signals on the microscope

To facilitate the discussion, we begin with a theoretical description of measured fluorescence signals from microscopic volumes, including a description of how finite excited state lifetimes of fluorescent molecules play a role in fluorescence measurements on the microscope. This derivation allows one to understand the role of excitation saturation in fluctuation measurements, useful for model discrimination when photophysical dynamics play an important role in FCS measurements (e.g., triplet state kinetics), and more importantly provides a common theoretical notation linking fluctuation measurements with fluorescence lifetime measurements. By defining fluorescence lifetime decays in terms of molecular concentrations and molecular brightness parameters rather than fractional intensities it becomes straight forward to globally link the physically fundamental fitting parameters across multiple data types.

Experimental measurement of fluctuations requires that the fluorescence signal must arise from a relatively small number of molecules, ranging from a single molecule to a few thousands. For typical concentrations of interest, this requires the signal to be measured from very small measurement volumes, typically on the scale of femtoliters or tenths of femtoliters, defined optically in three-dimensions using either two-photon or confocal microscopy (Fig. 7.1, left). We thus begin with a theoretical description of fluorescence signals measured from small volumes on the microscope. This chapter will focus exclusively on using two-photon excitation, although the theory below is easily generalized for one-photon confocal fluorescence measurements, as the same three-dimensional Gaussian profile is typically used to model one-photon confocal FCS experiments as well.

3.2. Time resolved two-photon fluorescence

The theory for two-photon excited fluorescence microscopy has been previously described (Denk, Strickler, & Webb, 1990; Wu & Berland, 2007; Xu & Webb, 1997). The instantaneous absorption rate of photon pairs follows the familiar intensity squared dependence, and is given by $W(\mathbf{r},t') = (\sigma_2 I^2(t') S^2(\mathbf{r}))/2$, where σ_2 is the two-photon absorption cross section, $I(t')$ is the laser flux (in photons $\mathrm{cm}^{-2}\,\mathrm{s}^{-1}$) at the center of the focused excitation source, and $S(\mathbf{r})$ is a dimensionless distribution function representing the spatial profile or point spread function (PSF) of the focused laser excitation. In the above notation, we have used the variable t' to denote the microtime, measured from the beginning of each excitation laser pulse. The variable t' resets to zero with each new laser pulse and has as a

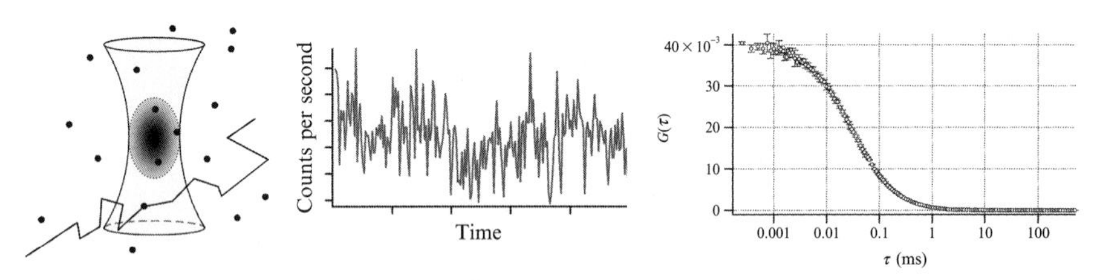

Figure 7.1 Fluorescent molecules diffusing under Brownian motion through an observation volume created at the focus of an objective lens (left) lead to fluctuations in the recorded fluorescence signal (middle). An autocorrelation function, $G(\tau)$, is computed to analyze the fluctuation statistics. The amplitude of the correlation function is related to the molecular concentration, and the temporal relaxation reveals the timescale and dynamic process responsible for the fluctuations.

maximum value the time between laser pulses. The microtime scale is used to describe the molecular excited state dynamics of the fluorescing molecules, including molecular excitation during each laser pulse and nanosecond scale excited state relaxation between laser pulses. We will later make use of a second macrotime variable, t, which does not reset with each laser pulse and instead and measures time continuously from the start of an experiment.

In the limit of low-excitation intensity, the excited state populations and corresponding fluorescence signals are linearly proportional to $W(\mathbf{r},t')$, which is consistent with the familiar "intensity-squared" dependence of the fluorescence signal on input laser power. The simple linear relationship between $W(\mathbf{r},t')$ and the measured fluorescence signals can be lost due to excitation saturation, and also influenced by the finite time molecules spend in the excited state if their fluorescence lifetime is a significant fraction of the time between laser pulses. Saturation and finite lifetime effects can be important in two photon microscopy due to the use of high peak power and high-repetition-rate lasers. We therefore describe below a complete model for the fluorescence signal that accurately captures the nanosecond excited state population dynamics in addition to the longer timescale fluctuation dynamics typically observed using FCS.

Fluorescent molecules can be reasonably modeled as two-state systems (Siegman, 1986). We use N_0 and N_1 to describe the probability that a particular molecule occupies the ground or first excited state, respectively, and $\Delta N = N_0 - N_1$ to describe the difference in these probabilities. Since molecules must be in either the ground or excited state, the sum of these probabilities is unity, that is, $N_0 + N_1 = 1$. The standard first-order differential equation describing the excited state dynamics for laser excitation rate, $W(\mathbf{r},t')$, and spontaneous excited state relaxation rate $\Gamma = 1/\tau$, where τ is the excited state lifetime, can be written as (Siegman, 1986):

$$\frac{\mathrm{d}\Delta N(r,t')}{\mathrm{d}t'} = -2W(r,t')\Delta N(r,t') + 2\Gamma N_1(r,t') \qquad [7.1]$$

There is a separation of timescales between the nanosecond excited state lifetimes of most fluorophores and the ~100 fs laser pulse widths used for two-photon excitation; therefore, we assume negligible spontaneous relaxation during the 100 fs incident laser pulse. As such, the excited state population change during a single laser pulse can be solved using $\mathrm{d}\Delta N(r,t')/\mathrm{d}t' = -2W(r,t')\Delta N(r,t')$. The temporal profile of the laser pulse will introduce a constant factor into the calibration of $W(\mathbf{r},t')$ in terms of the laser intensity (Xu & Webb, 1997), but otherwise is unimportant for the discussion here. We thus assume for simplicity that the laser pulse has a constant intensity during each laser pulse of pulse width α and solve for the change in excited state population during laser excitation. We use ΔN_b and ΔN_a to denote the probability difference between ground and excited states *b*efore and *a*fter the laser pulse, respectively, and find $\Delta N_a(r) = \Delta N_b(r)\exp[-2W(r)\alpha]$. For high repetition-rate lasers of pulse repetition rate f_p, the excited state populations will reach a steady state in which the number of molecules pumped from the ground to excited state during each excitation laser pulse exactly matches the number of spontaneous relaxation events between laser pulses. Solving for this steady state we find the excited state population *a*fter each laser pulse is given by:

$$N_{1a}(r) = \frac{1}{2}\left(\frac{1 - \exp(-2W(r)\alpha)}{1 - \exp\left(-2W(r)\alpha - \left(\Gamma/f_p\right)\right)}\right) \quad [7.2]$$

where $1/f_p$ is the time between the laser pulses (Wu & Berland, 2007). The probability of a molecule to remain in the excited state decays exponentially at a rate $N_1(r,t') = N_{1a}(r)e^{-\Gamma t'}$, where again the microtime t' represents the time measured from the start of each laser pulse. Thus, just *b*efore the next laser pulse the probability of a molecule occupying the excited state is $N_{1b}(r) = N_{1a}(r)e^{-\Gamma/f_p}$. We note that $N_1(r,t')$ does depend on molecular location, r, within the excitation beam, indicating that the fluorescence signal per molecule depends on where the molecule is located within the focused laser beam.

The number of molecular relaxation events per unit time (i.e., excited state to ground state transitions) is the product of the spontaneous relaxation rate, Γ, and the number of molecules in the excited state. The number of fluorescence photons generated per unit volume per unit time is thus $\kappa\Gamma e^{-\Gamma t'}C(r,t)N_{1a}(r)$ where $C(r,t)$ is the time and position dependent molecular concentration (number/volume) of the fluorophore and the constant κ accounts for the quantum yield and also the efficiency of the optical detection system for the fluorescent molecule being investigated. The concentration is written as a function of position, r, and the macrotime variable t (rather than t') as concentration only varies on timescales that are very long compared to the nanosecond scale excited state dynamics. The instantaneous time dependent fluorescence signal, $F_{inst}(t',t)$, is the integrated fluorescence signal over the sample volume:

$$F_{\mathrm{inst}}(t',t) = \kappa\Gamma \mathrm{e}^{-\Gamma t'} \int \mathrm{d}r\, C(r,t) N_{1a}(r) \qquad [7.3]$$

The integral in Eq. (7.3) is evaluated over all space, but for two-photon excitation the probability of occupying the excited state after each laser pulse, $N_{1a}(r)$, goes to zero far from the laser focus so the total integrated fluorescence signal is finite.

Equation (7.3) contains two distinct time variables that characterize the micro- and macrotime variations in the fluorescence signal. In practical fluorescence lifetime measurements using time-resolved photon counting methods, there are many laser pulses per detected photon, and to accurately measure the microtime dynamics the fluorescence signal is averaged over experimental acquisition times, T, that are sufficiently long that any spatial or temporal concentration fluctuations are averaged out. For this case, the time-averaged time-resolved fluorescence signal can be written in terms of the average concentration, with

$$\langle F_{\mathrm{inst}}(t')\rangle = \frac{1}{T}\int_0^T F_{\mathrm{inst}}(t',t)\mathrm{d}t = \kappa\Gamma \langle C\rangle \mathrm{e}^{-\Gamma t'} \int \mathrm{d}r N_{1a}(r) \qquad [7.4]$$

which depends only on the microtime. Alternatively, for fluorescence fluctuation measurements, we will be interested in macrotime variations in concentration and can average out the microtime behavior. The macrotime averaged fluorescence signal thus depends explicitly on the time dependent concentration and can be written as

$$F(t) = f_{\mathrm{p}} \int_0^{1/f_{\mathrm{p}}} F_{\mathrm{inst}}(t',t)\mathrm{d}t' = \kappa f_{\mathrm{p}}\left(1 - \mathrm{e}^{-\Gamma/f_{\mathrm{p}}}\right) \int \mathrm{d}r C(r,t) N_{1a}(r) \qquad [7.5]$$

Here, $F(t)$ is averaged over the microtime behavior but we omit the angular brackets indicating temporal averaging to avoid notational confusion when presenting FCS theory below. One can also find the average fluorescence intensity over long times by averaging over both macro- and microtimes to find:

$$\langle F\rangle = \kappa f_{\mathrm{p}}\left(1 - \mathrm{e}^{-\Gamma/f_{\mathrm{p}}}\right)\langle C\rangle \int \mathrm{d}r N_{1a}(r) \qquad [7.6]$$

For purposes of notational consistency with fluctuation spectroscopy theory, it is convenient to rewrite Eqs. (7.3)–(7.6) in terms of a position dependent "effective" molecular excitation rate (Nagy, Wu, & Berland,

2005; Wu & Berland, 2007; Xu & Webb, 1997), $\langle W_{\text{eff}}(r)\rangle$. This excitation rate indicates the average number of excitation events per molecule per second for molecules located at position r, including saturation and finite lifetime effects, and has the simple relationship with average intensity that $\langle F\rangle = \kappa\langle C\rangle \int \mathrm{d}r\langle W_{\text{eff}}(r)\rangle$. Comparing this expression with Eq. (7.6), we see that $\langle W_{\text{eff}}(r)\rangle = N_{1\text{a}}(r) f_{\text{p}}\left(1 - e^{-\Gamma/f_{\text{p}}}\right)$. In terms of this new notation, we can then rewrite Eqs. (7.4) and (7.5) as

$$\langle F_{\text{inst}}(t')\rangle = \frac{\kappa\Gamma \mathrm{e}^{-\Gamma t'}}{f_{\text{p}}\left(1-\mathrm{e}^{-\Gamma/f_{\text{p}}}\right)}\langle C\rangle \int \mathrm{d}r\langle W_{\text{eff}}(r)\rangle \quad [7.7]$$

and

$$F(t) = \kappa \int \mathrm{d}r C(r,t)\langle W_{\text{eff}}(r)\rangle \quad [7.8]$$

We will return to these expressions again after defining the observation volume and molecular brightness below.

3.3. Observation volumes and gamma factors

In the context of fluctuation spectroscopy, it is useful to define a quantity referred to as the "observation volume" that serves as an estimate of the size of the sample region from which most of the measured fluorescence signal arises. The observation volume is not a true volume or container size in that there are no well-defined boundaries for which the molecule is either "inside" or "outside" the volume. Nonetheless, it is possible to generate a quantity with dimensions of volume by dividing the total fluorescence signal of Eq. (7.8) by the fluorescence per unit volume generated by molecules located at the center of the focused laser beam (Nagy et al., 2005; Thompson, 1991). Specifically, the measurement volume is defined as

$$V = \frac{1}{\langle W_{\text{eff}}(0)\rangle}\int \langle W_{\text{eff}}(r)\rangle \mathrm{d}r \equiv \int \langle \hat{W}_{\text{eff}}(r)\rangle \quad [7.9]$$

where $\langle \hat{W}_{\text{eff}}(r)\rangle$ is the normalized fluorescence excitation probability that defines the profile of the observation volume and includes both saturation and finite lifetime effects. From this definition, it is easy to see that the observation volume is a normalized probability rather than a container size. Most fluctuation spectroscopy researchers make use of the convenient three-dimensional Gaussian (3DG) function, $S_{3\text{DG}}(r)$, as a sufficient if not physically precise description of the observation volume profile (Müller,

Chen, & Gratton, 2003). In cylindrical coordinates, the 3DG profile is defined as $S_{3DG}(\rho,z)=e^{-2(\rho^2/\omega_0^2)}e^{-2(z^2/z_0^2)}$ with $1/e^2$ radial and axial beam waists ω_0 and z_0, respectively (Qian & Elson, 1990; Rigler, Mets, Widengren, & Kask, 1993).

For fluctuation measurements, higher-order moments of $\langle \hat{W}_{\text{eff}}(r)\rangle$ are generally also needed to fully characterize the volume. For the specific case of FCS, the additional required parameter is referred to as the gamma factor (Elson & Magde, 1974; Mertz, Xu, & Webb, 1995; Thompson, 1991). The gamma factor is defined as

$$\gamma=\frac{\int \langle \hat{W}(r)\rangle^2 \mathrm{d}r}{\int \langle \hat{W}(r)\rangle \mathrm{d}r} \quad [7.10]$$

and its value (always less than unit value) characterizes the uniformity of the fluorescence signal from molecules located at various locations within the volume and the effective steepness of the boundary defining the volume. We note that some authors prefer to incorporate the gamma factor into their definition of the volume, defining an effective detection volume as $V_{\text{eff}}=V/\gamma$ (Mertz et al., 1995; Schwille, Haupts, Maiti, & Webb, 1999; Webb, 2001).

Using the volume notation, one can rewrite the average fluorescence signal as

$$\langle F\rangle=\kappa\langle C\rangle\langle W_{\text{eff}}(0)\rangle V\equiv\Psi\langle C\rangle V \quad [7.11]$$

where we have introduced the "molecular brightness" parameter $\Psi=\kappa\langle W_{\text{eff}}(0)\rangle$. The volume multiplied by the concentration estimates the number of molecules contributing to the fluorescence signal, and the molecular brightness is so named because it reports the average number of fluorescence photons measured per molecule per second (Chen, Müller, Berland, et al., 1999; Chen, Müller, Eid, & Gratton, 2001; Chen, Müller, So, et al., 1999). From its definition, one can see that the molecular brightness actually specifies the number of fluorescence photons measured per molecule per second when the molecule is located at the center of the focal volume within the excitation laser beam. It should be noted that molecular brightness is not a fundamental quantity, but rather depends explicitly on the molecular properties (cross section and quantum yield), the excitation conditions (laser power, pulse width, and beam waist), and the measurement instrumentation (detector and collection optics efficiencies). The molecular brightness is a very useful parameter for investigating molecular interactions as well as for predicting the signal-to-noise ratios of fluctuation spectroscopy measurements, with higher brightnesses leading to better signal statistics (Chen, Müller, Ruan, & Gratton, 2002; Chen, Müller, So, et al., 1999; Koppel, 1974; Müller, 2004; Saffarian & Elson, 2003; Wu & Müller, 2005).

Returning to Eqs. (7.7) and (7.8), and incorporating the observation volume and molecular brightness definitions we can now write an expression for the full time dependence of the fluorescence signal in terms of the molecular brightness as

$$\langle F_{\text{inst}}(t')\rangle = \frac{\Psi\langle C\rangle V}{\left(1-\mathrm{e}^{-\Gamma/f_{\mathrm{p}}}\right)}\frac{\Gamma}{f_{\mathrm{p}}}\mathrm{e}^{-\Gamma t'} \quad [7.12]$$

and

$$F(t) = \Psi\int \mathrm{d}r C(r,t)\left\langle \hat{W}_{\text{eff}}(r)\right\rangle \quad [7.13]$$

3.4. Fluorescence lifetime data

Experimental fluorescence lifetime data are measured for a total acquisition time, T, and histogrammed into time bins of width Δt. The time dependent decay for the full experiment is thus given by $F^{\text{data}}(t') = \langle F_{\text{inst}}(t')\rangle f_{\mathrm{p}} T$, that is, the instantaneous decay following each laser pulse multiplied by the total number of pulses for the experiment. The fully histogrammed data which specifies the number of counts per time bin, $F^{\text{data}}_{\text{hist}}(t')$ at time, t', is thus written:

$$F^{\text{data}}_{\text{hist}}(t') = \langle F_{\text{inst}}(t')\rangle f_{\mathrm{p}} T\Delta t = \frac{\Psi\langle C\rangle V}{1-\mathrm{e}^{-\Gamma/f_{\mathrm{p}}}}\Gamma\mathrm{e}^{-\Gamma t'} T\Delta t \quad [7.14]$$

For samples with multiple components, the total data signal in counts per bin are

$$F^{\text{data}}_{\text{hist}}(t') = \sum_i \frac{\Psi_i\langle C_i\rangle V}{1-\mathrm{e}^{-\Gamma_i/f_{\mathrm{p}}}}\Gamma_i\mathrm{e}^{-\Gamma_i t'} T\Delta t \quad [7.15]$$

The expression for the fluorescence lifetime data is now written fully in terms of the molecular brightness values and concentrations, that is,the same notation that is used for fluctuation spectroscopy. This will allow global fitting of fluorescence lifetime and FCS data sets, with fundamental molecular parameters such as brightness, Ψ, and concentration $\langle C\rangle$ linked across multiple experimental data acquisition and analysis methods as described below.

3.5. Fluorescence correlation spectroscopy

The basic theory of FCS, shown schematically in Fig. 7.1, is well described in the literature (Elson & Magde, 1974; Magde et al., 1972; Rigler & Elson, 2001; Thompson, 1991) and elsewhere within this volume, so we provide only a brief introduction with emphasis on notational consistency between FCS and fluorescence lifetime measurements. In fluctuation spectroscopy, measured fluorescence fluctuations serve as reporters of local concentration fluctuations within the observation volume (Fig. 7.1, left and middle). With the use of the volume and brightness notation from Eq. (7.13), the fluorescence fluctuations can be written as $\delta F(t) = \Psi \int \delta C(r,t)\hat{W}(r)\mathrm{d}r$, with $\delta F(r,t) = F(r,t) - \langle F \rangle$ and $\delta C(r,t) = C(r,t) - \langle C \rangle$ are the fluorescence and local concentration fluctuations respectively. Statistical analysis is needed to recover useful information from the random fluctuation measurements. A variety of statistical analysis methods have been introduced, but here we focus on the most commonly implemented auto- and cross-correlation analysis. The fluorescence autocorrelation function is generally defined as

$$G(\tau) = \frac{\langle \delta F(\mathrm{r},t)\delta F(\mathrm{r}',t+\tau)\rangle}{\langle F\rangle^2} \qquad [7.16]$$

which mathematically compares the signal with itself on various timescales, τ (Fig. 7.1, right). Theoretical forms for the correlation function are readily derived and can be used for curve fitting of measured FCS data to recover the desired experimental parameters. While the exact mathematical form of the correlation function depends on the profile of the observation volume as well as the underlying physical dynamics, it is generally possible to represent the normalized correlation function for a single molecular species as

$$G(\tau) = \frac{\gamma}{\langle C\rangle V} A(\tau), \qquad [7.17]$$

where $G(0) = \gamma/\langle C\rangle V$ represents the amplitude of the correlation function and $A(\tau)$ represents the temporal relaxation profile ($A(0) = 1$). The correlation amplitude is thus inversely proportional to the sample concentration.

Equation (7.17) can be further generalized to account for both multiple molecular species and multiple detection channels resulting in a correlation function that has a similar form although weighted by the molecular brightness of the individual species, written as

$$G_{mn}(\tau) = \frac{\gamma}{V} \frac{\sum_i \Psi_{i,m} \Psi_{i,n} \langle C_i \rangle A_i(\tau)}{\left(\sum_i \Psi_{i,m} \langle C_i \rangle\right)\left(\sum_i \Psi_{i,n} \langle C_i \rangle\right)} \qquad [7.18]$$

The indices m and n represent detector channels and the index i represents the various molecular species. The quantity $\Psi_{i,m}$ therefore represents the molecular brightness of molecular species i in detector channel m. Equation (7.18) is quite general, accurately describing the signal for both auto- and cross-correlation measurements with multiple molecular species. For autocorrelation measurements, the indices m and n refer to the same detector channel whereas for cross correlation they refer to two unique detector channels.

The specific mathematical form for the time dependence of the correlation function, $A(\tau)$, depends on the molecular processes that cause the fluctuations. For this work meant to illustrate the value of global fitting of lifetime and FCS data for correlation analysis, the functional form for pure Brownian diffusion is sufficient. It has been previously shown (Berland, So, & Gratton, 1995; Feder, BrustMascher, Slattery, Baird, & Webb, 1996; Kohler, Schwille, Webb, & Hanson, 2000; Krichevsky & Bonnet, 2002; Magde & Elson, 1978; Thompson, 1991) that for the 3DG volume profile and diffusion coefficient D, the temporal dependence of the correlation function is $A(\tau) = 1/((1+8D\tau/\omega_0{}^2)(1+8D\tau/z_0{}^2)^{1/2})$ which yields

$$G(\tau) = \frac{\gamma_{3\mathrm{DG}}}{\langle C \rangle V_{3\mathrm{DG}}} \frac{1}{(1+8D\tau/\omega_0^2)(1+8D\tau/z_0^2)^{1/2}} \qquad [7.19]$$

with $V_{3\mathrm{DG}} = \frac{\pi^{3/2}}{8}\omega_0^2 z_0$ and $\gamma_{3\mathrm{DG}} = \frac{1}{2\sqrt{2}}$ (Berland & Shen, 2003; Schwille, Haupts, et al., 1999). This solution is also sometimes written in terms of a diffusion time defined as $\tau_\mathrm{D} = \omega_0{}^2/8D$, which is related to the average time a molecule will reside within the observation volume before diffusing out. We note that the diffusion time for two-photon excitation has one-half the value of the equivalent expression for one-photon excitation.

4. METHODS

We illustrate the multimode global analysis approach using simulated FCS and fluorescence lifetime data for two-component mixtures. The authors have also demonstrated this method experimentally, with the results to appear in a forthcoming publication. Autocorrelation curves and fluorescence lifetime histograms where created using the two-component equations above (Eqs. (7.18) and (7.15)).

Autocorrelation lag times were chosen to match those used in experimental calculation of correlated data from time correlated single photon counting (TCSPC) data, ranging from 0.001 to 100 or 1000 ms depending on the magnitude of the diffusion coefficients. The beam waist parameter, ω_0, was chosen as 0.35 μm and the axial to radial beam waist ratio parameter, z_0/ω_0, was chosen as 5, typical values for our experimental setup. Noise added to each channel of the simulated autocorrelation functions, $G(\tau)$, was estimated from real experiments performed under comparable conditions.

All fluorescence lifetime decay histograms were calculated for an 80 MHz pulsed laser using a bin width of 32 ps with 320 data points to simulate tail fitting of experimental data, except for the reduced-χ^2 surface shown in Fig. 7.3, which used a bin width of 52 ps with 250 data points. Due to the photon counting nature of exponential decay data, noise added follows a Poissonian distribution such that

$$\sigma_{\text{decay}}(t'_i) = \sqrt{F_{\text{hist}}^{\text{data}}(t'_i)} \qquad [7.20]$$

where t'_i denotes the ith histogram bin of width Δt.

Once created, simulated data were fitted to Eqs. (7.18) and (7.15) using a Levenberg–Marquardt nonlinear least squares algorithm in Igor Pro (Wavemetrics, Inc., OR). The Igor Pro software minimizes and returns the χ^2 value, which describes the difference between the fitted function y and the corresponding simulated data point y_i having a standard deviation σ_i. This is then normalized by the number of degrees of freedom, $\nu = n - p - 1$, where n is the number of data points and p is the number of free parameters, to give the reduced-χ^2 (Straume & Johnson, 1992):

$$\chi_R^2 = \frac{1}{n-p-1}\sum\left(\frac{y-y_i}{\sigma_i}\right)^2 \qquad [7.21]$$

Given an appropriate measure of the standard deviations, a reduced-χ^2 value of 1 denotes an ideal fit. Readers interested in the fitting routines used here may contact the authors for additional information about the fitting software. A variety of global fitting packages are also available commercially.

When using global fitting for different types of data, certain fitting parameters are global (relating to and influenced by both data types), and others local (relevant to only one data type or the other). Table 7.1 lists all fitting parameters sorted as local or global parameters, the parameter values used for data simulations, and indicates which parameters are held fixed during global fitting and which are free fitting parameters. There are two main sets of simulated data, one in which the diffusion coefficient of each component is the same ($D_1 = D_2$) and a second in which the diffusion coefficients differ by a factor of two ($D_1 = 2 \times D_2$). For each set, data were simulated for a range of concentration values. Data points in the figures show the average returned values from fits to three individually simulated data sets. The total acquisition time for the TCSPC histograms was chosen to be 300 s. In the global fitting algorithm for τFCS, the calculated reduced-χ^2 value is explicitly

$$\chi_{\mathrm{R}}^2 = \frac{1}{k+l-p-1}\left(\sum_{i=0}^{k-1}\frac{\left(F_{\mathrm{hist}}^{\mathrm{data}}(x_i) - y_i\right)^2}{\sigma_{\mathrm{decay}}(x_i)} + \sum_{i=k}^{k+l-1}\frac{\left(G(x_i) - y_i\right)^2}{\sigma_{\mathrm{fcs}}(x_i)}\right) \qquad [7.22]$$

where k and l are the number of data points in the lifetime histogram and the calculated autocorrelation function, respectively. It is informative to consider that the number of data points as well as the standard deviations for each data type will affect the relative weighting of the minimized χ^2 value.

Table 7.1 The parameter values used to create simulated data

	Parameter	$D_1 = D_2$	Held/*free*	$D_1 = 2 \times D_2$	Held/*free*	Units
Global	C_1	0.01	Free	0.01	Free	μ*M*
	C_2	0.0001–1	Free	0.0001–1	Free	μ*M*
	Ψ_1	10	Held	10	Held	kcpsm
	Ψ_2	5, 10	Free	5	Free—τFCS Held—FCS only	kcpsm
Local	D_1	300	Held	300	Held	$\mu m^2\ s^{-1}$
	D_2	300	Free	150	Free	$\mu m^2\ s^{-1}$
	τ_1	4.0	Held	4.0	Held	ns
	τ_2	2.0	Free	2.0	Free	ns

Parameters are mapped in the table by how they are used for curve fitting, as either global or local parameters. The table also indicates which parameters are free fitting parameters and which are held fixed for curve fitting. The vertical column labels indicate the two data sets with equal diffusion coefficients, $\boldsymbol{D_1 = D_2}$, and different diffusion coefficients $\boldsymbol{D_1 = 2 \times D_2}$ corresponding to the data in Figs. 7.2 and 7.3, respectively.

5. RESULTS

We find the use of τFCS very effectively extends measurement capabilities of the use of FCS alone, with excellent resolution of molecular concentrations over a wide range of two-component mixture compositions. We illustrate the capability to resolve concentrations of multiple sample components even when their diffusion coefficients and molecular brightness are identical. We next demonstrate how τFCS can improve accuracy of curve fitting even when the two sample components would be resolvable by FCS alone. After discussing methods to reduce the number of free fitting parameters for τFCS analysis, we conclude with an analysis investigating the range of fluorescence lifetime ratios, and relative diffusion coefficients for a two-component sample for which it is possible to unambiguously identify the presence of multiple sample components using τFCS.

5.1. Resolution of concentration and diffusion coefficients when $D_1 = D_2$

When diffusion coefficients of multiple sample components have similar values, FCS measurements alone are unable to resolve the presence of multiple species. It is generally accepted that diffusion coefficients must differ by a factor of ~1.6–2 to be resolved by FCS (Meseth et al., 1999; Saffarian & Elson, 2003). This limitation can be overcome, provided an additional experimental parameter is able to resolve the component species. For example, we show here that τFCS can clearly recover the experimental parameters for a two component mixture even when the diffusion coefficients are identical as long as the fluorescence lifetimes of the two molecules are sufficiently different. To demonstrate, we simulated autocorrelation functions and lifetime decays of a system containing a mixture of two species at a variety of concentrations. The two species have the identical diffusion coefficient, $D = 300\ \mu m^2\ s^{-1}$, but distinct fluorescence lifetimes, $\tau_1 = 4$ ns and $\tau_2 = 2$ ns. When fit with FCS alone, there is no possibility to resolve multiple sample components. However, when analyzed using τFCS, the components of the sample are clearly resolved as shown in Fig. 7.2. Remarkably, even small fractions of one or the other component are clearly resolved with the τFCS analysis.

The identification of a second component is not unique to τFCS as fluorescence lifetime measurements alone can clearly resolve the second species. However, lifetime alone can only resolve fractional intensities of the two sample components and cannot determine concentrations. Serial fitting of lifetimes to first identify two species and then subsequent two-component FCS analysis is also unable to achieve this task. It is only via global fitting using the τFCS approach that we can achieve the demonstrated resolution.

In the example shown, the two components differ in molecular brightness by a factor of 2 and could thus also be easily resolved using brightness based assays such as PCH or Cumulant analysis (Chen, Müller, So, et al., 1999; Müller, 2004; Wu & Müller, 2005). Such measurements can be performed simultaneously with FCS and provide useful measurement capabilities. To illustrate how the τFCS approach adds a unique advantage for data analysis we simulated a second data set with the same parameters as above, with the exception that both the diffusion coefficients and molecular brightness values are identical for the two species. As seen in Fig. 7.2, τFCS still clearly succeeds in recovering the correct parameters that describe the sample composition. Given the equal brightness and diffusion coefficients, fluctuation methods alone would not be able to achieve this success.

We note that the fitting results shown in Fig. 7.2 were obtained under the assumption that we have independent knowledge of the diffusion coefficient, lifetime, and brightness for one of the two components. This is often the case in FCS applications. It is also possible to recover similar information without knowing the parameters for a single component *a priori*, although such curve fitting requires either excellent signal statistics or additional constraints. We discuss below some additional fitting constraints that can be implemented practically and allow full parameter recover even without assuming prior knowledge of the lifetime, brightness, and diffusion coefficient for one sample component. While not shown in the chapter, these approaches can successfully resolve the two-component mixtures shown here, although over a more limited range of relative concentrations and lifetime ratios.

5.2. Accurate concentration recovery with unknown molecular brightness

We demonstrated above that τFCS can dramatically enhance the resolution of fluctuation spectroscopy measurements, identifying a second sample component and correctly recovering the concentration of each component when FCS alone cannot achieve that result. Here, we illustrate that using τFCS to enhance resolution even for the case where FCS alone is able to

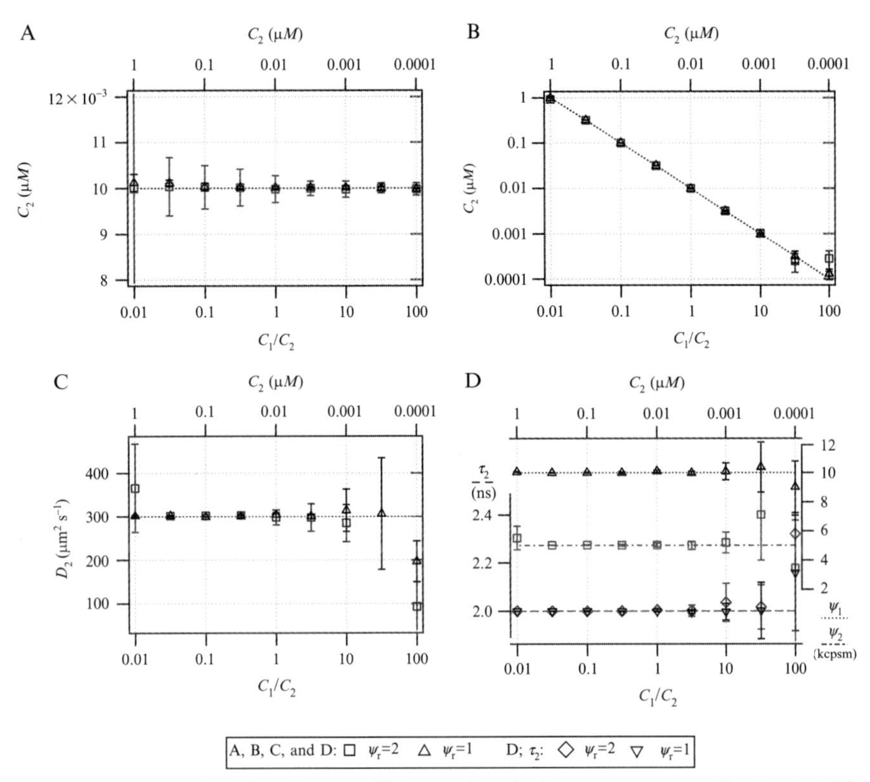

Figure 7.2 τFCS global fits of FCS and lifetime data for binary mixtures with identical diffusion coefficients ($D_1 = D_2 = 300$ μm^2 s^{-1}). Plots show data for mixtures of molecules with either different ($\Psi_r = 2$, squares) or identical ($\Psi_r = 1$, triangles) molecular brightness values. The concentration C_2 is titrated from 1 to 1×10^{-4} μ*M* (B) while C_1 is held fixed at 0.01 μ*M* (A). The lower horizontal axes indicate the concentration ratio and the upper axis is reported as the absolute concentration of C_2. The fluorescence lifetimes of the two components are $\tau_1 = 4$ and $\tau_2 = 2$ ns. The lifetime values for τ_2 are plotted as diamonds for $\Psi_r = 2$ and inverted triangles for $\Psi_r = 1$. D_1, Ψ_1, and τ_1 were fixed parameters in the analysis. FCS fits cannot be plotted as it is impossible to correctly analyze this data using FCS alone. However, these plots show that τFCS can recover concentrations C_1 (A) and C_2 (B) with remarkable accuracy across nearly four orders of magnitude. The recovery of D_2 (C), τ_2 (D, left axis) and Ψ_2 (D, right axis), is accurate at all but the very lowest C_2 concentrations. Dashed/dotted lines indicate the actual parameter value used to create the simulated data sets. Data points for the global fits are an average of three repeated simulations with global fitting and error bars indicate the standard deviation of the returned values.

correctly identify multiple sample components through diffusion analysis. To illustrate this, we have generated an additional simulated data set for a two-component mixture, in this case the diffusion coefficients are different by a factor of 2, enough to distinguish using FCS alone. However, inspection of Eq. (7.18) shows that in order to fit the data to a two-component model one needs to know the molecular brightness values for each component species. If the brightness of each species is known then full information recover is possible, but quite often this may not be the case and one needs to estimate the relative brightness values. For example, if one is using FCS for binding studies and a fluorescent ligand is partially quenched upon ligand binding its molecular brightness will be decreased by an unknown amount. If incorrect brightness values are used for curve fitting, then inaccurate concentration values are recovered from the FCS fits. This is shown in Fig. 7.3, where we have fit the FCS data using three different assumptions for the value of the molecular brightness ratio, $\Psi_r = \Psi_1/\Psi_2$. The brightness ratio is necessarily held fixed during FCS fitting—fits are otherwise unstable, as molecular brightness and concentration are covariant parameters—and one can clearly see the significant distortion and incorrect concentrations returned when incorrect brightness ratios are used (Fig. 7.3A and B). In contrast, by using τFCS to analyze the same data, the need to make assumptions about the molecular brightness ratio vanishes as Ψ_2 is easily resolvable from the τFCS analysis (Fig. 7.3D, horizontal diamonds).

Not only does τFCS accurately recover the concentrations without needing to estimate brightness ratios, but inspection of Fig. 7.3 also demonstrate that τFCS can resolve much smaller concentrations of the second species than FCS alone even for the case where the molecular brightness ratio is known accurately. The precise nature of the enhanced resolution will always depend on the specific experimental details, including brightness ratios, diffusion coefficients, lifetime differences, etc. For the example, we have shown here, with parameters $D_1 = 300\ \mu m^2\ s^{-1}$, $D_2 = 150\ \mu m^2\ s^{-1}$, $\tau_1 = 4$ ns, $\tau_2 = 2$ ns, $\Psi_1 = 10$, $\Psi_2 = 5$, we see that τFCS analysis can still resolve the concentration of the second component even when it is only 1% of the concentration of the first component that has double the molecular brightness. Standard FCS, even with the correct brightness ratio of 2, starts to be unable to resolve the concentration of the second component at much higher concentration. The concentration limit where τFCS begins to fail in recovering accurate concentrations understandably corresponds to the concentration where the signal from the second component becomes too small for the global analysis to correctly determine its lifetime, τ_2 (Fig. 7.3D; vertical diamonds). Nonetheless, we see a clear demonstration of how τFCS has dramatically enhanced the curve fitting resolution, and this enhancement requires only that there is a measurable difference in fluorescence lifetime values for the sample components.

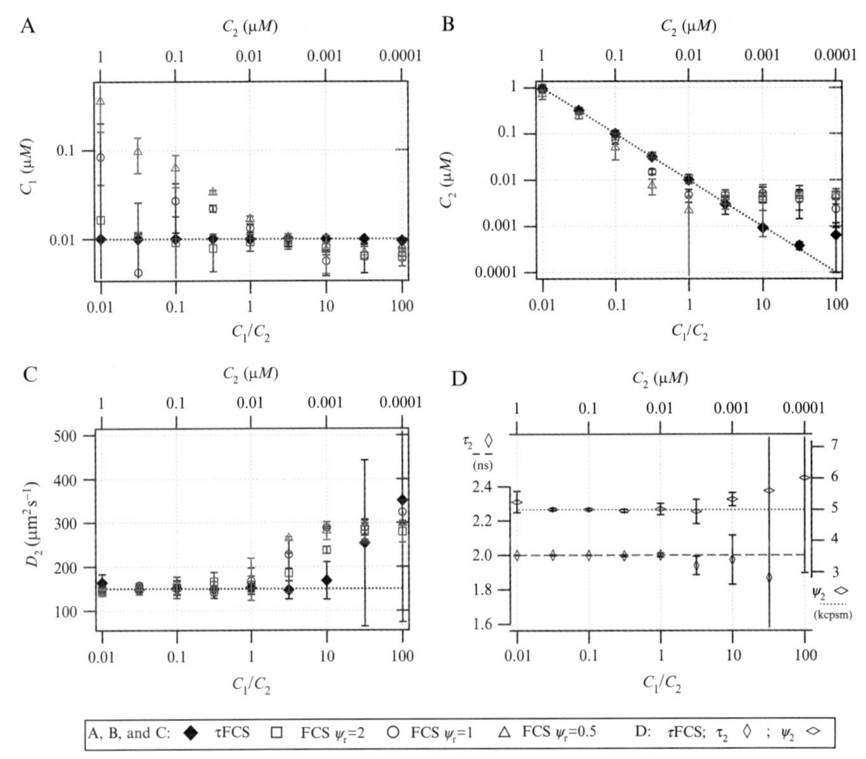

Figure 7.3 Improved accuracy of recovered concentrations achieved using τFCS. Data were simulated for a binary mixture with two resolvable diffusion coefficients ($D_1 = 300\ \mu m^2\ s^{-1}$, $D_2 = 150\ \mu m^2\ s^{-1}$) and a brightness ratio, $\Psi_r = \Psi_1/\Psi_2$, of 2. The concentration C_1 was held constant at 10 n*M*, while C_2 was sequentially diluted from 1 μ*M* to 100 p*M*. FCS curve fitting requires an estimated for Ψ_2 (or Ψ_r) in order to attain a stable fit (without fixing either *C* or Ψ the product $C\Psi^2$ is covariant). Data were thus fit with FCS assuming brightness ratios of $\Psi_r = 0.5$ (triangles), $\Psi_r = 1$ (circles), and $\Psi_r = 2$ (squares). We see that the recovered concentrations C_1 (A) and C_2 (B) are accurate when the correct brightness ratio is assumed, but badly distorted when incorrect brightness ratios are used for FCS fitting—even though stable fits with a misleading "low error" are achieved. In contrast, τFCS fits (diamonds) yield accurate concentrations for the full range of concentrations shown. Of significant interest is that τFCS is able to more accurately recover the concentration of C_2 than FCS alone even for the case where the correct molecular brightness ratio is known.

5.3. Additional curve fitting enhancements

We have shown above that using global analysis can enhance the capabilities of fluctuations methods, provided one finds at least one experimentally measurable parameter which clearly distinguishes the different molecular components within the sample, such as different fluorescent lifetimes in τFCS. Global analysis of data measuring the distinguishing parameter along with fluctuation data then provides excellent capabilities to characterize the properties of complex samples. We expect this approach can be widely applicable for many of the different fluctuation spectroscopy data acquisition and analysis methods. Despite the advantages offered by the global analysis approach, curve fitting of fluctuation data can still involve complicated procedures or require detailed knowledge about sample composition that is not easily accessible. For example, the data shown above were fit with the assumption that the properties of one sample component were known independently, effectively limiting the number of fitting parameters. In some cases, this independent information will not be available, yet for such cases accurate and stable curve fitting can often still be achieved, particularly if one can identify procedures to limit to total number of free fitting parameters. Specifically, we recognize that there is often information available from a fluorescence measurement that is unused by a particular analysis method and that the added information can be used to constrain fitting models, effectively reducing the number of fitting parameters. We provide two examples of how this can be achieved here.

First, we recognize that the time average fluorescence intensity is easily measured in most experiments, with no additional experimental requirements to do so. We know from Eq. (7.11) that the average intensity is given by $\langle F \rangle = \Psi \langle C \rangle V$, and thus by extension for multiple species we have

$$\langle F \rangle = \sum_i \Psi_i \langle C_i \rangle V \qquad [7.23]$$

For a sample with two components, we have two molecular brightness parameters and two concentrations, for a total of four fitting parameters. However, one can easily see that if the average fluorescence signal has been accurately measured, then only three of these parameters are truly unknown. Using the measured value of $\langle F \rangle$ in curve fitting thus effectively reduces the number of fitting parameters by one. We find that using this "average F" constraint in fitting models does significantly improve curve fitting stability in otherwise underdetermined systems, and can lead, for example, to stable curve fitting of the data shown in Fig. 7.3 without needing to know as much independent information about one sample component prior to global analysis.

We note that this constraint makes the assumption that all measured photons can be assigned to the molecular species of interest. In practice, all fluorescence measurements have some amount of background signal. Thus, to use this constraint it is important that either the background signal is very small when compared with the fluorescence signal of interest or alternatively that the background can be measured and corrected for. Background signals, typically constant over microtime scales, are easily subtracted from the lifetime data. Correlation amplitudes must also be corrected for background signal levels using

$$G(\tau)_{\text{corrected}} = G(\tau)_{\text{measured}}\left(\frac{\langle F\rangle + \langle B\rangle}{\langle F\rangle}\right)^2 \quad [7.24]$$

where the quantity $\langle F\rangle + \langle B\rangle$ is the actual measured signal level including fluorescence and background contributions, and $\langle B\rangle$ is the independently measured background signal level (Schwille, Korlach, & Webb, 1999; Thompson, 1991).

A second example of a useful fitting constraint is that for any particular fluorescent molecule there is a link between its excited state lifetime and its molecular brightness. For example, a dye labeled ligand may experience quenching upon binding of the ligand to its receptor molecule leading to a reduction in its molecular brightness. For such a case, there would be a corresponding reduction in the fluorescence lifetime. For the case of a dual-component system in which the two components represent the same fluorescent chromophore in two distinct environments (e.g., receptor bound and unbound), we expect the fluorescent lifetimes and molecular brightnesses are related by $\Psi_1/\tau_1 = \Psi_2/\tau_2$. For the experimental systems where this assumption is valid, it would also effectively reduce the number of fitting parameters by one. Making full use of the information encoded in the fluorescence signal thus offers significant opportunities to improve resolution and curve fitting stability in fluctuation measurements. The extent to which this can enhance measurement capabilities does depend on specific experimental circumstances and must be explored on a case-by-case basis.

5.4. Model discrimination and resolution requirements

We have so far demonstrated that τFCS can be extremely useful for accurate curve fitting in two-component systems, recovering molecular concentrations, lifetimes, diffusions coefficients, and molecular brightness values over a wide range of concentrations. For this demonstration, we began with the assumption that the sample was composed of a two-component mixture. There are many experiments one can envision for which it is not known whether a sample has single or multiple components, and for which data analysis procedures are needed to help determine the sample composition. We thus conclude with a computational analysis to investigate the conditions for which τFCS is able to unambiguously resolve the presence of a second sample component. This analysis also sets a benchmark for how different fluorescent lifetimes or diffusion coefficients must be in order to clearly resolve the second component.

For this resolution analysis, we create an array of two-component data sets. For each point in the 21×21 array, with logarithmically spaced abscissa that cover a range of diffusion and lifetime ratios, an autocorrelation function and lifetime decay are calculated as above. The simulated data assumed equal concentration and molecular brightness values for each of the two component species. We note that simulations here are especially convenient for this analysis, as it would be extremely impractical to cover the same parameter space experimentally. We then performed a one-component fit for each of the simulated data sets using a variety of fitting models including: (1) FCS, (2) fluorescence lifetime, (3) fluorescence lifetime with average fluorescence constraint (see above), and (4) τFCS. Diffusion coefficients, concentration, and lifetime were all treated as free parameters. Figure 7.4 depicts sections of the reduced-χ^2 surface for each fitting model. The regions in which the reduced-χ^2 values are closer to 1 denote parameter space regions for which the presence of a second component is not identifiable, that is, there is no reason from the fit alone to suspect more than a single component system. Reduced-χ^2 values of approximately 1.3 define the boundary at which there is justification for rejecting the one-component model.

As seen from the figure, fitting with FCS alone yields a horizontal χ^2 valley centered on the diffusion ratio of one (horizontal dot-dashed lines). This is consistent with the known requirements for differences in diffusion coefficients in order to resolve multiple components by FCS (Meseth et al., 1999). Similarly, fitting the lifetime data only with a single exponential function gives a vertical valley centered on the lifetime ratio of 1, also as expected (solid vertical lines). By incorporating the average fluorescence constraint (Eq. 7.23) into the amplitude of the single exponential decay (Eq. 7.15), it is possible to narrow the vertical valley, increasing the resolvability of a second component (vertical dashed lines). Data fitted using single

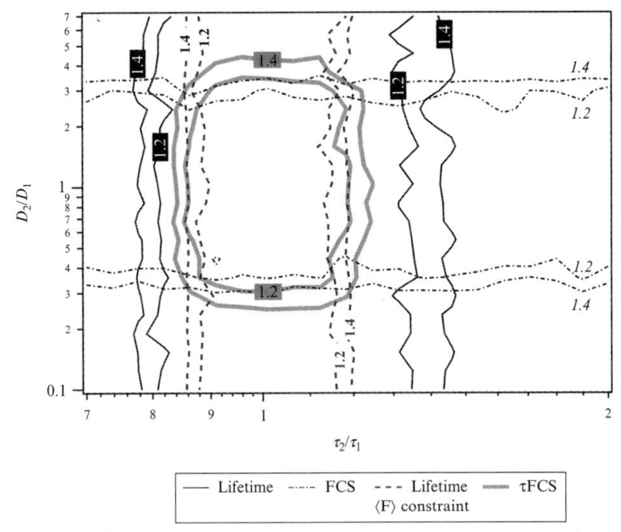

Figure 7.4 Reduced-χ^2 plots for curve fitting of binary mixtures with single component analysis. Curve fitting can indicate the single component model fails to fit the data if the reduced-χ^2 value is higher than ~1.3–1.5 when fitted using a one-component model. We show sections of four reduced-χ^2 surfaces demonstrating the regions in parameter space where: (1) FCS (horizontal dot-dashed lines), (2) fluorescence lifetime (vertical solid lines), (3) fluorescence lifetime with average fluorescence constraint (vertical dashed lines), and (4) τFCS (thick gray lines) successfully identify the presence of a second component. The central region of the plot indicates where none of these methods can resolve the second component. Moving to parameter space regions away from the center, where the reduced-χ^2 value becomes larger, indicate lifetime and/or diffusion coefficient ratios for which two components can be successfully identified by curve fitting. Using FCS alone, it is only possible to correctly identify a second component when diffusion ratios lie outside of the horizontal area centered on $D_2/D_1=1$ (bounded by horizontal dot-dashed lines). Similarly, fluorescence lifetime can only discern a second component outside a vertical area centered on $\tau_1/\tau_2=1$ (bounded by vertical solid lines), but cannot provide concentration or molecular brightness information. Introduction of the average fluorescent constraint into the amplitude of the fluorescence lifetime narrows this vertical region (bounded by vertical dashed lines), demonstrating the value of fitting model constraints as discussed in the text. The combined global analysis of τFCS reduces the area for which it is not possible to identify the presence of a second component, occupying a symmetric region centered on $\tau_1/\tau_2=D_2/D_1=1$ (thick gray lines), allowing for concentration, diffusion, and molecular brightness information to be identified over a much wider range of experimental parameters.

component τFCS, also incorporating the average fluorescence constraint (solid gray lines) allows us to combine the analyses and reduce the overall parameter space in which the second component is unidentifiable. Using τFCS, the overall parameter space in which the two components are clearly resolved is greatly increased. The central region of the plot, where the lifetimes and diffusion coefficients are still nearly identical does indicate that there are regions of parameter space for which τFCS cannot resolve the two components. This is as expected, as there is no difference in the physical parameters upon which to base the molecular discrimination. For such cases, resolution would require identification of an additional experiment parameter such as molecular brightness or rotational correlation times to be able to resolve the sample populations.

6. CONCLUSIONS

We have introduced τFCS as an excellent tool to enhance resolution and model discrimination capabilities in fluctuation spectroscopy measurements. By employing simultaneous acquisition of multiple modes of fluorescence data (e.g., fluorescence lifetime and FCS), at least one of which shows good contrast for the component species in the sample, we are then able to resolve signal contributions that would otherwise not be detectable via fluctuation measurements alone. Employing a global analysis curve fitting approach with global fitting parameters for the common parameters contributes to the success of this approach. We have demonstrated the capability to accurately resolve the molecular concentration of two sample components using τFCS, even for the case where the diffusion coefficients and molecular brightness values of the two components are identical and would therefore be undetectable by FCS alone. We have further demonstrated the utility of the τFCS fitting routine even for cases where FCS alone is capable of resolving multiple diffusing components. We expect the multimode global analysis approach can be widely useful in fluctuation spectroscopy applications, including τFCS but also extending to other acquisition modes such as anisotropy and energy transfer measurements.

ACKNOWLEDGMENTS

This work was partially supported by Emory University and NSF grants MCB0817966 and DMR0907435.

REFERENCES

Beechem, J. M. (1989). A second generation global analysis program for the recovery of complex inhomogeneous fluorescence decay kinetics. *Chemistry and Physics of Lipids*, *50*, 237–251.

Beechem, J. M. (1992). Global analysis of biochemical and biophysical data. *Methods in Enzymology*, *210*, 37–54.

Berland, K., & Shen, G. Q. (2003). Excitation saturation in two-photon fluorescence correlation spectroscopy. *Applied Optics*, *42*, 5566–5576.

Berland, K. M., So, P. T., Chen, Y., Mantulin, W. W., & Gratton, E. (1996). Scanning two-photon fluctuation correlation spectroscopy: Particle counting measurements for detection of molecular aggregation. *Biophysical Journal*, *71*, 410–420.

Berland, K. M., So, P. T. C., & Gratton, E. (1995). 2-Photon fluorescence correlation spectroscopy - Method and application to the intracellular environment. *Biophysical Journal*, *68*, 694–701.

Chen, Y., Müller, J. D., Berland, K. M., & Gratton, E. (1999). Fluorescence fluctuation spectroscopy. *Methods*, *19*, 234–252.

Chen, Y., Müller, J. D., Eid, J. S., & Gratton, E. (2001). Two-photon fluorescence fluctuation spectroscopy. In B. Valeur & J. C. Brochon (Eds.), *New trends in fluorescence spectroscopy: Applications to chemical and life sciences*. Berlin: Springer-verlag.

Chen, Y., Müller, J. D., Ruan, Q. Q., & Gratton, E. (2002). Molecular brightness characterization of EGFP in vivo by fluorescence fluctuation spectroscopy. *Biophysical Journal*, *82*, 133–144.

Chen, Y., Müller, J. D., So, P. T. C., & Gratton, E. (1999). The photon counting histogram in fluorescence fluctuation spectroscopy. *Biophysical Journal*, *77*, 553–567.

Denk, W., Strickler, J. H., & Webb, W. W. (1990). 2-Photon laser scanning fluorescence microscopy. *Science*, *248*, 73–76.

Ehrenberg, M., & Rigler, R. (1976). Fluorescence correlation spectroscopy applied to rotational diffusion of macromolecules. *Quarterly Reviews of Biophysics*, *9*, 69–81.

Eigen, M., & Rigler, R. (1994). Sorting single molecules: application to diagnostics and evolutionary biotechnology. *Proceedings of the National Academy of Sciences of the United States of America*, *91*, 5740–5747.

Elson, E. L. (2011). Fluorescence correlation spectroscopy: past, present, future. *Biophysical Journal*, *101*, 2855–2870.

Elson, E. L., & Magde, D. (1974). Fluorescence correlation spectroscopy. 1. Conceptual basis and theory. *Biopolymers*, *13*, 1–27.

Feder, T. J., BrustMascher, I., Slattery, J. P., Baird, B., & Webb, W. W. (1996). Constrained diffusion or immobile fraction on cell surfaces: A new interpretation. *Biophysical Journal*, *70*, 2767–2773.

Gregor, I., & Enderlein, J. (2007). Time-resolved methods in biophysics. 3. Fluorescence lifetime correlation spectroscopy. *Photochemical & Photobiological Sciences*, *6*, 13–18.

Kapusta, P., Wahl, M., Benda, A., Hof, M., & Enderlein, J. (2007). Fluorescence lifetime correlation spectroscopy. *Journal of Fluorescence*, *17*, 43–48.

Kask, P., Palo, K., Fay, N., Brand, L., Mets, U., Ullmann, D., et al. (2000). Two-dimensional fluorescence intensity distribution analysis: Theory and applications. *Biophysical Journal*, *78*, 1703–1713.

Knutson, J. R., Beechem, J. M., & Brand, L. (1983). Simultaneous analysis of multiple fluorescence decay curves—A global approach. *Chemical Physics Letters*, *102*, 501–507.

Kohler, R. H., Schwille, P., Webb, W. W., & Hanson, M. R. (2000). Active protein transport through plastid tubules: Velocity quantified by fluorescence correlation spectroscopy. *Journal of Cell Science*, *113*, 3921–3930.

Koppel, D. E. (1974). Statistical accuracy in fluorescence correlation spectroscopy. *Physical Review A*, *10*, 1938–1945.

Krichevsky, O., & Bonnet, G. (2002). Fluorescence correlation spectroscopy: The technique and its applications. *Reports on Progress in Physics*, *65*, 251–297.

Kudryavtsev, V., Felekyan, S., Wozniak, A. K., Konig, M., Sandhagen, C., Kuhnemuth, R., et al. (2007). Monitoring dynamic systems with multiparameter fluorescence imaging. *Analytical and Bioanalytical Chemistry*, *387*, 71–82.

Kuhnemuth, R., & Seidel, C. A. M. (2001). Principles of single molecule multiparameter fluorescence spectroscopy. *Single Molecules*, *2*, 251–254.

Lakowicz, J. R. (1999). *Principles of fluorescence spectroscopy* (2 ed.). New York: Plenum.

Magde, D., & Elson, E. L. (1978). Fluorescence correlation spectroscopy. 3. Uniform translation and laminar-flow. *Biopolymers*, *17*, 361–376.

Magde, D., Webb, W. W., & Elson, E. (1972). Thermodynamic fluctuations in a reacting system—Measurement by fluorescence correlation spectroscopy. *Physical Review Letters*, *29*, 705–708.

Mertz, J., Xu, C., & Webb, W. W. (1995). Single-molecule detection by two-photon-excited fluorescence. *Optics Letters*, *20*, 2532–2534.

Meseth, U., Wohland, T., Rigler, R., & Vogel, H. (1999). Resolution of fluorescence correlation measurements. *Biophysical Journal*, *76*, 1619–1631.

Müller, J. D. (2004). Cumulant analysis in fluorescence fluctuation spectroscopy. *Biophysical Journal*, *86*, 3981–3992.

Müller, J. D., Chen, Y., & Gratton, E. (2003). Fluorescence correlation spectroscopy. *Methods in Enzymology*, *361*, 69–92. http://www.sciencedirect.com/science/bookseries/00766879/361/supp/C.

Nagy, A., Wu, J., & Berland, K. M. (2005). Observation volumes and gamma factors in two-photon fluorescence fluctuation spectroscopy. *Biophysical Journal*, *89*, 2077–2090.

Nguyen, T. A., Sarkar, P., Veetil, J. V., Koushik, S. V., & Vogel, S. S. (2012). Fluorescence polarization and fluctuation analysis monitors subunit proximity, stoichiometry, and protein complex hydrodynamics. *PloS One*, *7*, e38209.

Qian, H., & Elson, E. L. (1990). Analysis of confocal laser-microscope optics for 3-D fluorescence correlation spectroscopy. *Applied Optics*, *30*, 1185–1195.

Rahim, N. A. A., Pelet, S., Kamm, R. D., & So, P. T. C. (2012). Methodological considerations for global analysis of cellular FLIM/FRET measurements. *Journal of Biomedical Optics*, *17*(2), 026013 [Mar 19]. http://dx.doi.org/10.1117/1.JBO.17.2.026013.

Rigler, R., & Elson, E. (2001). *Fluorescence correlation spectroscopy theory and applications, springer series in chemical physics*. New York: Springer.

Rigler, R., Mets, U., Widengren, J., & Kask, P. (1993). Fluorescence correlation spectroscopy with high count rate and low-background—Analysis of translational diffusion. *European Biophysics Journal with Biophysics Letters*, *22*, 169–175.

Saffarian, S., & Elson, E. L. (2003). Statistical analysis of fluorescence correlation spectroscopy: The standard deviation and bias. *Biophysical Journal*, *84*, 2030–2042.

Schwille, P., Haupts, U., Maiti, S., & Webb, W. W. (1999). Molecular dynamics in living cells observed by fluorescence correlation spectroscopy with one- and two-photon excitation. *Biophysical Journal*, *77*, 2251–2265.

Schwille, P., Korlach, J., & Webb, W. W. (1999). Fluorescence correlation spectroscopy with single-molecule sensitivity on cell and model membranes. *Cytometry*, *36*, 176–182.

Schwille, P., Meyer-Almes, F. J., & Rigler, R. (1997). Dual-color fluorescence cross-correlation spectroscopy for multicomponent diffusional analysis in solution. *Biophysical Journal*, *72*, 1878–1886.

Siegman, A. (1986). *Lasers*. University Science Books.

Sisamakis, E., Valeri, A., Kalinin, S., Rothwell, P. J., & Seidel, C. A. M. (2010). Accurate single-molecule fret studies using multiparameter fluorescence detection. *Methods in Enzymology*, *474*, 455–514.

Skakun, V. V., Hink, M. A., Digris, A. V., Engel, R., Novikov, E. G., Apanasovich, V. V., et al. (2005). Global analysis of fluorescence fluctuation data. *European Biophysics Journal*, *34*, 323–334.

Straume, M., & Johnson, M. L. (1992). Analysis of residuals—Criteria for determining goodness-of-fit. *Methods in Enzymology*, *210*, 87–105.

Thompson, N. L. (1991). Fluorescence correlation spectroscopy. In J. R. Lakowicz (Ed.), *Topics in fluorescence spectroscopy* (pp. 337–378). New York: Plenum.

Verveer, P. J., Squire, A., & Bastiaens, P. I. H. (2000). Global analysis of fluorescence lifetime imaging microscopy data. *Biophysical Journal, 78*, 2127–2137.

Verveer, P. J., Squire, A., & Bastiaens, P. I. H. (2001). Improved spatial discrimination of protein reaction states in cells by global analysis and deconvolution of fluorescence lifetime imaging microscopy data. *Journal of Microscopy, 202*, 451–456.

Webb, W. W. (2001). Fluorescence correlation spectroscopy: Inception, biophysical experimentations, and prospectus. *Applied Optics, 40*, 3969–3983.

Weidtkamp-Peters, S., Felekyan, S., Bleckmann, A., Simon, R., Becker, W., Kuhnemuth, R., et al. (2009). Multiparameter fluorescence image spectroscopy to study molecular interactions. *Photochemical & Photobiological Sciences, 8*, 470–480.

Wu, J., & Berland, K. (2007). Fluorescence Intensity is a poor predictor of saturation effects in two-photon microscopy: Artifacts in fluorescence correlation spectroscopy. *Microscopy Research and Technique, 70*, 682–686.

Wu, B., Chen, Y., & Müller, J. D. (2006). Dual-color time-integrated fluorescence cumulant analysis. *Biophysical Journal, 91*, 2687–2698.

Wu, B., & Müller, J. D. (2005). Time-integrated fluorescence cumulant analysis in fluorescence fluctuation spectroscopy. *Biophysical Journal, 89*, 2721–2735.

Xu, C., & Webb, W. W. (1997). Multiphoton excitation of molecular fluorophores and nonlinear laser microscopy. In J. Lakowicz (Ed.), *Topics in fluorescence spectroscopy* (pp. 471–540). New York: Plenum.

CHAPTER EIGHT

Dual-Focus Fluorescence Correlation Spectroscopy

Christoph Pieper, Kerstin Weiß, Ingo Gregor, Jörg Enderlein[1]
Third Institute of Physics–Biophysics, Georg-August-University Göttingen, Friedrich-Hund-Platz, Göttingen, Germany
[1]Corresponding author: e-mail address: enderlein@physik3.gwdg.de

Contents

Abstract

This chapter introduces into the technique of dual-focus fluorescence correlation spectroscopy or 2fFCS. In 2fFCS, the fluorescence signals generated in two laterally shifted but overlapping focal regions are auto- and crosscorrelated. The resulting correlation curves are then used to determine diffusion coefficients of fluorescent molecules or particles in solutions or membranes. Moreover, the technique can also be used for noninvasively measuring flow-velocity profiles in three dimensions. Because the distance between the focal regions is precisely known and not changed by most optical aberrations, this provides an accurate and immutable external length scale for determining diffusivities and velocities, making 2fFCS the method of choice for accurately measuring absolute values of these quantities at pico- to nanomolar concentration.

Methods in Enzymology, Volume 518
ISSN 0076-6879
http://dx.doi.org/10.1016/B978-0-12-388422-0.00008-X

1. INTRODUCTION

A fundamental property of particles in solution is their thermally induced random movement called translational diffusion. Translational diffusion is characterized by the diffusion coefficient D which is connected with the average hydrodynamic size of a molecule or particle via the Stokes–Einstein relation:

$$D = \frac{k_B T}{6\pi\eta r_H}, \quad [8.1]$$

where k_B denotes the Boltzmann constant, T the temperature, η the viscosity of the solution, and r_H the hydrodynamic radius. From the very start of fluorescence correlation spectroscopy (FCS; Elson & Magde, 1974; Magde, Elson, & Webb, 1972, 1974), it was clear that the method should be an ideal tool for measuring translational and rotational diffusion coefficients. The advantage of FCS over other methods, such as dynamic light scattering (DLS), pulsed-field gradient NMR, or size exclusion chromatography, is its unmatched sensitivity. This allows to measure sample concentrations from pico- to nanomolar. This is particularly important when dealing with proteins where the difficulty to produce sizable amounts of sample can be a serious limitation. However, for many years, FCS lacked the accuracy of DLS or NMR in measuring diffusion coefficients. Many biologically relevant conformational changes produce rather small changes in molecular size on the order of Ångstrøm (see e.g., Weljie, Yamniuk, Yoshino, Izumi, & Vogel, 2003). To monitor such small changes, it is necessary to measure the diffusion coefficient with an accuracy better than a few percent. The core problem with the accuracy of conventional FCS becomes obvious if one recalls that one needs a time and a length scale to measure a diffusion coefficient. The time scale of an FCS experiment is given by the clock of the detection system (timing of photon-detection events), and the length scale is given by the size of the detection volume. As a result, the method is enormously sensitive to the exact shape and size of the detection volume (subsequently referred to as molecule detection function or MDF) of the confocal microscope. Even the slightest changes of this volume will change the resulting correlation curve and thus the extracted value of diffusion coefficient. However, the precise size and shape of the MDF depends on many parameters that are difficult to control, see Enderlein, Gregor, Patra, Dertinger, and Kaupp (2005). We briefly discuss the most important ones here.

The first common problem is that most water-immersion objectives typically used in an FCS setup are designed to image through a cover slip of specific thickness. In this sense, the cover slip acts as the last optical element of the objective, and the optical quality of imaging (and laser focusing) critically depends on the exact matching between the actual cover slip thickness and the value the objective is adjusted to. Figure 8.1 illustrates what happens if the actual cover slip thickness differs from its design value by only a few micrometers. As a result, the size of the MDF becomes larger, shifting the temporal decay of the autocorrelation function toward longer lag times. Also, the autocorrelation amplitude becomes smaller and thus the apparent concentration larger (there are more molecules present in the detection volume because it has become larger). In general, increasing aberration of any kind leads to a larger detection volume and thus to an apparently lower diffusion coefficient and higher concentration. It should be noted that the errors shown here are largely unaffected when the position of the focus is varied.

This is in stark contrast to the second problem considered here, the effect of refractive index mismatch. An optical microscope using a

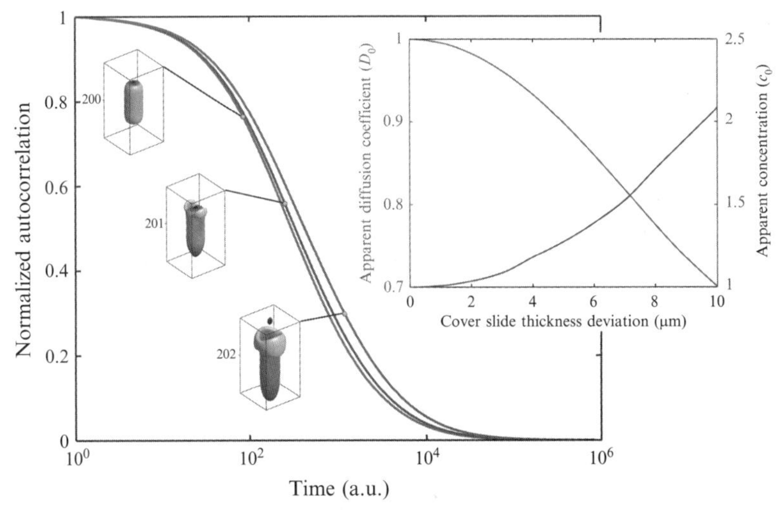

Figure 8.1 The main figure shows, from left to right, the MDF and autocorrelation function for three increasing values of cover slide thickness difference of 0, 5, and 10 μm compared to the design value. Box size of the MDF displays is 1 μm × 1 μm × 2 μm; the number next to the box gives the center position along the optical axis in μm. Note the shift of the center of the MDF along the optical axis for increasing values of thickness difference. The inset figure shows the dependence of apparent diffusion coefficient and the concentration on thickness difference. (For color version of this figure, the reader is referred to the online version of this chapter.)

water-immersion objective is optimally corrected for imaging in water. However, in many biophysical applications, buffer solutions are used with slightly different refractive indices, typically in the range between 1.333 and 1.360. Figure 8.2 shows the impact of refractive index mismatch on the MDF, the autocorrelation function, and on the apparent diffusion coefficient and concentration. In contrast to the cover slip thickness, the aberrations introduced by refractive index mismatch accumulate with increasing distance of the focus from the cover slide surface, because a thicker layer of solution with mismatched refractive index lies between the optics and the detection volume. The effect of refractive index mismatch can be much reduced when positioning the detection volume closer to the surface.

A third and particularly intriguing problem in FCS measurements is the dependence of the autocorrelation function on excitation intensity due to optical saturation. Optical saturation means that increasing excitation intensity does not lead to a proportional increase in emitted fluorescence

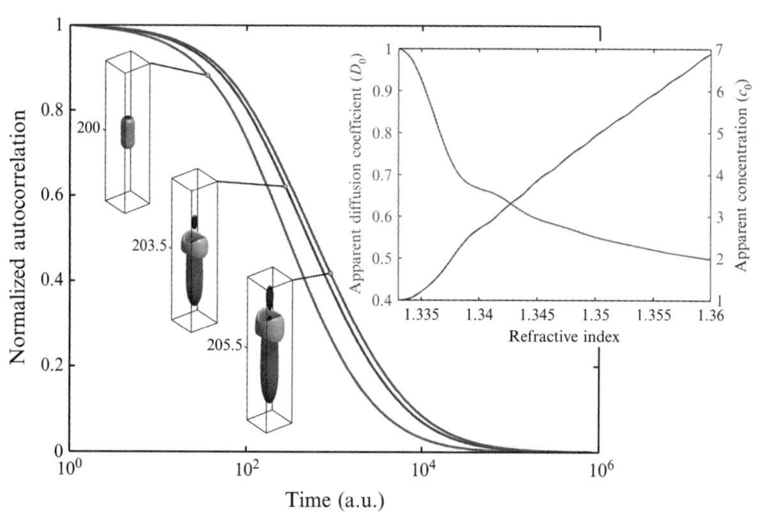

Figure 8.2 The main figure shows, from left to right, the MDF and ACF for three increasing values of refractive index of the sample solution of 1.333, 1.346, and 1.360. Box size of the MDF displays is 1 μm × 1 μm × 5 μm. Note again the shift of the center of the MDF along the optical axis for increasing values of refractive index. The inset figure shows the dependence of apparent diffusion coefficient and the concentration on refractive index. The impact of even slight refractive index mismatch is much more dramatic than that of cover slip thickness. This is mostly due to the large distance of the focus position from the cover slide surface (200 μm, the default value of the commercial instrument Confocor I by Carl Zeiss). (For color version of this figure, the reader is referred to the online version of this chapter.)

intensity. This is due to the fact that a molecule spends more and more time in a nonexcitable state when increasing the excitation rate. The most common sources of optical saturation are (i) excited state saturation, that is, the molecule is still in the excited state when the next absorbable photon arrives; (ii) triplet state saturation, that is, the molecule undergoes intersystem-crossing from the excited to the triplet state so that it can no longer become excited until it returns back to the ground state; (iii) other photo-induced transitions into a nonfluorescing state, such as the photo-induced *cis–trans* isomerization in cyanine dyes, or the optically induced dark states in quantum dots. The detailed relationship between fluorescence emission intensity and excitation intensity can be very complex (Enderlein, 2005) and even dependent on the excitation mode (pulsed or continuous wave) (Gregor, Patra, & Enderlein, 2005). Figure 8.3 shows how optical saturation changes the shape of the MDF and autocorrelation function and the apparent diffusion coefficient and concentration. An important feature of optical saturation is the behavior of apparent diffusion and concentration in the limit of vanishing excitation intensity: the slopes of the apparent diffusion and concentration curves have their largest absolute value at zero saturation. Whereas for cover slip thickness difference or refractive index mismatch the slope of the curves tend to zero for vanishing aberration.

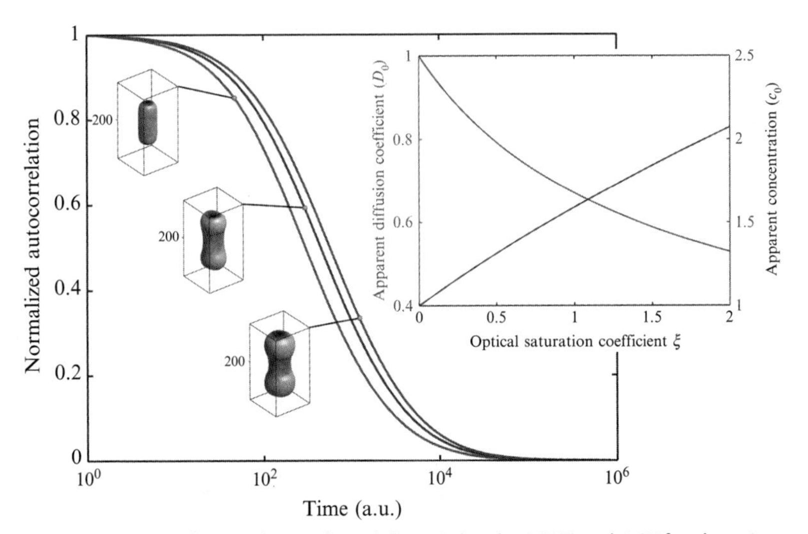

Figure 8.3 The main figure shows, from left to right, the MDF and ACF for three increasing values of optical saturation. Box size of the MDF displays is 1 μm × 1 μm × 2 μm. The inset figure shows the dependence of apparent diffusion coefficient and concentration on optical saturation, that is, excitation intensity. (For color version of this figure, the reader is referred to the online version of this chapter.)

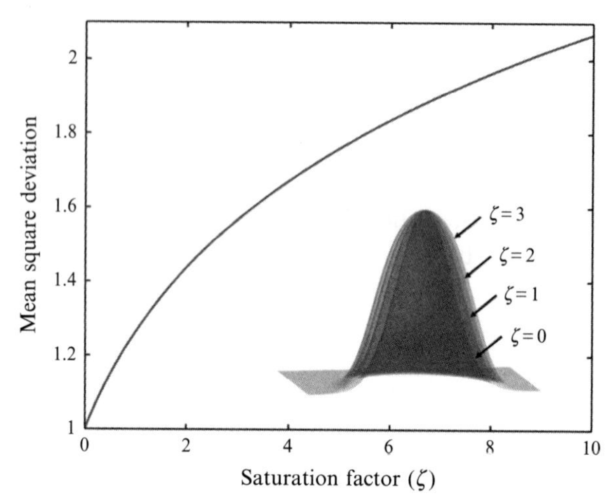

Figure 8.4 Change of the mean square deviation of the distribution $I/(1+I/I_{sat})$ with increasing optical saturation $\zeta = \max(I/I_{sat})$. (For color version of this figure, the reader is referred to the online version of this chapter.)

To better understand the reason for such behavior, we consider an ideal Gaussian excitation profile $I = I_0 \exp(-x^2/2)$. To simulate saturation, we transform this distribution using $I^* = I/(1 + I/I_{sat})$ for different values of the so-called saturation intensity I_{sat}. A measure of the width of the distribution is its mean square deviation. Figure 8.4 shows the widening of such a profile due to saturation. As can be seen, the relative change in profile width is largest in the limit of zero intensity $I_0 \rightarrow 0$, explaining why one sees the biggest change of an autocorrelation function at low saturation levels.

All these effects make a quantitative evaluation of conventional FCS measurements unreliable. Even relative measurements are problematic due to the strong dependence of a FCS result on optical saturation, which is itself determined, in a complex manner, by the photophysics of the dye. Even more, the photophysical properties of one and the same dye can change upon binding to a protein or to another target molecule.

2. DUAL-FOCUS FLUORESCENCE CORRELATION SPECTROSCOPY AND TRANSLATIONAL DIFFUSION

As we have seen, the core problem of conventional FCS is the absence of a reliable and unchangeable length scale. This problem is solved by dual-focus Fluorescence Correlation Spectroscopy or 2fFCS (Dertinger et al., 2007). This technique satisfies two requirements: (i) it introduces an external ruler into the measurement by generating two overlapping laser foci of precisely known and fixed distance, (ii) it generates the two foci and corresponding detection regions in such a way that the corresponding MDFs are sufficiently well described by a simple two-parameter model yielding accurate diffusion coefficients when applied to 2fFCS data analysis. Both these properties allow for measuring absolute values of the diffusion coefficient with an accuracy of a few percent. Moreover, the new technique is robust against all the above-mentioned optical problems that are interfering in conventional FCS measurements.

2.1. Experimental setup

The 2fFCS setup, as shown in Fig. 8.5, is based on a confocal epifluorescence microscope. However, instead of using a single excitation laser, the light of two identical, linearly polarized pulsed lasers is combined by a polarizing beam splitter. Both lasers are pulsed *alternately* with a high repetition rate (typically 10–80 MHz), an excitation scheme called pulsed interleaved excitation (Müller, Zaychikov, Bräuchle, & Lamb, 2005). Thus, the combined light consists of a train of laser pulses with alternating orthogonal polarization. The beam is then coupled into a polarization-maintaining single-mode fiber. At the output, the light is collimated, and a dichroic mirror reflects the beam toward the microscope's water-immersion objective. Before entering the objective, the light beam is passed through a Nomarski prism that is normally used in differential interference contrast (DIC) microscopy (Microscopy, n.d.). The Nomarski prism is an optical element that deflects the laser pulses into two different directions according to their polarization. The deflection axes are collinear with the laser polarization. After focusing the light through the objective, two overlapping excitation foci are generated with a small lateral shift between them. The distance between the foci is uniquely defined by the chosen Nomarski prism and is independent on the sample's refractive index, cover slide thickness, or optical saturation, because all these effects introduce aberrations but will not change the distance between the axes of propagation of both focused laser beams. An important detail is the diameter of the collimated laser beam which is sent into the objective. As we have shown in Dertinger et al. (2008), 2fFCS is most accurate for relaxed focusing, where the resulting diameter of the laser foci in the sample solution is larger than 300 nm.

As in conventional FCS, the generated fluorescence is collected by the same objective, passed through the Nomarski prism and the dichroic mirror,

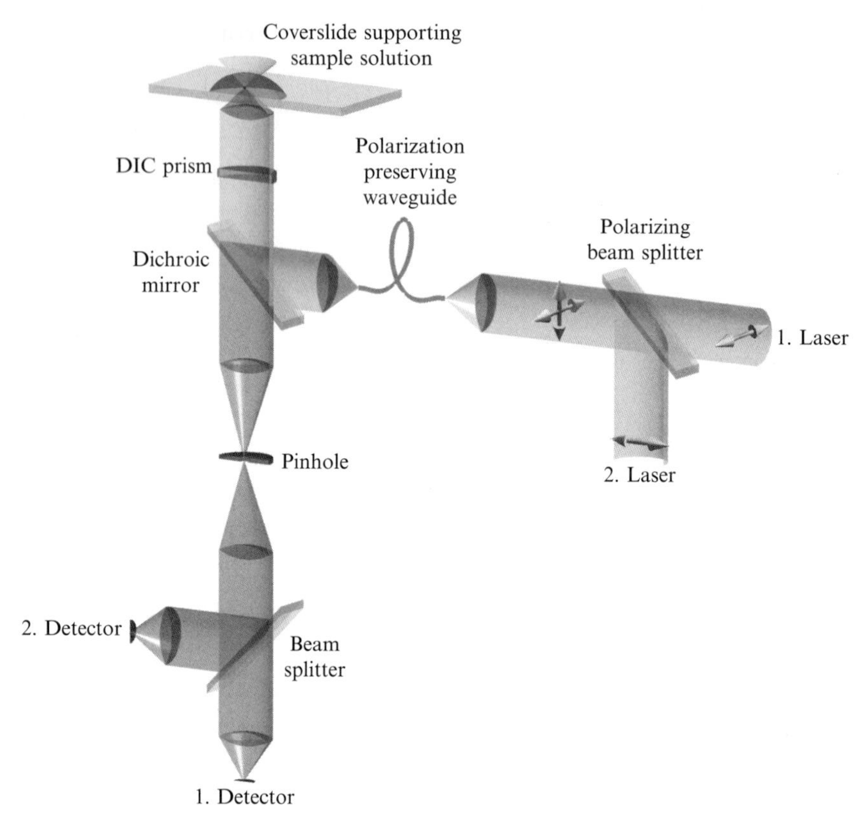

Figure 8.5 Schematic of the 2fFCS setup. Excitation is done by two interleaved pulsed lasers of the same wavelength. The polarizations of both lasers are linear and orthogonal to each other. Light is then combined by a polarizing beam splitter and coupled into a polarization-maintaining optical single-mode fiber. After exiting the fiber, the laser light is collimated by an appropriate lens and reflected by a dichroic beam splitter through a DIC prism. The DIC prism deflects the laser light into two beams according to the polarization of the incoming laser pulses. The microscope objective focuses the two beams into two laterally shifted foci. Fluorescence is collected by the same objective. The tube lens focuses the detected fluorescence from both excitation foci on a single pinhole. Subsequently, the fluorescence light is split by a 50/50 beam splitter and detected by two single-photon avalanche diodes. (See Color Insert.)

and focused onto a single circular aperture which is positioned symmetrically with respect to the positions of both foci. The aperture is chosen large enough to allow the light generated in both foci to pass through. A recommended value of pinhole size is about 150 μm for a magnification of 60×. After the pinhole, the light is collimated, split by a nonpolarizing beam splitter cube and focused onto two single-photon counting avalanche diodes. A dedicated single-photon counting electronics is used to record the detected photons from both detectors with picosecond temporal resolution. Nowadays, turn-key commercial 2fFCS systems are available, such as the MicroTime 200 from PicoQuant (Berlin, Germany).

The picosecond temporal resolution is now used to decide which laser has generated which fluorescence photon, that is, in which laser focus/ detection volume the light was generated. This is done by correlating the detection time of each photon with the time of the last preceding laser pulse. A typical histogram of these time correlations, a so-called time-correlated single-photon counting histogram, is shown in Fig. 8.6. There, one sees two fluorescence decay curves that correspond to the two lasers. In the data analysis, all photons that fall into the first time window are associated with the first laser, and all photons that fall into the second time window with the second laser. A successful application of this method requires that the time between the laser pulses is significantly larger than the fluorescence lifetime of the fluorescent molecules.

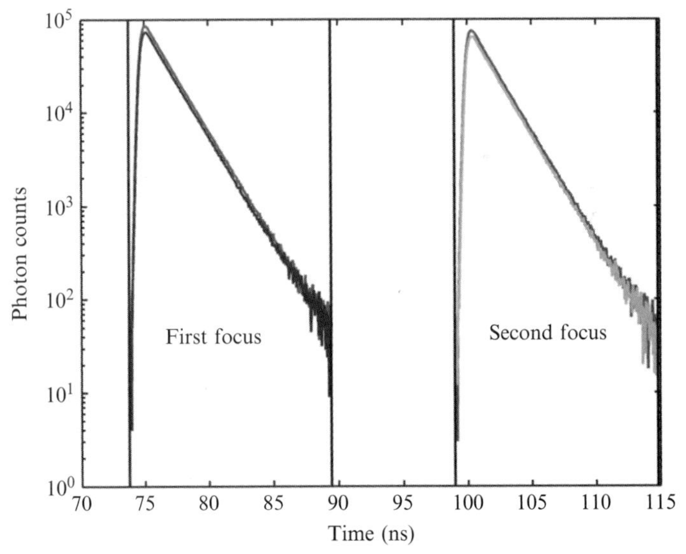

Figure 8.6 TCSPC histograms measured on an aqueous solution of Atto655. The photon counts in the left time window are generated by the first laser, that is, first focus, the photon counts in the second time window are generated by the second laser, that is, second focus. In both time windows, there are two curves corresponding to the two SPAD detectors, respectively. (For color version of this figure, the reader is referred to the online version of this chapter.)

2.2. Calculation of correlation curves

Once each photon is assigned to its respective detection volume, the autocorrelation functions for each detection volume as well as the cross-correlation function between the two detection volumes can be calculated. This can be done with a software algorithm that allows for converting the asynchronously measured single-photon counting data into a correlation curve. Let us assume that the detection times of photons are given by the linear file $\{t_1, t_2, \ldots, t_N\}$, where N is the total number of detected photons. A special feature of these detection times is that they are integer multiples of some minimal time δt, determined by the temporal resolution of the detection electronics. Without restriction of generality, it can be assumed that all times are measured in units of δt, so that all the numbers t_j take integer values. The value $g(\tau)$ of the autocorrelation function for a given lag time τ is defined by

$$g(\tau) = \langle I(t)I(t+\tau)\rangle_t. \qquad [8.2]$$

For a photon-detection measurement with temporal resolution δt, the intensity values $I(t)$ within consecutive time intervals can only take the values $1/\delta t$ or 0, depending on whether there was a photon-detection event during a time interval of width δt or not. The average in Eq. (8.2) is then calculated as the sum over all consecutive time intervals of width δt, divided by the total number of summed intervals. A straightforward way of calculating the autocorrelation function is to divide the total measurement time $t_N - t_1$ into intervals of unit length δt, and to sort the detected photons into these intervals corresponding to their arrival times t_j. The result is a synchronous photon-detection intensity file I_j with j running from 1 through $t_N - t_1$, where the I_j can only adopt the values one or zero. The fluorescence autocorrelation can then be calculated as given by Eq. (8.2). In practice, such an approach is prohibitively demanding of memory and computationally expensive. An alternative, and much more efficient FCS algorithm works directly on the arrival times $\{t_1, t_2, \ldots, t_N\}$, without converting them into time-binned data (Wahl, Gregor, Patting, & Enderlein, 2003). For a given lag time τ, the algorithm counts all photon pairs in the data stream that have this temporal distance τ apart from each other. The number of these photon pairs is directly proportional to the autocorrelation value at lag time τ.

Typically, one does not compute the autocorrelation function for all possible values of lag time τ, but at increasingly spaced lag time values. If the temporal resolution of the photon detection is, for example, 100 ns, and one desires to follow correlation processes up to a minute, possible values of lag time τ are any value between 100 and 60 s in intervals of 100 ns, resulting in 6×10^8 possible lag time values. Calculation of $g(\tau)$ for all of these values would be an enormously time-consuming numerical effort. Instead, the autocorrelation is calculated for only few, approximately logarithmically spaced values of τ, which is sufficient for most FCS applications. A convenient choice for these values of τ is

$$\tau_j = \theta \times \begin{cases} 1 & \text{if } j = 1 \\ \tau_{j-1} + 2^{\lfloor (j-1)/B \rfloor} & \text{if } j > 1 \end{cases}, \quad [8.3]$$

where θ is the base time resolution, and j takes integer values starting with one and running up to some maximum number $j_{\max} = n_{\text{casc}}\, B$. The integer number B defines how many τ-values have the same temporal spacing before doubling it, and the bracket $\lfloor\rfloor$ gives the integer part of the enclosed expression. The resulting lag times are grouped into n_{casc} cascades with equal spacing of $2^{\lfloor (j-1)/B \rfloor}$. The advantage of such a choice of lag times is that all τ_j have integer values so that fast integer arithmetic can be used in subsequent computations. For example, when using values of $B = 10$, $n_{\text{casc}} = 3$, and $\theta = 1$, one obtains the lag time sequence

$$\{\tau_j\} = \{1, 2, \ldots, 9, 10, 12, \ldots, 28, 30, 34, \ldots, 70\}.$$

The correlation algorithm is rather straightforward and visualized in Fig. 8.7. For a given lag time τ, a second vector of arrival times $\{t'_1, t'_2, \ldots, t'_N\}$ is generated, containing the time values $t'_j = t_j + \tau$. In the beginning, the value of the autocorrelation at lag time τ is set to zero. The algorithm starts with the time t_1 in the original vector and moves to consecutive time entries in that vector until it encounters a value t_j that is equal to some t'_k, in which case the value of the autocorrelation at lag time τ is increased by one.

Next, the algorithm switches to the entries of the second vector and, starting with t'_k, moves to consecutive time entries in that vector until it encounters a value t'_l that is equal to some value t_p in the first vector, in which case the value of the autocorrelation at lag time τ is again increased by one. The algorithm switches back to the first vector and, starting with t_p, moves to consecutive time entries in that vector until it encounters a value t_q that is equal to some value t'_r in the second vector, and so on until the last entry in one of both vectors is reached. This correlation algorithm calculates, up to some constant factor, the probability of detecting a photon at some time $t + \tau$ if there was a photon-detection event at time t.

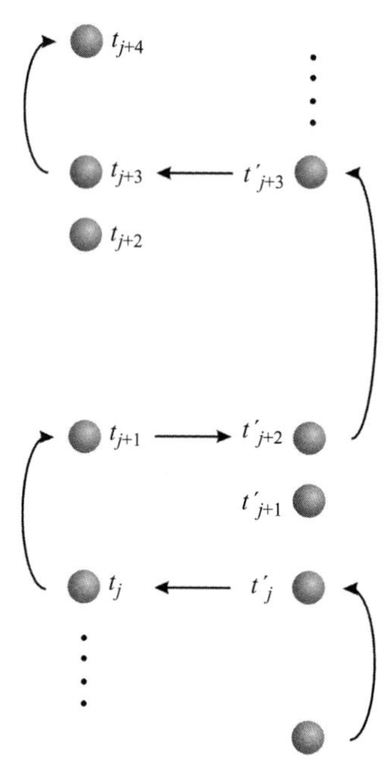

Figure 8.7 Visualization of the software correlation for asynchronous photon-detection data. For details see main text. (For color version of this figure, the reader is referred to the online version of this chapter.)

When applying the algorithm only at the increasingly spaced lag times as given by Eq. (8.3), it will miss, for example, the autocorrelation of any periodic signal with a repetition time not included within the vector of used lag times. To avoid that, one applies an averaging procedure by coarsening the time resolution of the photon-detection times t_j when coming to the calculation of the correlation function at increasingly larger lag times. This is done by modifying the algorithm in the following way. Besides working only with the original and shifted vectors of arrival times, $\{t'_1, t'_2, \ldots, t'_N\}$ and $\{t'_1, t'_2, \ldots, t'_N\}$, all time entries t_j and t'_j are associated with weight values w_j and w'_j that are all set to one at the start of the algorithm. In case of equal $t_j = t'_k$, the autocorrelation is increased not by one but by the product $w_j w'_k$. A time coarsening step is inserted each time when finishing the calculations for one cascade of B lag times τ_j with equal spacing and before starting with the next cascade of B lag times with doubled spacing: All values $\{t'_1, t'_2, \ldots, t'_N\}$ used in the previous cascade are divided by two and rounded to the nearest lower integer value, which will occasionally lead to the occurrence of consecutive entries with the same time value. Such double entries are reduced to one entry, and the corresponding weight of that remaining entry is increased by the weight of the eliminated one. Thus, with increasing lag time τ_j, the time scale underlying the correlation calculation is increased, and the total number of time entries to be processed becomes smaller. To correct for the varying time scale of the correlation calculation, one has finally to divide, at each lag time τ_j, the calculated autocorrelation value by the corresponding scale factor $2^{\lfloor (j-1)/B \rfloor}$.

The correlation algorithm is easily generalized to calculations of cross-correlation functions between two detection channels (e.g., the signal form the first against the signal from the second focus). This is done by using the time values of the first channel as the entries for the first vector $\{t'_1, t'_2, \ldots, t'_N\}$, and the time values of the second channel for calculating the lag time shifted vector $\{t'_1, t'_2, \ldots, t'_N\}$. In that case, one obtains the cross-correlation of the second channel with positive lag times against the first channel. By reversing the order, that is, assigning the time values of the second channel to $\{t'_1, t'_2, \ldots, t'_N\}$ and using the time values of the first channel for $\{t'_1, t'_2, \ldots, t'_N\}$, one obtains the crosscorrelation of the first channel with positive lag times against the second channel.

2.3. Data fitting

A crucial point for a successful 2fFCS data analysis is to have an appropriate model function for the MDF. A suitable expression is given by

$$U(\mathbf{r}) = \frac{\kappa(z)}{w^2(z)} \exp\left[-\frac{2}{w^2(z)}\left(x^2 + y^2\right)\right], \qquad [8.4]$$

where $\kappa(z)$ and $w(z)$ are functions of the axial coordinate z (optical axis) defined by

$$w(z) = w_0 \left[1 + \left(\frac{\lambda_{\mathrm{ex}} z}{\pi w_0^2 n}\right)^2\right]^{1/2} \qquad [8.5]$$

and

$$\kappa_z = 2\int_0^{\alpha} \frac{\mathrm{d}\rho\rho}{R^2(z)} \exp\left(-\frac{2\rho^2}{R^2(z)}\right) = 1 - \exp\left(-\frac{2a^2}{R^2(z)}\right), \qquad [8.6]$$

where the $R(z)$ itself is defined by an expression similar to Eq. (8.5):

$$R(z) = R_0 \left[1 + \left(\frac{\lambda_{\mathrm{em}} z}{\pi R_0^2 n}\right)^2\right]^{1/2}. \qquad [8.7]$$

In the above equations, λ_{ex} is the excitation wavelength and λ_{em} the center emission wavelength, n is the refractive index of the immersion medium (water), a is the radius of the confocal aperture divided by magnification, and w_0 and R_0 are two (generally unknown) model parameters. Equation (8.5) is nothing else than the scalar approximation for the radius of a diverging laser beam with beam waist radius w_0. Equation (8.4) is a modification of the three-dimensional Gaussian. It says that in each plane perpendicular to the optical axis, the MDF is approximated by a Gaussian distribution having width $w(z)$ and amplitude $\kappa(z)/w^2(z)$. Using the parameterization of the MDF as given by Eqs. (8.4–8.7), the auto- and crosscorrelation curves of the dual-focus setup can be calculated as

$$g(t,\delta) = g_\infty(\delta) + \frac{\varepsilon_1 \varepsilon_2 c}{4}\sqrt{\frac{\pi}{Dt}}\int_{-\infty}^{\infty} dz_1 \int_{-\infty}^{\infty} dz_2 \frac{\kappa(z_1)\kappa(z_2)}{8Dt + w^2(z_1) + w^2(z_2)} \exp\left[-\frac{(z_2 - z_1)^2}{4Dt} - \frac{2\delta^2}{8Dt + w^2(z_1) + w^2(z_2)}\right]. \quad [8.8]$$

Here, we have taken into account that the MDFs of both detection volumes are identical but shifted by a distance δ, and we have denoted the overall detection efficiencies from both volumes by ε_1 and ε_2.

Data fitting is performed with least-square fitting of the model curve, Eq. (8.8), against the measured autocorrelation functions ($\delta = 0$, $\varepsilon_1\varepsilon_2$ replaced by either $\varepsilon_1{}^2$ or $\varepsilon_2{}^2$) and crosscorrelation functions *simultaneously* in a global fit. As fitting parameters, we need to use $\varepsilon_1 c^{1/2}$, $\varepsilon_2 c^{1/2}$, D, w_0, and R_0, as well as three offset values g_∞. The distance δ between the detection regions is determined by the properties of the Nomarski prism and has to be exactly known *a priori*, thus introducing an external length scale into data evaluation. It is important to notice that a crucial criterion of fit quality is not only to simultaneously reproduce the temporal shape of both autocorrelation functions and the crosscorrelation function but also to reproduce their three amplitudes $g_{t\to 0} - g_\infty$ using only the two parameters $\varepsilon_1 c^{1/2}$ and $\varepsilon_2 c^{1/2}$. The ratio between the amplitudes of the crosscorrelation and autocorrelation functions is determined by the overlap between the two MDFs, and thus by the shape parameters w_0 and R_0. Thus, the fit quality of this amplitude ratio is an additional constraint that increases the accuracy of the determined diffusion coefficient.

Due to the external length scale as given by the distance δ between the detection volumes and the reasonably accurate model of the MDF, 2fFCS gives superior accuracy and stability when measuring diffusion. An optimal distance between the two foci is ca. 400 nm, giving a sufficiently large overlap between detection volumes so that the amplitude of the cross-correlation function between both detection volumes is roughly one half of the amplitude of each autocorrelation function. Larger distances will lead to significantly longer measurement times for accumulating a sufficiently good crosscorrelation, smaller distances will lead to a crosscorrelation function too similar to the autocorrelation functions so that fitting becomes increasingly unstable.

2.4. Calibration

A crucial point for precise absolute 2fFCS measurements is the exact knowledge of the center distance δ between foci. This has to be determined once and does typically not change if one does not modify the optical setup of the confocal microscope.

In Müller, Weiss, Richtering, Loman, and Enderlein (2008), we used fluorescent 0.1 μm TetraSpeck™ microspheres (T-7279, Life Technologies GmbH, Darmstadt, Germany) for calibrating a 2fFCS system. In a first step, we measured with DLS the hydrodynamic size of fluorescently labeled TetraSpeck™ spheres at 298.15 ± 0.1 K using a detection angle of 90°. A typical set of measured ACFs is presented in the inset of Fig. 8.8. For each wavelength, measurements were repeated over 50 times to obtain a sufficiently small standard deviation.

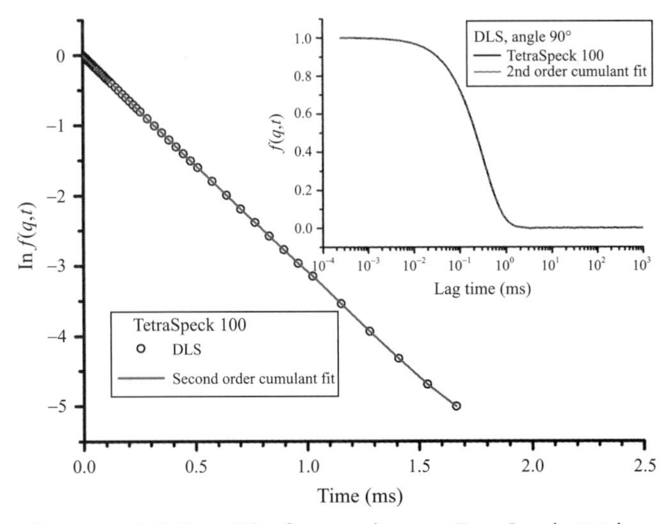

Figure 8.8 Main panel: DLS at 90° of mono disperse TetraSpeck 100 latex particles, fitted with a second-order cumulant fit. Inset: standard plot of ACF. (For color version of this figure, the reader is referred to the online version of this chapter.)

A semilogarithmic plot of the data is shown in the main panel of Fig. 8.8, together with a second-order cumulant fit. The good fit quality proves the monodispersity of the bead sample. The hydrodynamic radius R_h of the beads was determined to be 55.6 ± 0.6 nm.

In a second step, 2fFCS measurements were performed at the three excitation wavelengths of $\lambda_{ex} = 470$, 532, and 637 nm, respectively. Due to the high label density of the beads, total excitation power was reduced to less than 0.1 μW within each detection volume. During each measurement, fluorescence was collected for 5 min, and each measurement was repeated more than 80 times per excitation wavelength. A typical measurement result is shown in the inset of Fig. 8.9.

To obtain the distance between the overlapping detection volumes, each set of ACFs and CCF was globally fitted by the model function of Eq. (8.8) to obtain a value of the diffusion coefficient D and thus hydrodynamic radius R_h, assuming that the distance δ between the detection volumes had a certain value between 360 and 416 nm. The obtained hydrodynamic radius R_h as a function of assumed distance δ is shown in Fig. 8.9. The intersection of this curve with the actual value of the hydrodynamic radius as obtained from

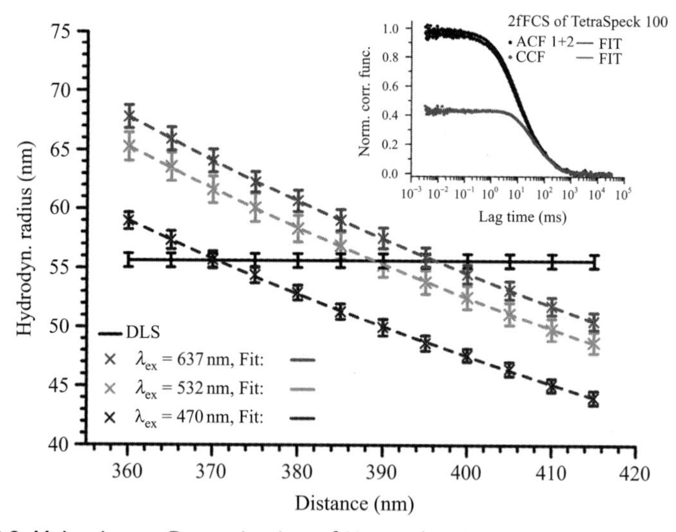

Figure 8.9 Main picture: Determination of Nomarski—DIC—prism shear distance, by comparison of DLS and 2fFCS measurements at different wavelengths. Inset: 2fFCS measurement of 0.1 μm TetraSpeck™ microspheres. Autocorrelation function (ACF) and crosscorrelation (CCF) function were fitted with the 2fFCS model. (For color version of this figure, the reader is referred to the online version of this chapter.)

Table 8.1 Reference diffusion coefficients of various dyes in aqueous solution

Dye	λ_{abs} (nm)	λ_{em} (nm)	$D_{25°C}$ (cm^2/s)
Cy5	650	670	$3.7 \pm 0.15 \times 10^{-6}$
Atto655-COOH	665	690	$4.26 \pm 0.08 \times 10^{-6}$
Atto655-maleimide	665	690	$4.07 \pm 0.1 \times 10^{-6}$
Rhodamine 6G	530	560	$4.14 \pm 0.05 \times 10^{-6}$
Oregon Green 488	488	540	$4.11 \pm 0.06 \times 10^{-6}$

the DLS measurements (horizontal line in Fig. 8.9) yields the actual distance between the detection volumes and thus the shear distance of the DIC prism.

The observed wavelength dependence of the shear distance is due to the dispersion of the Nomarski prism. The result emphasizes how important it is to measure the shear distance of DIC prisms as function of wavelength.

As an alternative to fluorescent beads, one can also use fluorescent dyes with known diffusion coefficient as reference for calibration. In Müller, Loman, et al. (2008), we have measured the diffusion coefficients of several dyes in aqueous solution, for excitation/emission wavelengths across the visible spectrum. The values are listed in Table 8.1.

It is important to notice that the values shown are measured for a sample temperature of 25 °C (298.15 K), and for dye solutions in pure water. If one uses a different solvent with a viscosity different from water, or if one measures at a different temperature, one can use the Stokes–Einstein relation, Eq. (8.1), to calculate the corresponding new diffusion coefficient values from the ones presented in Table 8.1.

2.5. Example

An example for auto- and crosscorrelation curves of a 2fFCS measurement with the corresponding fit curves is presented in Fig. 8.10.

The superior performance of 2fFCS is shown in Fig. 8.11, presenting 2fFCS results for the diffusion coefficient of the dye Atto655 as a function of viscosity of aqueous solutions of guanidine hydrochloride (GdHCl). The concentration of GdHCl was varied from zero and 6 M between different measurements, leading not only to a strong increase in solution viscosity but also to a significant increase in the solution's refractive index. According to the Stokes–Einstein relation, one expects to find a linear relationship between diffusion coefficient and inverse of the viscosity. Indeed, 2fFCS precisely yields these results, despite increasing aberrations introduced by

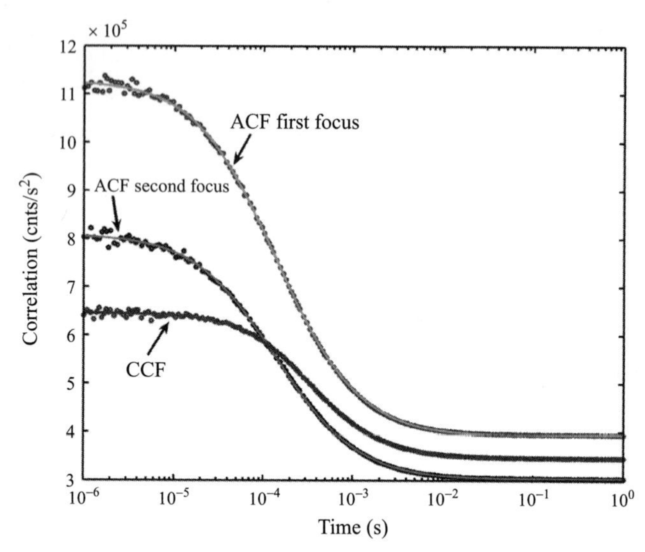

Figure 8.10 2fFCS measurement on a nanomolar aqueous solution of Atto655. Shown are the autocorrelation functions for the first focus, the second focus, and the crosscorrelation between both foci (CCF). The *shape* of both ACFs is virtually identical. Circles are experimental values and solid lines are global fits as described below. (For color version of this figure, the reader is referred to the online version of this chapter.)

the refractive index mismatch. Moreover, as is also shown in Fig. 8.11, the *absolute* values of the diffusion coefficient perfectly match the one determined by pulsed-field gradient NMR, an independent spectroscopic method capable of measuring absolute diffusion coefficients. The achievable accuracy of 2fFCS in diffusion coefficient measurements was estimated to be smaller than 5%, opening the possibility to measure changes in the hydrodynamic radius of nanometer-sized molecules on the order of one Ångstrøm.

3. FLOW MEASUREMENTS WITH 2fFCS

An interesting application of dual-focus FCS is the measurement of directed flow in solution (Arbour & Enderlein, 2010). Due to the asymmetry of the dual-focus geometry, one can simultaneously measure flow velocities both along and perpendicular to the connecting line of the two foci, as well as along the optical axis. For a spatially homogenous (over the size of the detection volume) constant flow-velocity vector $\mathbf{v} = \{v_x, v_y, v_z\}$, Eq. (8.8) is modified to

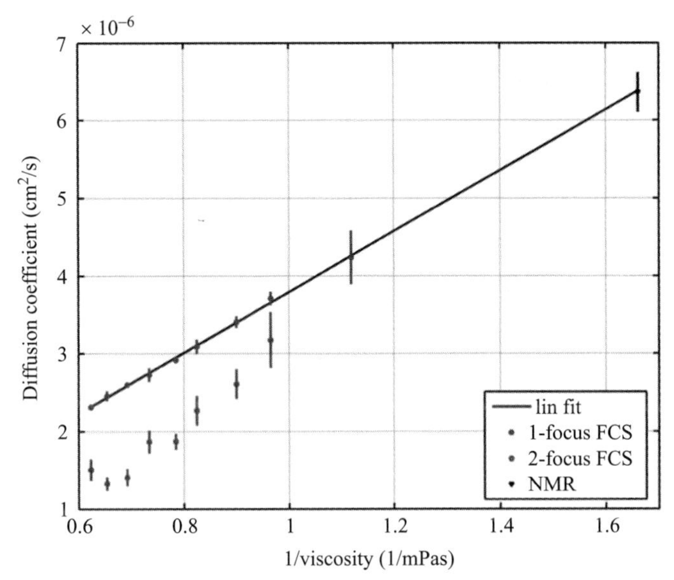

Figure 8.11 Dependence of the diffusion coefficient of Atto655 in aqueous GdHCl solutions and in d_4-deuterized methanol at 25 °C as a function of solvent viscosity. Solid line is linear least-square fit to all data. Standard deviations are shown as error bars and are each derived from 10 repeated measurements. For comparison, the results of single-focus FCS using a standard model that assumes a three-dimensional Gaussian MDF are also shown. Because single-focus FCS can only measure relative values of diffusion coefficient, the value for pure water was used as reference value. (For color version of this figure, the reader is referred to the online version of this chapter.)

$$g(t,\delta,\mathbf{v}) = g_\infty + \frac{\varepsilon_j\varepsilon_k c\sqrt{\pi}}{2}\int d\xi \int d\eta \frac{\kappa(\zeta_-)\kappa(\zeta_+)}{8Dt + w^2(\zeta_-) + w^2(\zeta_+)} \exp\left[-\left(\xi - \frac{v_z}{2}\sqrt{\frac{t}{D}}\right)^2 - \frac{2(\delta - v_x t)^2 + 2(v_y t)^2}{8Dt + w^2(\zeta_-) + w^2(\zeta_+)}\right]. \quad [8.9]$$

where ζ ± = η ± (Dt)1/2 ξ Data evaluation is done as before by fitting this model curve against an experimentally measured correlation function, but now with the velocity vector giving three additional fit parameters. If one knows *a priori* that the flow velocity is zero along the optical axis (which is usually the case when measuring fluid flows in microfluidic channels), one can set v_z to zero which improves fit robustness. A typical measurement result is shown in Fig. 8.12 for a fluid flow within a square-bore capillary.

By scanning across the capillary, a line profile of the flow velocity can be recorded, as shown in Fig. 8.13. For each flow-profile acquisition, the

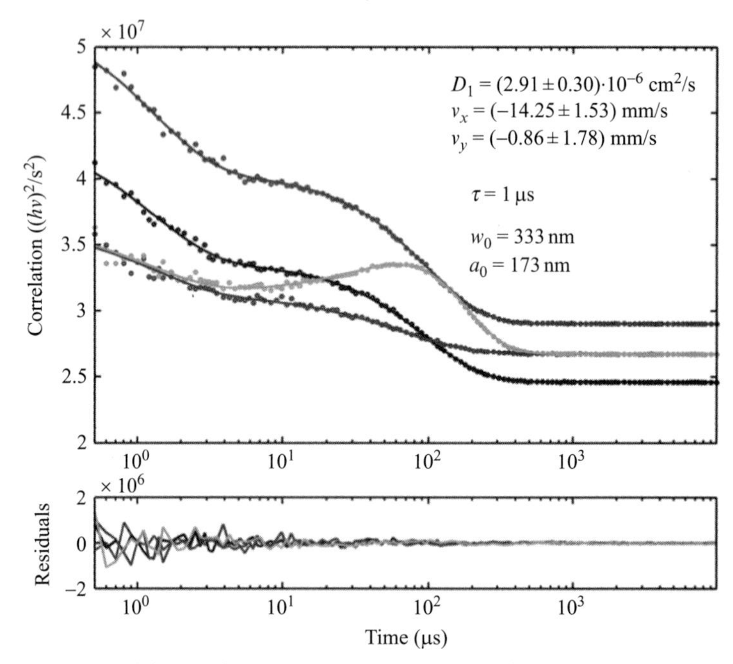

Figure 8.12 Typical fitted 2fFCS correlation curves. The four curves are comprised of two autocorrelations (one from each focus) and the forward and reverse crosscorrelations. Shown are also the fitted parameters: diffusion coefficient (D_1), flow velocity along (v_x) and perpendicular to (v_y) the foci-connecting line, triplet state time (τ), focus diameter (w_0), and confocal aperture parameter (a_0). (For color version of this figure, the reader is referred to the online version of this chapter.)

confocal spot was manually scanned in the direction of the optical axis using the microscope's vertical adjustment knob. The position of the inner capillary surface nearest the objective (designated as the zero position) was found by observing the Airy disks at the outer and then inner glass surfaces. Measurement at each position was continued until at least seven million photon-detection events had been recorded (as determined directly from file size), typically requiring 5–10 min. The input volumetric flow rate was 38 μl/min for the flow-profile scans, corresponding to a bulk linear fluid velocity of ~7 mm/s. Laser excitation intensity was 50 μW for the flow-profile scans and 10 μW for the no-flow reference measurements.

An outstanding feature of FCS applied to flow measurement is the wide dynamic which can be covered. A linearity check is shown in Fig. 8.14. The measured velocities, when plotted against the volumetric flow rate input to the syringe pump, display highly linear behavior down to the lowest velocities measured (ca. 200 μm/s).

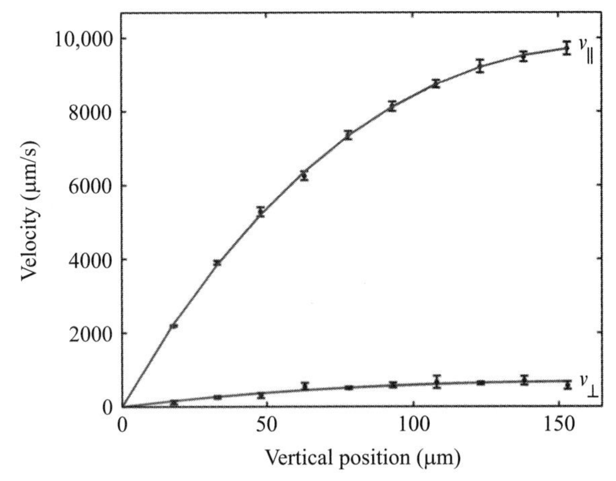

Figure 8.13 Measured flow-velocity profiles in square-bore capillary with 300 μm inner edge length. Solid lines are fitted exact analytical results for a laminar flow profile. The resulting ratio of $v_{\|}$ to $v_{\perp}$ indicates a 4° inclination of the microcapillary axis with respect to the foci-connecting line. (For color version of this figure, the reader is referred to the online version of this chapter.)

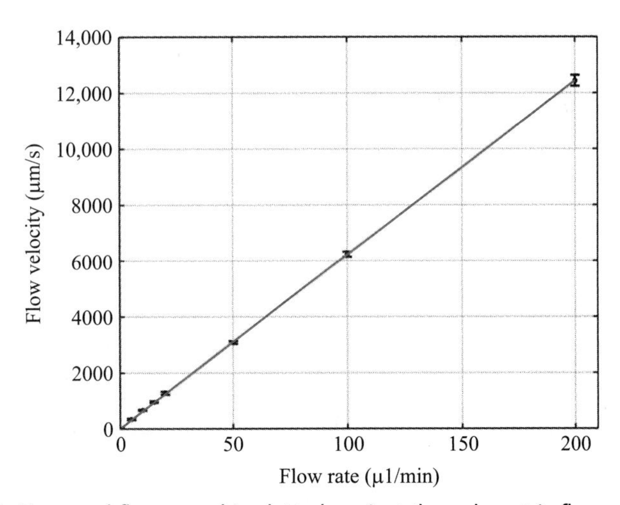

Figure 8.14 Measured flow speed is plotted against the volumetric flow rate input to the syringe pump. The high degree of linearity and small error bars demonstrate 2fFCS's flow measurement precision over a wide dynamic range of flow rates. (For color version of this figure, the reader is referred to the online version of this chapter.)

4. 2fFCS MEASUREMENTS IN LIPID MEMBRANES

An important application of 2fFCS is diffusion measurements in lipid membranes (lipid bilayers). In that case, it is essential to precisely position the foci with respect to the plane of the membrane (which is here supposed to be oriented perpendicular to the optical axis). One could think that the position of maximum fluorescence intensity would be the perfect choice. However, it was shown by Benda et al. (2003) that this does not necessarily coincide with the accurate focus position with respect to the bilayer. Instead, they employed so-called z-scan FCS, where the sample is scanned in 0.2 μm steps along the optical axis (z-axis) perpendicular to the bilayer plane. The particle number and the diffusion time τ_D are both dependent on the focus diameter (and thus on the intersection between diverging laser beam and membrane). With increasing vertical distance between laser beam waist and bilayer, the observed diffusion time increases. In the z-scan method, the observed dependence of diffusion time on vertical distance is fitted by a parabolic profile to determine the beam waist diameter and relative position between laser focus and membrane. This can then be used for calculating the lateral diffusion coefficient D. The disadvantage of this method is that it requires recording multiple ACFs at different vertical focus positions and is thus time-consuming and sensitive to mechanical drift of the setup. In this respect, 2fFCS is simpler because only one measurement is needed (instead of a complete z-stack of measurements). For 2fFCS measurement of diffusion within a membrane, the fit function simplifies to

$$g(t,\delta) = g_\infty + \frac{\varepsilon_1 \varepsilon_2 c}{2w^2 + 8Dt} \exp\left(-\frac{2\delta^2}{2w^2 + 8Dt}\right) \qquad [8.10]$$

with the same meaning of the symbols as in Eq. (8.8), except that w^2 here is the mean square beam waist of both foci within the plane of the membrane.

As already mentioned, accurate positioning of the foci is important also for 2fFCS. In what follows, we use model calculations for demonstrating that for different experimental situations without and with optical aberration. In these calculations, the following parameters are used: the numerical aperture of the objective is 1.2 (water-immersion objective), the principal focal distance of the objective is assumed to be 3 mm, and the focal distance of the tube lens 180 mm, so that image magnification at the confocal aperture is 60×. The aperture diameter is set to 150 μm. The excitation wavelength is 640 nm, and the peak emission wavelength is 670 nm. Furthermore, we assume that the laser beam which is focused through the objective into the sample has a Gaussian intensity profile. To investigate how this focusing affects a 2fFCS measurement of diffusion in a membrane, we have varied the $1/e^2$-radius of this Gaussian profile from 1.25 to 3.5 mm, thus covering the range from relaxed to nearly diffraction-limited focusing. Figure 8.15 shows how the laser-beam radius correlates to the waist radius of the resulting focus in sample space.

Figure 8.16 presents the computational results for a 2fFCS measurement on molecules diffusing within a plane under ideal optical conditions (no aberrations). Shown are the "fitted" values of the diffusion coefficients as a function of the position of the membrane with respect to the beam waist and for different degrees of focusing. As can be seen, for laser-beam radius values between 1.25 and ca. 2.25 mm, one has a range of ca. ± 0.5 μm around the focal plane where the average systematic error between fitted and actual value of the diffusion coefficient remains below 5%. For tighter focusing, this range narrows and one quickly obtains large systematic errors

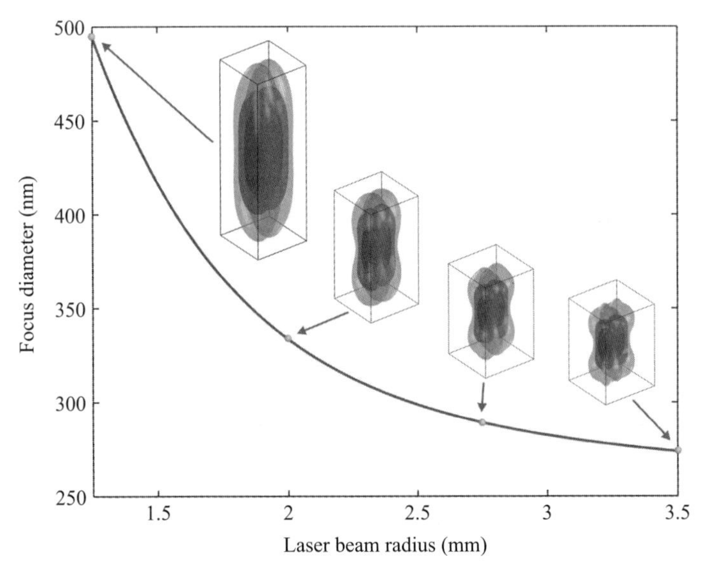

Figure 8.15 Correlation between the radius of the laser beam that is coupled into the objective and the radius of the resulting focus in sample space. The insets show the shape of the two overlapping MDFs for selected laser-beam radius values, each box has the same transversal size of 1.2 μm × 1.6 μm. Shown are the isosurfaces, for both foci, where the combined excitation and detection efficiency for a fluorescence photon has fallen off to $1/e^2 \sim 13\%$ of its maximum value in the very center of the focus. (For color version of this figure, the reader is referred to the online version of this chapter.)

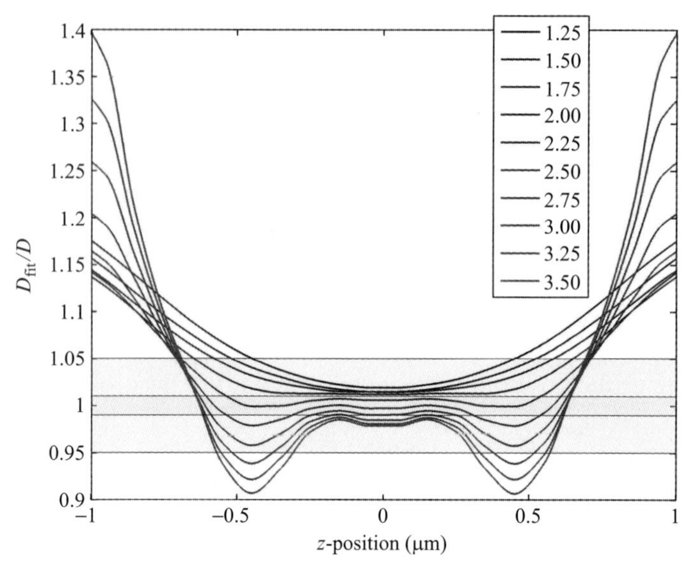

Figure 8.16 Result of modeling a 2fFCS measurement of molecular diffusion within a plane under ideal optical conditions (no aberrations). Shown is the ratio of "fitted" values to "actual" values of the diffusion coefficients as a function of the position of the plane with respect to the beam waist, and for different degrees of focusing. Here and in the following figures, the legend indicates the radius values of laser beam coupled into the objective (in mm). Shaded areas indicate 1% and 5% error margins. (For color version of this figure, the reader is referred to the online version of this chapter.)

when the plane of the diffusing molecules moves farther away from the focal plane. The main reason for this is that with tighter focusing, the transversal excitation intensity profile can no longer be well approximated with a Gaussian distribution, which is, however, the basis of the correlation fit curves. Thus, when aiming at good accuracy and least sensitivity to focus placement, it is recommended to work with relaxed focusing, corresponding to focus beam waists of around 300–400 nm. The obtained results indicate also that 2fFCS will be robust against vertical membrane fluctuations having amplitudes below a couple of 100 nm. This can be important when applying 2fFCS to diffusion measurements in GUVs, where even slight differences in osmolarity between the in- and outside of the vesicle can cause non-negligible membrane fluctuations.

As already mentioned in Section 1, two of the most common and nearly unavoidable origins of optical aberrations are deviation of the cover slide thickness from its design value to which the objective is adjusted, and refractive index mismatch between sample solution and the objective's immersion medium. When working with free-standing lipid membranes, the distance between membrane and cover slide surface is typically rather large (on the order of 100 μm). Thus, even slight mismatches in refractive indices accumulate over the long optical path length and cause large aberrations. First, we consider the situation of refractive index mismatch. As a typical example, we assume that the sample solution has a refractive index of 1.36 instead of 1.33 for pure water, and that the focal plane of the objective is located 100 μm above the cover slide.

For this situation, the top panel of Fig. 8.17 depicts the computational result of the model 2fFCS measurements, again showing the "fitted" diffusion coefficient as a function of the z-position along the optical axis (position of the membrane, i.e., the plane of diffusing molecules). The graphs purposefully start at a z-position larger than 100 μm because positioning the foci 100 μm above the cover slide by moving the objective by that distance actually places the focus deeper in solution due to refractive index mismatch. Here and hereafter, we restrict ourselves to cases of relaxed focusing (laser-beam radius below 2 mm).

In a real experiment, one still has to find the correct position of the laser focus with respect to the membrane. One option would be to adjust the focal plane to the maximum of fluorescence intensity. The middle panel of Fig. 8.17 shows the fluorescence intensity (integrated over the whole plane of the membrane) as a function of focus position. As can be seen, the position of maximum intensity is shifted to larger z-values compared to the region of least systematic error of the 2fFCS measurement given in Fig. 8.17 (top panel). Thus, choosing the position of maximum intensity for the measurement would result in a systematic overestimation of the diffusion coefficient. The reason for this shift is the not so widely recognized fact that spherical aberrations always place the maximum of the detection efficiency distribution to a different position along the optical axis than the minimum beam waist of the focused laser beam. This has the effect that the transversal profile of the MDF quickly becomes non-Gaussian when moving away from the plane of tightest focus diameter. This is not a specific problem of 2fFCS, but will also affect line-scan or scanning focus FCS. The only FCS method which does not suffer from this problem is z-scan FCS, but for the price of ca. 10 times longer measurement times for the whole z-stack of FCS measurements. Another option is to search for the position of maximum molecular brightness. The calculated (normalized) brightness values as a function of position z are shown in Fig. 8.17 (lower panel). As can be seen, the

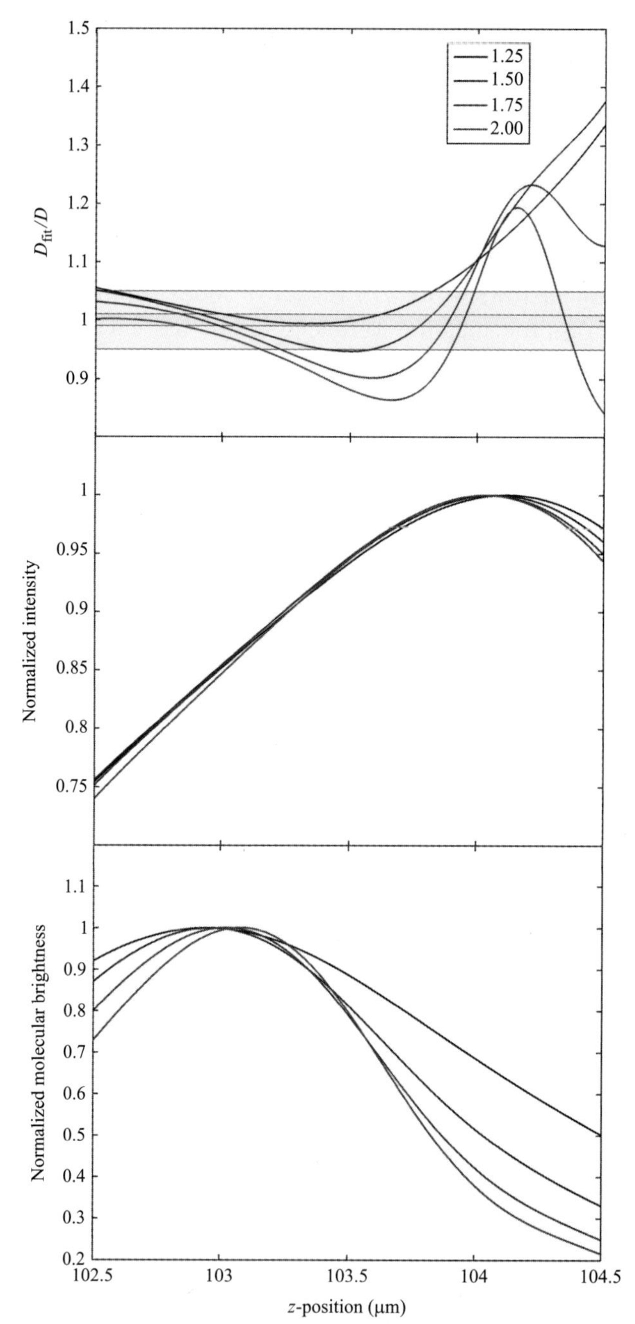

Figure 8.17 Results of modeling a 2fFCS measurement with aberrations caused by refractive index mismatch. Top panel: Fitted diffusion coefficient relative to its exact value. Shaded areas indicate 1% and 5% error margins. Middle panel: Mean fluorescence intensity (average over plane). Bottom panel: Molecular brightness. All curves are given as a function of position along the optical axis. Legend indicates the radius values of laser beam coupled into the objective (in mm). (For color version of this figure, the reader is referred to the online version of this chapter.)

position of maximum molecular brightness is a much better choice for obtaining correct diffusion coefficients from a 2fFCS measurement.

A second important origin of aberrations are cover slide thickness deviations. State-of-the-art water-immersion objectives are designed to take into account the presence of a glass cover slide of specific thickness between the objective and the sample. Most of these objectives also have adjustment rings for matching them to a specific cover slide thickness. In practice, however, cover slides rarely have the exact thickness value as indicated by the supplier, and correct positioning of an objective's adjustment ring is far from trivial.

Thus, deviations of the order of 10 μm between the actual cover slide thickness and the value an objective is adjusted to are mostly unavoidable. Although this is close to undetectable in typical imaging applications of a microscope, FCS experiments are extremely sensitive to size and shape changes of the MDF. As an example, we consider here a cover slide thickness deviation of only 10 μm.

The top panel of Fig. 8.18 again shows the values of "fitted" diffusion coefficients as a function of focal plane position, and the middle and lower panel of the same figure depict the corresponding profile of the fluorescence intensity and molecular brightness, respectively. Again, we see a shift between the position of maximum intensity and the region of least systematic error, although not as dramatic as caused by the refractive index mismatch. And again, the position of maximum molecular brightness is a better choice for minimizing systematic errors in diffusion coefficient determination.

Thus, when applying 2fFCS for diffusion measurements in membranes, short FCS measurements around the position of maximum intensity are performed to find the position of maximum molecular brightness, which should then be used for the diffusion measurements. When performing FCS measurements on membranes, one concern is that the membrane may move out-of-focus during the measurement (mechanical drift). This is avoided by choosing short measurement times of 10 min and by readjusting the foci before each measurement. Moreover, membrane movement out-of-focus or rupture of the membrane during the measurement lead to a pronounced decrease in the count rate and loss of correlation. Therefore, meaningful correlation curves with an intact membrane and correct foci position can be distinguished from flawed ones. Finally, when measuring in membranes, the excitation power should be in the lower μW range to avoid artifacts due to photobleaching.

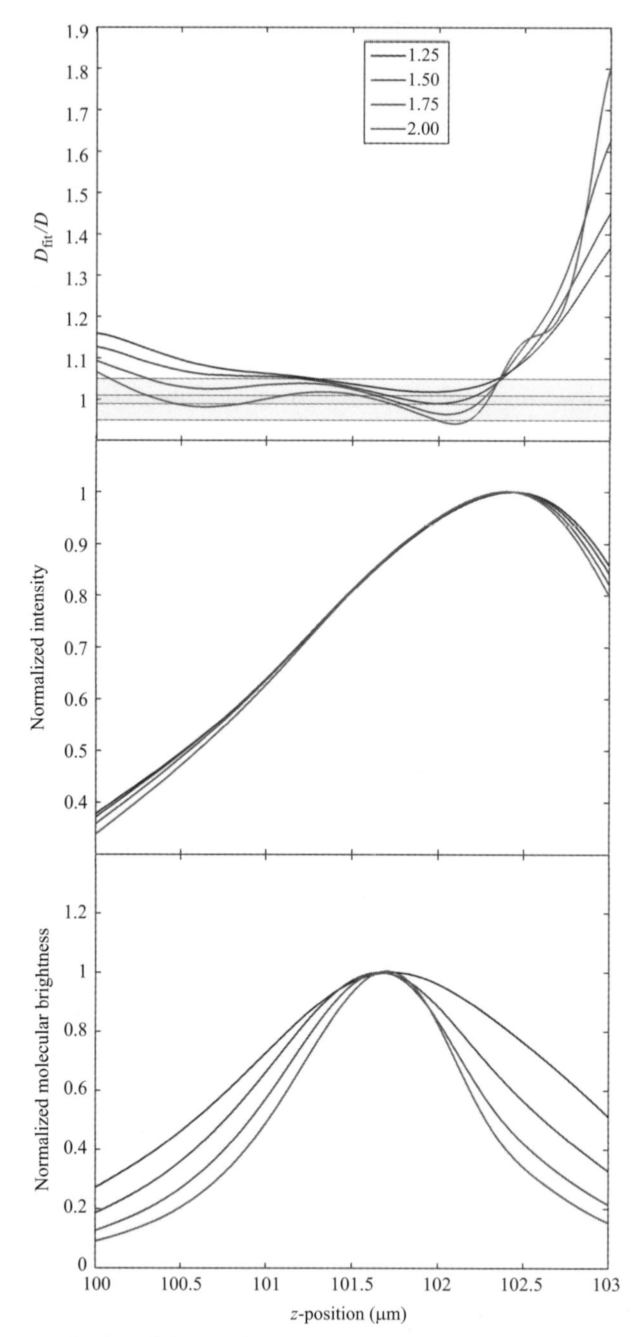

Figure 8.18 Result of modeling a 2fFCS measurement with aberrations caused by cover slide thickness deviation. Top panel: Fitted diffusion coefficient relative to its exact value. Shaded areas indicate 1% and 5% error margins. Middle panel: Mean fluorescence intensity (average over plane). Bottom panel: Molecular brightness. All curves are given as a function of position along the optical axis. Legend indicates the radius values of laser beam coupled into the objective (in mm). (For color version of this figure, the reader is referred to the online version of this chapter.)

5. SUMMARY

We have presented a description of 2fFCS with strong emphasis on its application for precise diffusion measurements. Along the way, we have also given a thorough analysis of the optical performance of a confocal microscope. Of course, there are other important applications of FCS which are covered in the present book, for which the precise shape and size of the detection volume is much less important. However, whenever the length scale of the FCS measurement becomes important for the data evaluation, 2fFCS is a robust and easy to set up alternative to conventional FCS.

REFERENCES

Arbour, T. J., & Enderlein, J. (2010). Application of dual-focus fluorescence correlation spectroscopy to microfluidic flow-velocity measurement. *Lab on a Chip*, *10*, 1286–1292.

Benda, A., Beneš, M., Mareček, V., Lhotský, A., Hermens, W. T., & Hof, M. (2003). How to determine diffusion coefficients in planar phospholipid systems by confocal fluorescence correlation spectroscopy. *Langmuir*, *19*, 4120–4126.

Dertinger, T., Loman, A., Ewers, B., Müller, C. B., Krämer, B., & Enderlein, J. (2008). The optics and performance of dual-focus fluorescence correlation spectroscopy. *Optics Express*, *16*, 14353–14368.

Dertinger, T., Pacheco, V., von der Hocht, I., Hartmann, R., Gregor, I., & Enderlein, J. (2007). Two-focus fluorescence correlation spectroscopy: A new tool for accurate and absolute diffusion measurements. *ChemPhysChem*, *8*, 433–443.

Elson, E. L., & Magde, D. (1974). Fluorescence correlation spectroscopy. I. Conceptual basis and theory. *Biopolymers*, *13*, 1–27.

Enderlein, J. (2005). Dependence of the optical saturation of fluorescence on rotational diffusion. *Chemical Physics Letters*, *410*, 452–456.

Enderlein, J., Gregor, I., Patra, D., Dertinger, T., & Kaupp, B. (2005). Performance of fluorescence correlation spectroscopy for measuring diffusion and concentration. *ChemPhysChem*, *6*, 2324–2336.

Gregor, I., Patra, D., & Enderlein, J. (2005). Optical saturation in fluorescence correlation spectroscopy under continuous-wave and pulsed excitation. *ChemPhysChem*, *6*, 164–170.

Magde, D., Elson, E., & Webb, W. W. (1972). Thermodynamic fluctuations in a reacting system-measurement by fluorescence correlation spectroscopy. *Physical Review Letters*, *29*, 705–708.

Magde, D., Elson, E., & Webb, W. W. (1974). Fluorescence correlation spectroscopy. II. An experimental realization. *Biopolymers*, *13*, 29–61.

Microscopy, n.d. http://www.microscopyu.com/articles/dic/dicindex.html.

Müller, B., Loman, A., Pacheco, V., Koberling, F., Willbold, D., Richtering, W., et al. (2008). Precise measurement of diffusion by multi-color dual-focus fluorescence correlation spectroscopy. *Europhysics Letters*, *83*, 46001.

Müller, C. B., Weiss, K., Richtering, W., Loman, A., & Enderlein, J. (2008). Calibrating differential interference contrast microscopy with dual-focus fluorescence correlation spectroscopy. *Optics Express*, *16*, 4322–4329.

Müller, B. K., Zaychikov, E., Bräuchle, C., & Lamb, D. (2005). Pulsed interleaved excitation. *Biophysical Journal*, *89*, 3508–3522.

Wahl, M., Gregor, I., Patting, M., & Enderlein, J. (2003). Fast calculation of fluorescence correlation data with asynchronous time-correlated single-photon counting. *Optics Express*, *11*, 3583–3591.

Weljie, A. M., Yamniuk, A. P., Yoshino, H., Izumi, Y., & Vogel, H. J. (2003). Protein conformational changes studied by diffusion NMR spectroscopy: Application to helix-loop-helix calcium binding proteins. *Protein Science*, *12*, 228–236.

CHAPTER NINE

Pulsed Interleaved Excitation: Principles and Applications

Jelle Hendrix*,†, Don C. Lamb*,‡,1

*Physical Chemistry, Department of Chemistry, Munich Center for Integrated Protein Science (CiPSM) and Center for Nanoscience (CeNS), Ludwig-Maximilians-Universität München, Munich, Germany
†Laboratory of Photochemistry and Spectroscopy, Department of Chemistry, University of Leuven, Celestijnenlaan, Leuven, Belgium
‡Department of Physics, University of Illinois at Urbana-Champaign, Urbana, Illinois, USA
[1]Corresponding author: e-mail address: d.lamb@lmu.de

Contents

Abstract

Pulsed interleaved excitation (PIE) is the methodology of interleaved or alternating excitation of different fluorophores on the nanosecond timescale, which allows quasi-simultaneous, yet independent measurements to be performed. PIE simplifies quantification of several fluorescence techniques such as FCCS and FRET experiments. Foremost, it allows to specifically filter out spectral emission bleedthrough (crosstalk) and direct excitation without a decrease in the signal-to-noise ratio (SNR) of the experiment. Next, PIE allows determination of the absolute FRET efficiency from FCCS experiments in the case of non-perfect labeling. In recent years, PIE has been utilized in many different advanced FFS techniques. Combining MFD with PIE allows highly accurate and species-specific spFRET analyses to be performed. The combination of scanning FFS techniques with PIE combines the best of both techniques and allows for false-positive free measurements of molecular

Methods in Enzymology, Volume 518
ISSN 0076-6879
http://dx.doi.org/10.1016/B978-0-12-388422-0.00009-1

interactions *in vitro* and in living cells. In succession, a comprehensive overview of the principle and versatility of the PIE technique is discussed, the theory for analysis with PIE is outlined by comparing CW– and PIE–FCCS and finally, some of the most important applications of the PIE technique in literature are reviewed.

1. INTRODUCTION: THE NEED FOR PIE

When expanding fluorescence fluctuation spectroscopy techniques for analysis of multiply colored molecules, one will be faced with different challenges related to the properties of fluorophores. For example, the excitation and emission spectra of fluorophores have a finite width that limits how well one can separate the fluorescence coming from different fluorophores. Photophysical interactions such as Förster resonance energy transfer (FRET) between different fluorophores may also occur (Förster, 1948), which can be used to quantify their interaction but can also lead to complications in other spectroscopical applications. For quantitative fluorescence cross-correlation spectroscopy (FCCS) (Schwille, Meyer-Almes, & Rigler, 1997), for example, it is crucial to have control over, or at least good knowledge of, the rate of photon emission of the fluorophores in an experiment. In general, different complications in analysis of fluorescence experiments can be described as "misguided" photon detection events inherent to multicolor fluorescence spectroscopy. While it may be possible to correct for misguided photons in some dual-color analyses, these corrections become more and more complicated making a quantitative analysis near to impossible for three-color or multicolor spectroscopy.

A straightforward method for avoiding the detection of misguided photons is to excite (and detect) different fluorophores at different points in time. This has been done for years, and is still done, in sequential multicolor fluorescence imaging experiments. Sequential (or alternated) excitation creates an extra dimension for separating fluorophores: time. Crosstalk of a fluorophore in another fluorophores' detection channel, for example, will then only occur at times where the other fluorophore is not being excited. Misguided photons can thus be specifically "mistimed" allowing them to be filtered from fluorophore-specific photons. In general, by taking advantage of the differences in the extinction and emission spectrum of different fluorophores and by exciting different fluorophores sequentially in time, their emission can be specifically separated. Importantly, such *modus operandi* is not limited to two-color analysis: fluorophores spanning the ultraviolet, visible, and near-infrared spectrum can be imaged independently by doing sequential excitation and detection.

To independently study single molecules and/or complexes labeled with different fluorophores that are freely diffusing in solution with a confocal microscope, the rate of alternating excitation has to be significantly faster than the time the molecules or complexes need to diffuse through the observation volume. Alternating laser excitation (ALEX) was originally developed in the group of Shimon Weiss by Achillefs Kapanidis and coworkers utilizing electrooptical modulators that controlled the illumination source and could be alternated on the timescale of a few microseconds (Kapanidis et al., 2005, 2004). However, for many quantitative and/or single-molecule fluorescence applications, it is desirable to image different fluorophores independently on the nanosecond timescale of fluorescence emission and this ultimately is how the idea for the pulsed interleaved excitation (PIE) technique came about. PIE is the methodology of interleaved or alternating excitation of different fluorophores on the nanosecond timescale (Müller, Zaychikov, Bräuchle, & Lamb, 2005). The method can be seen as a modern multicolor extension of the time-correlated single-photon counting (TCSPC) technique (Becker et al., 1999), which originally employs pulsed lasers and detection hardware with picosecond resolution and single-photon sensitivity to measure the fluorescence lifetime and anisotropy accurately.

In this chapter, a comprehensive overview of the PIE technique is given. First, the principle is explained and illustrated, a standard PIE microscope is presented and simple formulas for the independent photon streams from a typical experiment are described. Second, the extraordinary versatility of the technique is briefly overviewed. Third, the theoretical foundations for analysis of FCCS and FRET experiments are described in detail. Some interesting applications are reviewed, and finally, the important role that fast alternating excitation techniques fulfill in present and in future multicolor spectroscopy is discussed.

2. HOW DOES PIE WORK?

The principle of PIE is simple and similar to that of ALEX (Kapanidis et al., 2005, 2004): in addition to separating the emission of different fluorophores spatially and spectrally as in standard multicolor fluorescence spectroscopy, the excitation sources for different probes, and thus also the fluorescence emissions of the probes, are temporally alternated. PIE differs from microsecond ALEX (μsALEX) in that picosecond to

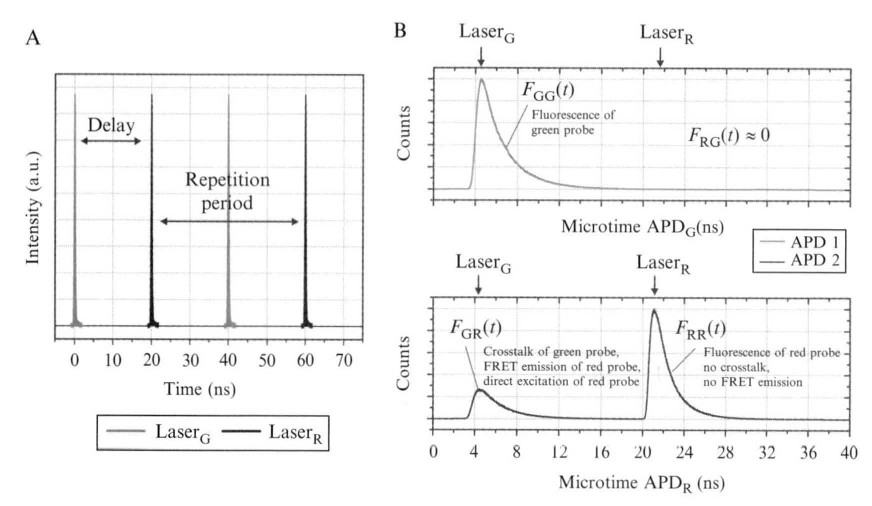

Figure 9.1 Schematic illustration of the principle for excitation and detection using PIE on a dual-color microscope. (A) The train of excitation pulses consists of alternating green and red picosecond laser pulses. (B) The photons are registered with picosecond time resolution by two single-photon counting devices used to create two microtime histograms. $F_{ij}(t)$ refers to the four PIE detection channels in Eq. (9.3). APD, avalanche photodiode. The electronic or optical delay between the lasers and the overall repetition rate/period needs to be adjusted to allow sufficient relaxation of one fluorophore before arrival of the next excitation pulse.

nanosecond pulsed lasers with a high (>1 MHz) repetition rate are used (Fig. 9.1A). By using TCSPC in combination with alternating excitation, the emission of different fluorophores is separated not only spectrally but also temporally. Moreover, fluorescence lifetime information for each of the fluorophores is available (Fig. 9.1B). In other words, the fluorescence from different probes is detected independently, yet quasi-simultaneously. This provides the necessary sub-microsecond time resolution for multicolor auto- and cross-correlation analysis of translational, rotational, and conformational dynamics of biomolecules. The nanosecond ALEX technique is similar in terms of instrumentation but originally focused on the analysis of FRET experiments only (Laurence, Kong, Jager, & Weiss, 2005).

An illustration of a typical dual-color excitation, dual-color detection setup is given in Fig. 9.2. Two picosecond pulsed lasers are synchronized by a master clock in the TCSPC hardware. One laser is then delayed electronically with respect to the other to ensure temporal separation of excitation of all fluorophores. Importantly, the repetition rate of and the delay

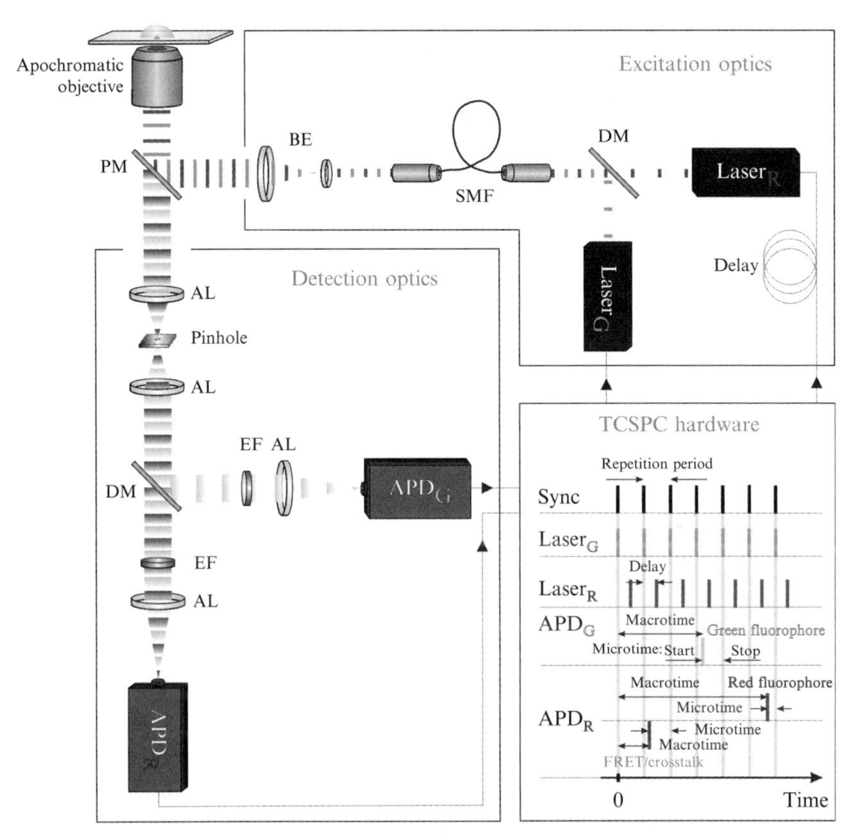

Figure 9.2 Schematic illustration of a dual-color excitation dual-color detection PIE–FCCS microscope. (*Excitation optics*) The emission of the lasers is synchronized by the TCSPC hardware that sends electrical pulses with a fixed repetition rate. The red pulses interleave the green ones because of electronic delay (shown in the image) or optical delaying (not shown) of the red laser. The lasers are combined with a dichroic mirror (DM). The two beams are made concentric and their beam profile is cleaned up by coupling into a single-mode optical fiber (SMF) with an achromatic collimator. A Keplerian beam expander (BE) finally expands the beam to ensure overfilling of the objective back aperture, if desired. (*Detection optics*) A polychroic mirror (PM) spectrally and spatially separates excitation from emission light. The latter is focused through a confocal pinhole and collimated again by achromatic lenses (AL) to limit chromatic aberrations. The emission dichroic mirror splits up the emission light into two spectral ranges. An emission filter (EF) transmits the correct spectral band while blocking the lasers and a lens finally focuses the emission light on the active area of an avalanche photodiode (APD). (*TCSPC hardware*) The counting electronics register three parameters per photon: the detector, the microtime and the macrotime. The microtime is the arrival time of the photon with respect to the laser pulse and is measured with picosecond resolution in the reverse start–stop principle, irrespective of the laser that triggered its emission. The macrotime is the arrival time of the photon with respect to the start of the measurement and is measured as an integer multiple of clock ticks with respect to the previously detected photon. (See Color Insert.)

between the excitation sources needs to be set according to the fluorescence lifetime of the fluorophores under investigation to avoid temporal crosstalk of the fluorophores in the different PIE channels. The different laser lines are coupled into a single-mode fiber to clean up the beam profile, collimated with a broadband apochromatic collimator to ensure beam diameters of comparable size and finally expanded to ensure a maximally focused diffraction limited probe volume. Excitation and emission light are spatially and spectrally separated by a polychroic mirror. The emission is collected via the same objective, passes through the polychroic and is focused through a pinhole by achromatic lenses and split into the two emission wavelength ranges by a dichroic mirror. Finally, two single-photon counting devices register the photons. The TCSPC hardware stores at least four parameters per photon: (i) the origin of the photon (green or red excitation), (ii) the detection channel (green or red detection), (iii) the macrotime, which is the arrival time with respect to the start of the measurement, stored as clock ticks since the previous photon detection, and (iv) the microtime, which is the arrival time with respect to the sync pulse (reverse start–stop), stored with picosecond resolution.

From the TCSPC data, microtime histograms for all the detected photons can be constructed. Instead of histograms for just the green and red detection channels, a dual-color PIE setup has four different microtime data channels determined from the laser that illuminated the sample and the detector that registered the emitted photon (Fig. 9.1). The $F_{GG}(t)$ channel displays the fluorescence decay of the green fluorophore after green excitation. Likewise, the $F_{RR}(t)$ channel contains the decay of the red fluorophore after red excitation. It is immediately clear that the relative delay of the two lasers and the overall repetition rate needs to be chosen to avoid a significant contribution of photons from one decay channel into the adjacent PIE channel. Importantly, red fluorescence emission originating from crosstalk and FRET-sensitized acceptor emission is now temporally separated (and observed in the red channel after green excitation, $F_{GR}(t)$) from the $F_{RR}(t)$ channel.

Analysis models for FCCS and/or FRET with two-color interleaved excitation and two-color detection PIE will be outlined later in this chapter. For now, we describe the four PIE channels of Fig. 9.1B with simple mathematical formulas. In general, when fluorescence is measured on a confocal microscope, the fluorescence intensity $F(t)$ registered on the detector is given by

$$F(t) = \int_V \mathrm{d}\mathbf{r}\, W(\mathbf{r}) \varepsilon C(\mathbf{r}, t) \quad [9.1]$$

where V is the size of the probe volume, $\mathbf{r}$ is a variable describing the position in three-dimensional space, and $W(\mathbf{r})$ is the point-spread function (PSF) that describes the fluorescence signal generated on the detector by a point emitter in the sample at location $\mathbf{r}$. The PSF is the product of the normalized excitation profile with the collection efficiency of the detection optics. The ε and $C(\mathbf{r},t)$ are the molecular brightness and number concentration of the fluorophore under investigation, respectively. The molecular brightness of the fluorophore at the center of the volume V is defined as

$$\varepsilon = \kappa\sigma\Phi I_0 \qquad [9.2]$$

where κ is the efficiency by which a specific fluorophore is detected in the system, σ and Φ are the absorption cross section at the wavelength of excitation and quantum yield of the fluorophore, respectively, and I_0 is the amplitude of the excitation intensity.

With PIE, the number of data channels in general equals the number of different excitation sources times the number of detectors. As such, $\varepsilon_{i,j,k}$ can be defined as the molecular brightness of species i with concentration C_i after excitation j and registered by detector k. Due to the shape of excitation and emission spectra, the red laser generally does not excite the green fluorophore and the red fluorophore emission is generally not detected in the green channel > That is, $\varepsilon_{\mathrm{G,R,G}} = \varepsilon_{\mathrm{G,R,R}} = \varepsilon_{\mathrm{R,G,G}} = \varepsilon_{\mathrm{R,R,G}} = 0$. The four PIE channels $F_{jk}(t)$ illustrated in Fig. 9.1B are then defined as follows:

$$\begin{aligned}
F_{\mathrm{GG}}(t) &= \int_V \mathrm{d}\mathbf{r}\, W(\mathbf{r}) \left(\varepsilon_{\mathrm{G,G,G}} C_{\mathrm{G}}(\mathbf{r},t) + \varepsilon_{\mathrm{GR,G,G}} C_{\mathrm{GR}}(\mathbf{r},t)\right) \\
F_{\mathrm{RG}}(t) &\approx 0 \\
F_{\mathrm{GR}}(t) &= \int_V \mathrm{d}\mathbf{r}\, W(\mathbf{r}) \left(\varepsilon_{\mathrm{G,G,R}} C_{\mathrm{G}}(\mathbf{r},t) + \varepsilon_{\mathrm{GR,G,R}} C_{\mathrm{GR}}(\mathbf{r},t) + \varepsilon_{\mathrm{R,G,R}} C_{\mathrm{R}}(\mathbf{r},t)\right) \\
F_{\mathrm{RR}}(t) &= \int_V \mathrm{d}\mathbf{r}\, W(\mathbf{r}) \left(\varepsilon_{\mathrm{R,R,R}} C_{\mathrm{R}}(\mathbf{r},t) + \varepsilon_{\mathrm{GR,R,R}} C_{\mathrm{GR}}(\mathbf{r},t)\right)
\end{aligned} \qquad [9.3]$$

When quenching of the green fluorophore occurs in the double-labeled species, then $\varepsilon_{\mathrm{GR,G,G}} < \varepsilon_{\mathrm{G,G,G}}$. When crosstalk of the green fluorophore in the red detection channel occurs, then $\varepsilon_{\mathrm{G,G,R}}$ and $\varepsilon_{\mathrm{GR,G,R}} > 0$. When the red fluorophore is directly excited by the green laser, $\varepsilon_{\mathrm{GR,G,R}}$ and $\varepsilon_{\mathrm{R,G,R}} > 0$. Finally, when the red fluorophore is quenched in the double-labeled species,

$\varepsilon_{GR,R,R} < \varepsilon_{R,R,R}$. A detailed interpretation of Eq. (9.3) will be given later on in this chapter.

3. JUGGLING PHOTONS WITH PIE

On the excitation side, the simplest realization uses two pulsed interleaved lasers (Fig. 9.3A, i). This concept is easy to expand to three or even multicolor experiments by alternating multiple pulsed lasers. This opens a new world for specific fluorophore imaging in a remarkably simple way. However, the principle is also amenable to more complex excitation schemes, specifically suited for certain applications. For example, in most

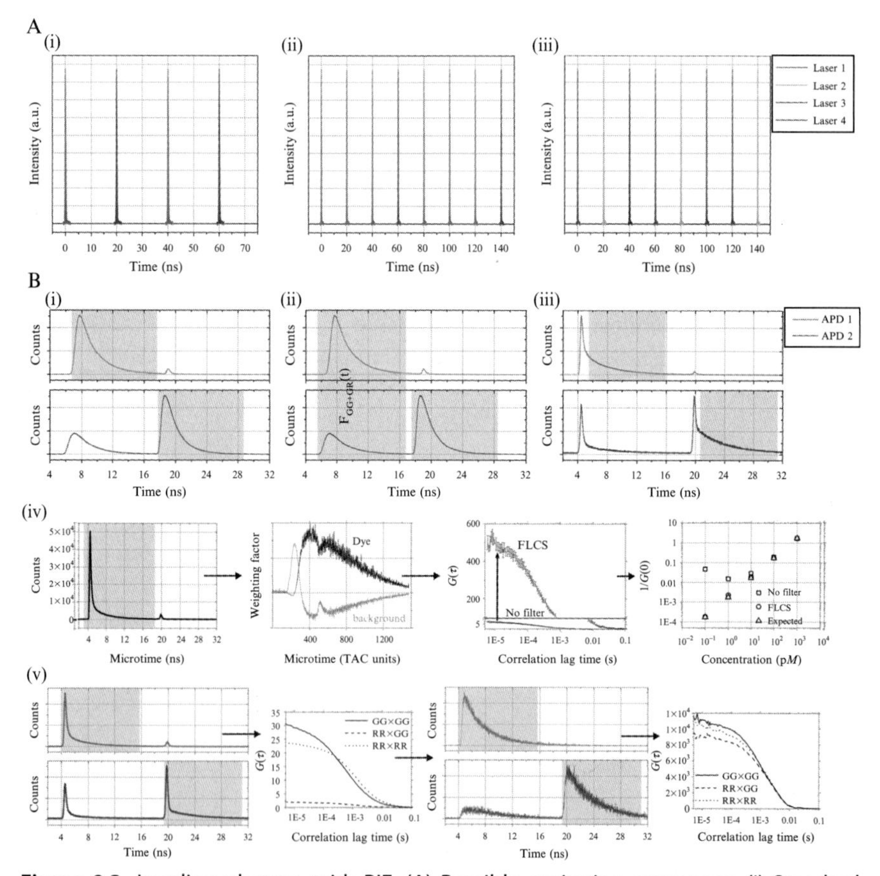

Figure 9.3 Juggling photons with PIE. (A) Possible excitation sequences. (i) Standard dual-color PIE where green and red excitation alternate each other with a fixed

FRET experiments, the photons collected after donor excitation are used to quantify the energy transfer, so it is desirable to probe more often with donor excitation than to probe the acceptor. This was done in duty cycle-optimized ALEX (Zarrabi et al., 2009) (Fig. 9.3A, ii) and is easily implementable with modern laser devices and TCSPC. When performing multicolor FRET experiments with PIE, further excitation schemes are possible for optimal signal detection of each of the fluorophores and an increase in the duty cycle for the fluorophores involved in multicolor FRET (Fig. 9.3A, iii). An alternative excitation scheme, where not the color but the polarization is alternated, is routinely used in dual-focus FCS (2fFCS) (Dertinger et al., 2007) and even a combination of color and polarization

---

repetition rate and fixed time delay between the lasers, (ii) FRET donor optimized excitation.where the green laser is allowed to fire more often than the red laser, (iii) multicolor excitation, where the green laser has a fixed repetition rate that determines the master clock of the whole system, the blue laser precedes the green laser at the same frequency and the red and infrared lasers alternate and are pulsed at half the frequency of the blue and green laser. In this way, the photons needed to determine FRET are probed more often. (B) Possible macrotime cross-correlation strategies for dual-color PIE–FCCS. (i) Artifact-free and species-specific analysis. The macrotimes corresponding to $F_{GG}(t)$ and $F_{RR}(t)$ do not contain misguided photons and can be cross-correlated and analyzed with Eqs. (9.14) and (9.23) to verify if double-labeled species exist. (ii) Addition of microtime channels. The macrotimes corresponding to $F_{GG+GR}(t)$ and $F_{RR}(t)$ do not contain contributions from FRET and can be cross-correlated and analyzed with Eqs. (9.15), (9.21), and (9.16) to allow for absolute calculation of concentrations of single- and double-labeled species. (iii) Time gating. At low concentrations, time gating of the microtime histograms allows reduction of the background and scattering-related artifacts. (iv) FLCS. Lifetime filtering allows further reduction of background and scattering artifacts, making it possible to perform quantitative FCS/FCCS analyses at ultra-low concentrations. Illustrated is an actual measurement of ATTO 532 at different concentrations. The raw microtime data is shown in the left panel. The weighting functions used to separate scattered light form ATTO 532 fluorescence is shown in the second panel. The third panel shows the autocorrelation analysis of the same data once as normal FCS data and once using FLCS. With FLCS, the measured concentration remains inversely proportional to the correlation function amplitude, even at a concentration of 100-fM (fourth panel). (v) Burstwise filtering. The left two panels illustrate unfiltered burst analysis data measured with MFD–PIE. The CCF has a very low amplitude due to the presence of an excess of single-labeled species. An *E–S* plot (Fig. 9.8A) allows sorting of fluorescent species according to the label content. This subsequently allows for a species-specific macrotime cross-correlation analysis. The right two panels illustrate the same data, now filtered according to stoichiometry. The near absence of scattering light in the microtime histograms and very high CCF amplitude is obvious. (See Color Insert.)

alternation for crosstalk-free dual-color 2fFCS is possible for certain applications (Garcia-Saez, Ries, Orzaez, Perez-Paya, & Schwille, 2009; Müller et al., 2008).

On the detection side, data channels are not limited to the four channels mentioned previously in Fig. 9.1B. Even in the case of dual-color excitation with dual-color detection, every part of the microtime histogram from each detector can be used for analysis. For example, a species-specific analysis can be done simply by selecting those photons corresponding to a specific fluorophore in the microtime histogram and ignoring all misguided photons (Fig. 9.3B, i). In some cases, as we see later on, it is advantageous to group the photons from different channels (Fig. 9.3B, ii). When measuring at low concentration, selecting only those subsets of the detection channels that do not contain scattered light, so-called time gating (Lamb, Schenk, Rocker, Scalfi-Happ, & Nienhaus, 2000), increases the signal-to-noise ratio (SNR) for further analysis considerably (Fig. 9.3B, iii). When lifetime filtering approaches can be used (Böhmer et al., 2002; Kapusta, Wahl, Benda, Hof, & Enderlein, 2007), this allows for a larger increase in the SNR and in some cases allows separation of different lifetime species with the same color (Fig. 9.3B, iv). Finally, by doing burst analysis with PIE, as will be discussed later, microtime channels can even be filtered according to the presence of a specific species, such as single- or double-labeled molecules (Margittai et al., 2003) (Fig. 9.3B, v) and even allows filtering of experimental data via any molecular property, such as anisotropy, color, or lifetime (Felekyan, Kalinin, Sanabria, Valeri, & Seidel, 2012).

In summary, by using a specific fine-tuned illumination scheme, PIE provides the necessary freedom and control over which photons, and thus which species, are included into a specific analysis, be it FRET or FCCS. Moreover, one can perform different kinds of analyses in parallel on the same (macrotime) dataset by first filtering according to the microtime data. This unique property, in a way, creates an extra dimension for data analysis and is ultimately the strength of the technique.

4. SIMPLE AND QUANTITATIVE CORRELATION ANALYSIS WITH PIE

4.1. Primer on correlation analysis

In general, the normalized autocorrelation function (ACF) expressed in terms of the deviation $\delta F(t)$ from the mean fluorescence intensity $\langle F(t) \rangle$ is given by (Elson & Magde, 1974; Magde, Webb, & Elson, 1972).

$$G(\tau) = \frac{\langle \delta F(t) \delta F(t+\tau) \rangle}{\langle F(t) \rangle^2} \tag{9.4}$$

where $\langle\rangle$ denotes a time-averaged value. When $W(\mathbf{r})$ in Eq. (9.1) is approximated by a three-dimensional Gaussian, the probe volume can be defined as a cuboid of homogeneous illumination with the fluorescence intensity inside equal to the peak intensity of the actual PSF (Ivanchenko & Lamb, 2011):

$$V = \left(\frac{\pi}{2}\right)^{3/2} \omega_r^2 \omega_z \tag{9.5}$$

Using this definition for V, $G(\tau)$ can be solved analytically for freely diffusing, noninteracting fluorophore of a single species (Aragon & Pecora, 1975; Elson & Magde, 1974; Ivanchenko & Lamb, 2011):

$$G(\tau) = \frac{2^{-3/2}}{\langle N \rangle} \left(\frac{1}{1+4D\tau/\omega_r^2}\right) \left(\frac{1}{1+4D\tau/\omega_z^2}\right)^{1/2} \equiv \frac{1}{\langle N \rangle} G_{\mathrm{diff}}(D,\tau) \tag{9.6}$$

where $\langle N \rangle$ is the average number concentration, ω_r and ω_z are the lateral and axial distance, respectively, from the center of the probe volume to the position where the maximum excitation intensity has decayed to $1/e^2$ (0.1353), D is the translational diffusion coefficient, and τ is the correlation lag time. The notation $G_{\mathrm{diff}}(D,\tau)$ will be used in the rest of the chapter as a short notation for the diffusion term in Eq. (9.6) containing the geometrical factor $2^{-3/2}$ (for a 3D Gaussian PSF). Importantly, using this definition of $\langle N \rangle$, $\varepsilon = \langle F(t) \rangle / \langle N \rangle$ represents the peak brightness of the fluorophore in the center of the probe volume (Thompson, 1991). For comparison of brightnesses between different FFS techniques such as FCS, PCH (Chen, Müller, So, & Gratton, 1999), FIDA (Kask, Palo, Ullmann, & Gall, 1999), or FCA (Müller, 2004), it is advised to verify that the same definitions for V, ε, and $\langle N \rangle$ are used. In the case of a mixture of species with unequal brightness, the autocorrelation amplitude is a brightness-weighted sum of all species present in solution (Lamb et al., 2000; Schwille, 2001).

When a dual-color excitation dual-color detection microscope is used, two fluorescence signals $F_G(t)$ and $F_R(t)$ are recorded, ideally corresponding to each fluorophore. In addition to the ACFs of the individual detection channels, the normalized cross-correlation function (CCF), $G_{GR}(\tau)$, reveals the similarity of the two signals (Rigler et al., 1998; Schwille et al., 1997):

$$G_{G\times R}(\tau) = \frac{\langle \delta F_G(t)\delta F_R(t+\tau)\rangle}{\langle F_G(t)\rangle\langle F_R(t)\rangle} \quad [9.7]$$

A symmetrical CCF calculation is typically assumed, meaning there is no difference between $G_{G\times R}(\tau)$ and $G_{R\times G}(\tau)$, and the two PSFs are assumed to be identical. A detection correction factor between the green and red channel can be defined, to take the differential detection efficiencies of the two detection channels for the given fluorophores into account:

$$\gamma = \frac{\kappa_{R,R}\Phi_R}{\kappa_{G,G}\Phi_G} \quad [9.8]$$

where $\kappa_{i,k}$ and Φ_i are the detection efficiency and quantum yield, respectively, of species i in detection channel k. With the same assumptions used for autocorrelation analysis, the CCF can be solved analytically:

$$G_{G\times R}(\tau) = \frac{N_{GR}}{(N_G + N_{GR})(N_R + N_{GR})} G_{diff}(D_{GR}, \tau) \quad [9.9]$$

where D_{GR} is the diffusion coefficient of the double-labeled species and the cross-correlation amplitude is a brightness-weighted sum of the green and red species (Schwille, 2001), with N_G being the average number of green-only particles, N_R is the average number of red-only particles, and N_{GR} is the number of double-labeled particles. For clarity, the average notation $\langle\rangle$ has been omitted from the equations. For concentric, yet differently sized green and red PSFs, the effective probe volume V_{CC} where double-labeled particles are observed is given by the root mean square of the radial and axial radii of the green and red probe volumes (Schwille et al., 1997):

$$\omega_{r,CC} = \sqrt{\frac{\left(\omega_{r,G}^2 + \omega_{r,R}^2\right)}{2}}$$

$$\omega_{z,CC} = \sqrt{\frac{\left(\omega_{z,G}^2 + \omega_{z,R}^2\right)}{2}} \quad [9.10]$$

$$V_{CC} = \left(\frac{\pi}{2}\right)^{3/2} \left(\frac{\omega_{r,G}^2 + \omega_{r,R}^2}{2}\right) \sqrt{\frac{\left(\omega_{z,G}^2 + \omega_{z,R}^2\right)}{2}}$$

The average molecule numbers from the CCF can then be converted into absolute concentrations:

$$\frac{N_{GR}/V_{GR}}{(N_{G,\,total}/V_G)(N_{R,\,total}/V_R)}N_A^{-1} = \frac{[GR]}{[G_{total}][R_{total}]} \tag{9.11}$$

where $N_{k,total}$ is the total number of particles detected in channel k, G_{total} and R_{total} are the total concentration of molecules carrying a green and red fluorophore, respectively, and N_A is Avogadro's number.

4.2. How PIE–FCCS deals with misguided detection

Due to the asymmetrical shape of most fluorescence spectra, the green fluorophore often exhibits emission bleedthrough or crosstalk into the red detection channel. With normal CW–FCCS, the $F_R(t)$ detection channel is thus contaminated with crosstalk of the green fluorophore. Assuming no brightness changes occur in the double-labeled species, crosstalk can be taken into account in normal cross-correlation analysis:

$$G_{G\times R}(\tau) = \frac{\beta' N_G G_{diff}(D_G,\tau) + (\beta'+1)N_{GR} G_{diff}(D_{GR},\tau)}{(N_G+N_{GR})(N_R+N_{GR}+\beta'(N_G+N_{GR}))} \tag{9.12}$$

where $\beta' = \varepsilon_{G,R}/\varepsilon_{R,R}$ is the relative brightness of the green fluorophore versus the red fluorophore in the red detection channel. This definition of the crosstalk (β') differs from that used in single-pair FRET (spFRET) experiments (β) that will be discussed later in the chapter. Thus, when the brightness of the fluorophores in the double- and single-labeled species is not significantly different (i.e., negligible FRET and/or quenching) and when the crosstalk parameter β' is determined, one can correct the obtained FFS data for crosstalk *a posteriori*. This has been successfully done in the early reports of FCCS (Foldes-Papp, 2005; Rigler et al., 1998) as well as in more recent *in vivo* measurements of protein interactions (Hwang & Wohland, 2005). In the absence of crosstalk ($\beta'=0$), Eq. (9.12) reduces to Eq. (9.9) and, in the absence of a double-labeled species, Eq. (9.12) reduces to

$$G_{G\times R}(\tau) = \frac{\beta'}{N_R+\beta' N_G} G_{diff}(D_G,\tau) \tag{9.13}$$

where the amplitude of the CCF then represents the so-called false-positive cross-correlation amplitude that is observed with CW–FCCS. The lower the number of red-labeled molecules, N_R, the more the CCF amplitude increases toward the $G_G(\tau)$ ACF amplitude. Finally, in the presence of green and red fluorophores under dual-color excitation conditions, the ACF $G_R(\tau)$ also has to be corrected for crosstalk.

With PIE, green fluorophore crosstalk and red fluorophore emission are detected in separate data channels on the red detector. Cross-correlating the photons of $F_{GG}(t)$ with $F_{RR}(t)$ results in

$$G_{GG\times RR}(\tau)_{\text{no FRET}} = \frac{N_{GR}}{(N_G + N_{GR})(N_R + N_{GR})} G_{\text{diff}}(D_{GR}, \tau) \quad [9.14]$$

Importantly, there is no correction for crosstalk necessary since both correlation channels are devoid of crosstalk. Second, when direct excitation of the red fluorophore by the green laser can be ignored and if γ is unity, then the summed signal $F_{GG}(t) + F_{GR}(t)$ can be cross-correlated with $F_{RR}(t)$, which again results in

$$G_{(GG+GR)\times RR}(\tau) = \frac{N_{GR}}{(N_G + N_{GR})(N_R + N_{GR})} G_{\text{diff}}(D_{GR}, \tau) \quad [9.15]$$

This CCF is devoid of crosstalk, and moreover, since all photons generated by the green fluorophore are used for the analysis, the SNR of the experiment is maximized.

As discussed before, whereas classical FCCS analysis works well for the qualitative proof-of-high-affinity interactions, the technique fails for studying low-affinity interactions that are usually masked by the high false-positive crosstalk amplitude of the CCF (Eq. (9.12)). *In extremis*, proving the absence of interaction with classical FCCS is impossible. In addition to being the technique of choice for quantitative FCCS analysis of molecular affinities, PIE–FCCS can also be used to prove the absence of molecular interactions. As an example, we discuss the interactions of the GroEL chaperonin system with substrates (Chakraborty et al., 2010; Sharma et al., 2008). In this study, we investigated whether the substrates bound to the chaperonin one at a time or if both rings of GroEL could simultaneously bind substrates and whether the denatured substrates formed transient aggregates. Hence, we fluorescently label the substrate, the double-mutant maltose-binding protein (DM-MBP) and proved with FCCS that denatured DM-MBP did not oligomerize. Therefore, the basis for accelerated folding of the DM-MBP substrate by GroEL cannot be due to encapsulation of DM-MBP in GroEL preventing transient aggregation (Chakraborty et al., 2010). To investigate whether it is possible for more than one substrate to bind to a single-GroEL molecule, we performed PIE–FCCS experiments (Fig. 9.4). As a negative control, a mixture of the two fluorophores was measured and no cross-correlation amplitude

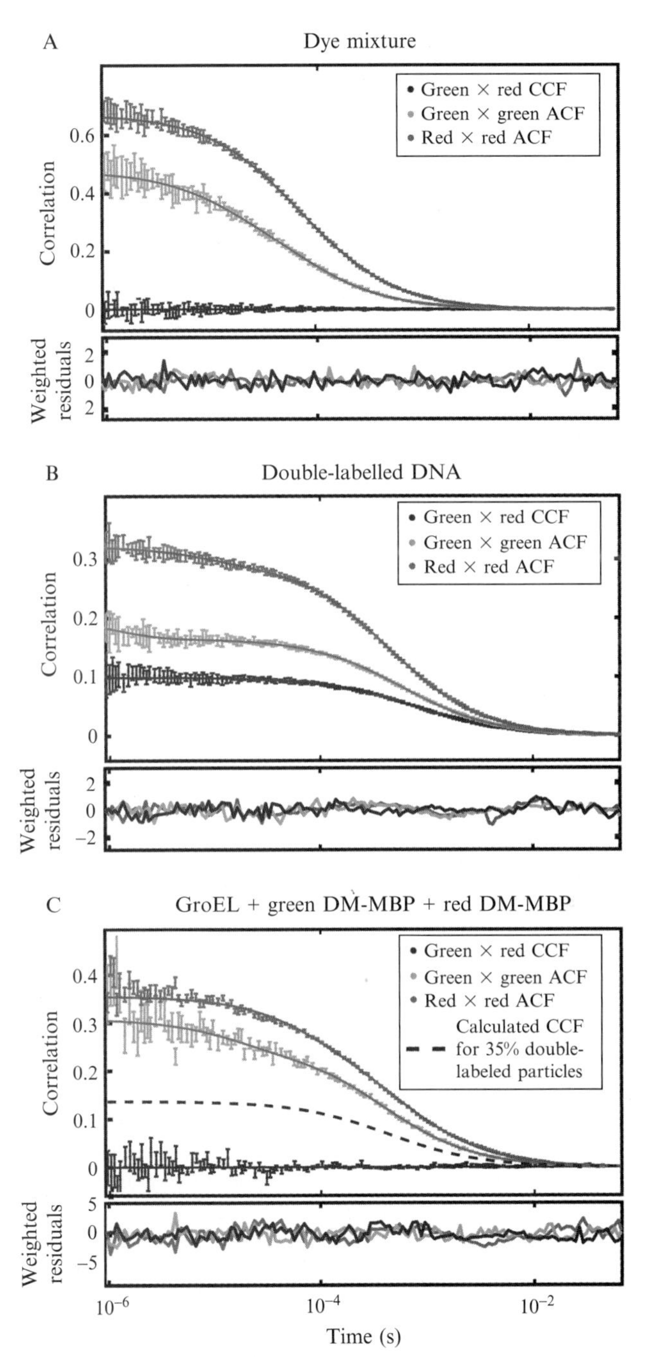

Figure 9.4 FCCS measurements investigating the role of GroEL in the refolding of DM-MBP. (A) ACFs and CCF of a mixture of ATTO 532 and ATTO 647N as a negative control. (B) ACFs and CCF of dsDNA labeled with ATTO 532 and ATTO 647N as positive

was observed after calculation with Eq. (9.14) (Fig. 9.4A). As a positive control for FCCS, a double-stranded DNA molecules labeled with the same fluorophores was measured and a high cross-correlation amplitude was observed (Fig. 9.4B). Finally, DM-MBP labeled either with ATTO 532 or with ATTO 647N was mixed together with GroEL. The PIE–FCCS experiments showed no significant cross-correlation (Fig. 9.4C). Thus, we could confidently conclude that the GroEL does not bind more than one substrate at a time at nanomolar concentrations (Sikor, 2011).

Finally, crosstalk-free, FRET-free PIE–FCCS can also be performed on a single detector (Müller, 2007), similar to a μsALEX single-detector approach (Thews et al., 2005). However, when using a single detector for PIE–FCCS, a residual cross-correlation amplitude might still be detected for noninteracting species due to direct excitation of the red fluorophores by the green laser (see Section 4.3). A μsALEX crosstalk-free FCCS approach with two detectors has also been shown recently (Takahashi et al., 2008).

4.3. How PIE–FCCS deals with misguided excitation

Due to the asymmetrical shape of an excitation spectrum, the red fluorophore can usually be excited by the green excitation light. In the presence of significant direct excitation, PIE–FCCS can still be quantified in a straightforward manner. If the detection correction factor, γ, is assumed to be unity, the green fluorophore is not quenched by any process other than FRET (which will be discussed in Section 4.4) and the red fluorophore is assumed to have a homogeneous brightness (which are reasonable assumptions for most FCCS experiments), then the total brightness (from both the green and red detection channels) of the green fluorophore after green excitation will be the same for the single-labeled and the double-labeled species. Hence, only direct excitation of the red fluorophore increases

control. (C) ACFs and CCF of a mixture of DM-MBP single-labeled with ATTO 532 or with ATTO 647N. As well as the calculated CCF assuming 35% double-labeled particles. If the accelerated folding observed in the presence of GroEL were due to inhibition of reversible dimerization of DM-MBP by encapsulation of the substrate within the GroEL cavity, a cross-correlation amplitude of 35% (shown as a dashed line) would be observable. (A-C) The ACF of the signal in the detection channel for ATTO 532 is shown in light gray, the ACF of the signal in the detection channel for ATTO 647N is shown in dark gray, and the CCF of the signals of both dyes is shown in black. The solid lines overlaying the data are fits using the normal FCS model, Eq. (6), including a triplet component. The lower panels show the weighted residuals of the fits. *Adapted from Sikor, 2011.* (For interpretation of the references to color in this figure legend, the reader is referred to the online version of this chapter.)

the total brightness of the double-labeled species, that is, $\varepsilon_{G,G,G}+\varepsilon_{G,G,R}=\varepsilon_{GR,G,G}+\varepsilon_{GR,G,R}-\varepsilon_{R,G,R}$. The CCF can then be shown to take the following form:

$$G_{(GG+GR)\times RR}(\tau)=\frac{\alpha' N_R G_{diff}(D_R,\tau)+(\alpha'+1)N_{GR}G_{diff}(D_{GR},\tau)}{(N_R+N_{GR})(N_G+N_{GR}+\alpha'(N_R+N_{GR}))} \quad [9.16]$$

where $\alpha'=\varepsilon_{R,G,R}/\varepsilon_{G,G,G+R}$ is the relative brightness of the red fluorophore versus the green fluorophore in the $F_{GG+GR}(t)$ combined detection channel. This definition of the direct excitation (α') differs from the one used for spFRET (α), discussed later in this chapter. In the absence of direct excitation, Eq. (9.16) reduces to Eq. (9.9). In the presence of direct excitation but in the absence of double-labeled species, a residual cross-correlation due to direct excitation of the red fluorophore will be observed:

$$G_{(GG+GR)\times RR}(\tau)=\frac{\alpha'}{N_G+\alpha' N_R}G_{diff}(D_R,\tau) \quad [9.17]$$

Importantly, Eq. (9.16) is valid even with significant amounts of crosstalk and/or FRET (as will be discussed next), in contrast to CW–FCCS experiments.

4.4. How PIE–FCCS deals with FRET

When two fluorophores are in close proximity (within 10 nm), energy can be transferred from the donor fluorophore to the acceptor fluorophore via FRET. The quantum yield for energy transfer, also known as the FRET efficiency (E), is given Eq. (9.18). When the fluorescence of the donor is not quenched by any process other than FRET, E can be expressed in terms of the fluorescence intensity of the donor in the presence (F_{DA}) or absence of the acceptor (F_D), as is done in many FRET assays. It is also possible to define the FRET efficiency over the molecular brightness ($\varepsilon_{i,k}$) of the donor in the presence and the absence of the acceptor, as we will demonstrate later:

$$E=\frac{k_T}{k_T+k_D+k_r}=1-\frac{F_{DA}}{F_D}=1-\frac{\varepsilon_{GR,G}}{\varepsilon_{G,G}} \quad [9.18]$$

where k_T and k_D are the rate constants for energy transfer and donor fluorescence, respectively, k_r is the summed rate constant for all other processes that de-excite the donor, and $\varepsilon_{i,k}$ is defined as before. This definition for E as a function of the measured fluorescence intensity holds true for a single D–A complex or for a population of 100% acceptor-labeled complexes. In a typical FRET measurement, however, the labeling is hardly ever complete. When the fraction of complexes with an active acceptor, f_{GR}, is known,

Eq. (9.18) can be expressed as a function of the donor brightness under conditions of incomplete donor labeling (Cheung, 1991):

$$E = 1 - \frac{F_{\mathrm{DA,\,measured}} - F_{\mathrm{D,\,measured}}(1 - f_{\mathrm{GR}})}{F_{\mathrm{D,\,measured}} f_{\mathrm{GR}}} = \left(1 - \frac{F_{\mathrm{DA,\,measured}}}{F_{\mathrm{D,\,measured}}}\right) \frac{1}{f_{\mathrm{GR}}} \quad [9.19]$$

The FRET efficiency can also be defined from sensitized acceptor emission (Clegg, 1992; Müller et al., 2005). Obtaining a correct value for f_{GR} is often difficult and moreover, when the donor is strongly quenched by the acceptor, a small, unknown percentage of species lacking an acceptor molecule can lead to a large error in E. This makes absolute FRET quantifications with standard assays difficult. When the quantum yield of the acceptor is assumed to be independent of the presence of the donor, the CCF for normal CW–FCCS is given by

$$G_{\mathrm{G \times R}}(\tau) = \frac{(1-E)(1+E')N_{\mathrm{GR}}}{(N_{\mathrm{G}} + (1-E)N_{\mathrm{GR}})(N_{\mathrm{R}} + (1+E')N_{\mathrm{GR}})} G_{\mathrm{diff}}(D_{\mathrm{GR}}, \tau) \quad [9.20]$$

where $E = 1 - \varepsilon_{\mathrm{GR,G,G}}/\varepsilon_{\mathrm{G,G,G}}$ is the FRET efficiency as defined in Eq. (9.18) and $E' = \varepsilon_{\mathrm{GR,G,R}}/\varepsilon_{\mathrm{R,R,R}}$ is also a measure for energy transfer. When the laser powers are adjusted such that $\varepsilon_{\mathrm{G,G,G}} = \varepsilon_{\mathrm{R,R,R}}$ and the detection efficiencies of the donor and acceptor are similar, then $E = E'$. To extract the absolute FRET efficiency from Eq. (9.20), the labeling degree has to be known *a priori*. As the labeling degree is often unknown, FRET quantification with CW–FCCS is in practice difficult. Likewise, a quantitative FCCS analysis in the presence of FRET is only possible when the FRET efficiency is known, making quantitative analyses in many cases extremely difficult.

Provided the sensitivity of the apparatus is similar for fluorescence from the donor and from the acceptor, then the decrease of brightness of the donor in the $F_{\mathrm{GG}}(t)$ channel is exactly compensated by the FRET-sensitized emission of the acceptor in the $F_{\mathrm{GR}}(t)$ channel and Eq. (9.15) can be used for the analysis. A quantitative FCCS analysis is thus possible without the need for knowing the FRET efficiency. Moreover, the labeling degree can be calculated from the following ACF and CCF:

$$\frac{G_{\mathrm{(GG+GR) \times RR}}(0)}{G_{\mathrm{RR \times RR}}(0)} = f_{\mathrm{GR}} \quad [9.21]$$

where $f_{\mathrm{GR}} = N_{\mathrm{GR}}/(N_{\mathrm{G}} + N_{\mathrm{GR}})$ is the fraction of double-labeled species relative to the total amount of molecules carrying a green label. Unlike normal CW–FCCS, this ratio is independent of the number of acceptor-only molecules, even in the presence of FRET, as we have shown experimentally in the original PIE publication (Müller et al., 2005).

In a recent work, parameters affecting FCCS experiments were analyzed in detail inside living cells (Foo, Naredi-Rainer, Lamb, Ahmed, & Wohland, 2012). In this study, PIE–FCCS was used to specifically investigate the effect of FRET on the different correlation functions of a tandem dimer of eGFP and mCherry fluorescent proteins (FPs). Both inside CHO cells (Fig. 9.5A) and in cell lysates (Fig. 9.5B), the amplitudes of the correlation functions were lowered in the presence of FRET as predicted by the equations. With the aid of PIE, a FRET-free FCCS analysis could be performed. Importantly, since the concentration ratio between green and red proteins in this study was similar, effects of crosstalk in the red detection channel were minimal. When the appropriate correction factors are known, a quantitative analysis in the presence of FRET and crosstalk can still be performed, although the needed analysis models were considerably more complicated as for PIE–FCCS.

5. QUANTIFYING FRET WITH PIE–FRET–FCCS

FRET can also be quantified with PIE–FCCS. In the presence of FRET, the molecular brightness of the donor fluorophore detected in the green channel after green excitation will decrease for the double-labeled species. In this case, the molecular brightness of the double-labeled species in Eq. (9.3) can be replaced with the FRET efficiency:

$$F_{\mathrm{GG}}(t)=\int_V \mathrm{d}\mathbf{r}\,W(\mathbf{r})\varepsilon_{\mathrm{G,G,G}}\left(C_{\mathrm{G}}(\mathbf{r},t)+(1-E)C_{\mathrm{GR}}(\mathbf{r},t)\right) \qquad [9.22]$$

Provided the quantum yield of the acceptor is assumed to be independent of the presence of the donor, the CCF, $G_{\mathrm{GGRR}}(\tau)$, is given by

$$G_{\mathrm{GG\times RR}}(\tau)_{\mathrm{FRET}}=\left(\frac{1-E}{1-Ef_{\mathrm{GR}}}\right)\frac{N_{\mathrm{GR}}}{(N_{\mathrm{G}}+N_{\mathrm{GR}})(N_{\mathrm{R}}+N_{\mathrm{GR}})}G_{\mathrm{diff}}(D_{\mathrm{GR}},\tau) \qquad [9.23]$$

Assuming negligible direct excitation of the acceptor, the FRET efficiency can be determined using Eq. (9.15), and the ACF amplitude of the red channel after red excitation, $G_{\mathrm{RR}}(0)$:

$$E=\frac{\left(1-\frac{G_{\mathrm{GG\times RR}}(0)_{\mathrm{FRET}}}{G_{\mathrm{(GG+GR)RR}}(0)}\right)}{\left(1-\frac{G_{\mathrm{GG\times RR}}(0)_{\mathrm{FRET}}}{G_{\mathrm{RR}}(0)}\right)} \qquad [9.24]$$

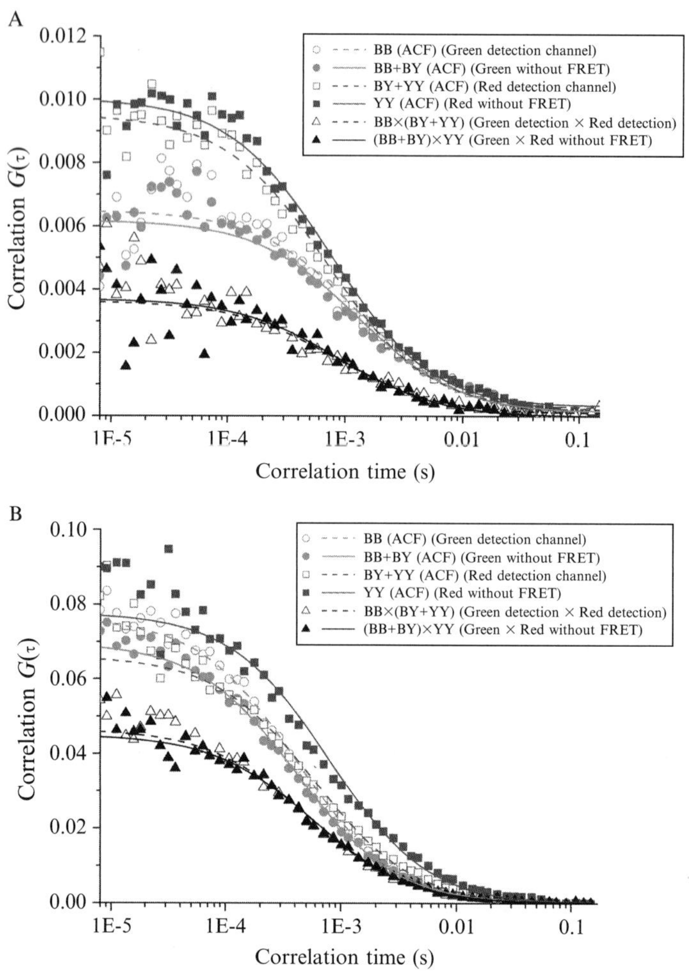

Figure 9.5 Quantitative FCCS analysis independent of FRET using PIE in live-cells and cell lysates. FCCS experiments with (a) mCherry-14-EGFP in live CHO cells and (b) mCherry-7-EGFP in cell lysate are shown. The correlation functions in the presence of FRET are represented by open circles (green ACF in green), open squares (red ACF in red), and open triangles (CCF in black). The correlation functions after correcting the PIE-FCCS data for FRET are represented by solid circles (green ACF in green), solid squares (red ACF in red), and solid triangles (CCF in black). The lines are the fits with Eq. (6) and (9). The influence of FRET is clear in both experiments. *Adapted from Foo et al. (2012).* (For interpretation of the references to color in this figure legend, the reader is referred to the online version of this chapter.)

In contrast with the analysis model for CW–FCCS in the presence of FRET, Eq. (9.20), prior knowledge regarding the fraction of complexes containing a functional FRET acceptor is not necessary. As a proof of principle that PIE–FRET–FCCS works, measurements were performed on a dsDNA

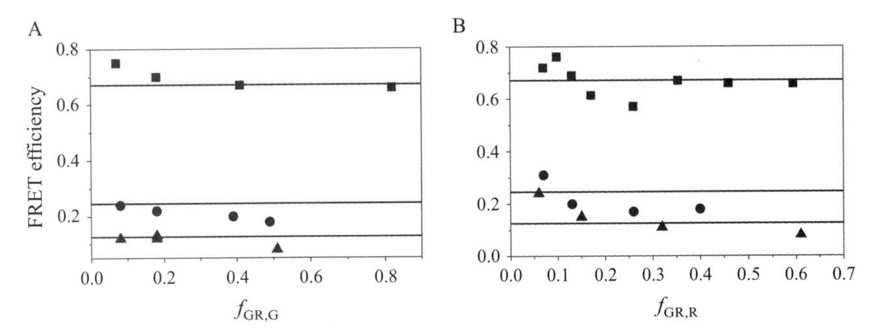

Figure 9.6 PIE allows a quantitative cross-correlation analysis in the presence of FRET. The FRET efficiency of the DNA with donor and acceptor separated by 10 bp (squares), 15 bp (circles), and 20 bp (triangles) is plotted as a function of (A) $f_{GR,G}$ and (B) $f_{GR,R}$ where $f_{GR,G}$ and $f_{GR,R}$ represent the fraction of double-labeled species relative to the total number of green and red bearing species, respectively. The corresponding FRET efficiency determined from spFRET measurements are shown as solid lines for comparison. *Adapted from Müller et al. (2005).*

molecule fluorescently labeled on each strand with a FRET donor (ATTO 532) and FRET acceptor (ATTO 647), spaced by 10, 15, or 20 bp (Müller et al., 2005). Different amounts of donor-only (Fig. 9.6A) or acceptor-only (Fig. 9.6B) labeled DNA were added to the solution. The absolute FRET efficiency that could be determined was, within a certain range, independent of the fraction of the donor-only or acceptor-only species. Moreover, the FRET efficiency was verified to be correct with spFRET, a technique that measures FRET on a single-molecule basis (discussed in more detail below).

Generally, there are two important guidelines for PIE–FRET–FCCS: on the one hand, one needs to make sure that f_{GR} is not too low, as otherwise the amplitude of $G_{GG \times RR}(\tau)$ will be too low to be accurately determined. On the other hand, when f_{GR} approaches one, there is only one species present in the sample and the numerator in Eq. (9.24) becomes invalid. An absolute FRET efficiency can also be calculated directly from measured FCS data without using PIE, for example, by investigating the photoinduced *trans–cis* isomerization of Cy5, which will depend upon the FRET efficiency as shown by Widengren, Schweinberger, Berger, and Seidel (2001). A proper calibration of the system is necessary for this approach. The group of Johnson used a global analysis of the green and red ACFs and the green–red and red–green CCFs to estimate the FRET efficiency under static conditions (Price, DeVore, & Johnson, 2010). This method relied on the preparation of a purely double-labeled complex, which allowed for quantitative interpretation of the correlation function amplitudes. The advantage of PIE–FRET–FCCS with respect to the latter two reports is that (i) it is a calibration-free method,

(ii) it does not require a pure double-labeled sample or a sample with known fractional labeling, and (iii) it does not suffer from crosstalk.

In the case of a mixture of FRET species with different FRET efficiencies and a donor-only species, the average FRET efficiency $\bar{E}$ measured with PIE–FRET–FCCS is given by

$$\bar{E} = \frac{E_1 N_{\mathrm{GR1}} + E_2 N_{\mathrm{GR2}}}{N_{\mathrm{GR1}} + N_{\mathrm{GR2}}} \qquad [9.25]$$

When the amount of direct excitation, α', is known, Eqs. (9.16) and (9.23) together with the $G_{\mathrm{GG}}(\tau)$ and $G_{\mathrm{RR}}(\tau)$ ACFs can still be used to extract E quantitatively. Hence, even in the presence of FRET, donor crosstalk and direct excitation of the acceptor, an FCCS experiment using PIE can be fully quantified.

Other methods for quantifying the absolute FRET efficiency are based on burst analysis, such as ALEX (Kapanidis et al., 2004) and/or spFRET–MFD (Widengren et al., 2006). ALEX- and MFD-based approaches have many advantages, as we discuss below, but are based upon the detection of individual bursts. Typical concentrations used in burst analysis measurements are well below 1 n*M*. For studying low-affinity interactions or for cellular applications, such concentrations are difficult to achieve. One advantage of PIE–FRET–FCCS approach is that absolute FRET efficiencies can be calculated at any concentration accessible to an FCCS analysis, which is roughly 1 p*M* to 1 μ*M* depending on the experimental setup.

Equations (9.22)–(9.25) hold true only when FRET is assumed to be static. Hence, for this approach to work, any fluctuations that alter the FRET efficiency must occur on a timescale significantly longer than the average diffusion time of the molecule through the probe volume in the FCCS experiment. However, conformational dynamics can occur on a timescale that is similar to the FCS timescale or faster. Conformational fluctuations on the microsecond timescale can also be studied with FRET using FRET-FCCS. Analysis of the ACFs and CCFs allows the rate constants to be extracted from FCCS experiments (Ferreon, Gambin, Lemke, & Deniz, 2009; Margittai et al., 2003; Price et al., 2010; Torres & Levitus, 2007). In such studies, PIE–FRET–FCCS would be useful as the FRET contributions to the correlation functions can also be removed by correlating $G_{(\mathrm{GG+GR})\times\mathrm{RR}}$, Eq. (9.15). This can be beneficial when the dynamics occurs on a similar timescale to diffusion. With the aid of PIE, not only is the correlated FRET dynamics available from the analysis but also the FRET signal can be removed from the same FCCS data and the timescale of diffusion determined independently or globally.

6. ACCURATE SINGLE-PAIR FRET WITH MFD–PIE

Burst analysis is a special fluorescence fluctuation technique in which the concentration of fluorophore is kept so low that the probability of finding more than one molecule in the probe volume is virtually zero. In doing so, the fluorescence properties of an individual molecule such as its fluorescence lifetime can be measured (Zander et al., 1996). SpFRET is a multicolor extension of burst analysis in which interfluorophore distances can be measured at the single-molecule level (Deniz et al., 1999). SpFRET burst analysis experiments with PIE have already been used to study of conformational states and transitions in several other biological systems, for which the reader is referred to the literature (Chakraborty et al., 2010; Hillger et al., 2008; Hofmann et al., 2010; Kobitski, Hengesbach, Helm, & Nienhaus, 2008; Kukolka et al., 2006; Mapa et al., 2010; Marcinowski et al., 2011; Müller, Reuter, Simmel, & Lamb, 2006; Rosenkranz, Schlesinger, Gabba, & Fitter, 2011; Sharma et al., 2008). Recently, we have combined the advantages of PIE with multiparameter fluorescence detection (MFD).

MFD extends the capabilities of burst analysis even further by extracting the maximum amount of information from the bursts, such as intensity, color, lifetime, quantum yield, and anisotropy (Eggeling et al., 2001) and, combined with spFRET, provides a powerful tool to study conformational dynamics (Rothwell et al., 2003). MFD traditionally employs linearly polarized excitation of the donor with two-color and two-polarization detection to monitor donor emission and FRET-sensitized acceptor emission independently in the parallel and orthogonal polarizations. To detect a high number of photons from the individual bursts, MFD is typically performed on a confocal microscope at relatively high excitation intensities (5–25 kW/cm^2) with a large probe volume (>1 fL), which increases the observation time per molecule. MFD with PIE (MFD–PIE) is the dual-color excitation variant of MFD. As in PIE, both the donor and acceptor can be excited and detected independently. A "classical" MFD setup can be converted into an MFD–PIE setup in a straightforward manner by adding a second picosecond pulsed laser with a wavelength appropriate for acceptor excitation (Fig. 9.7). This laser has to be synchronized with respect to the donor excitation laser and delayed so that the fluorescence decay is complete before the next laser pulse arrives. In a MFD–PIE setup, there are, in principle, eight independent PIE detection channels: each detector contains two time windows corresponding to the excitation laser, and the four detection channels registering either a green or red photons with parallel or perpendicular polarization.

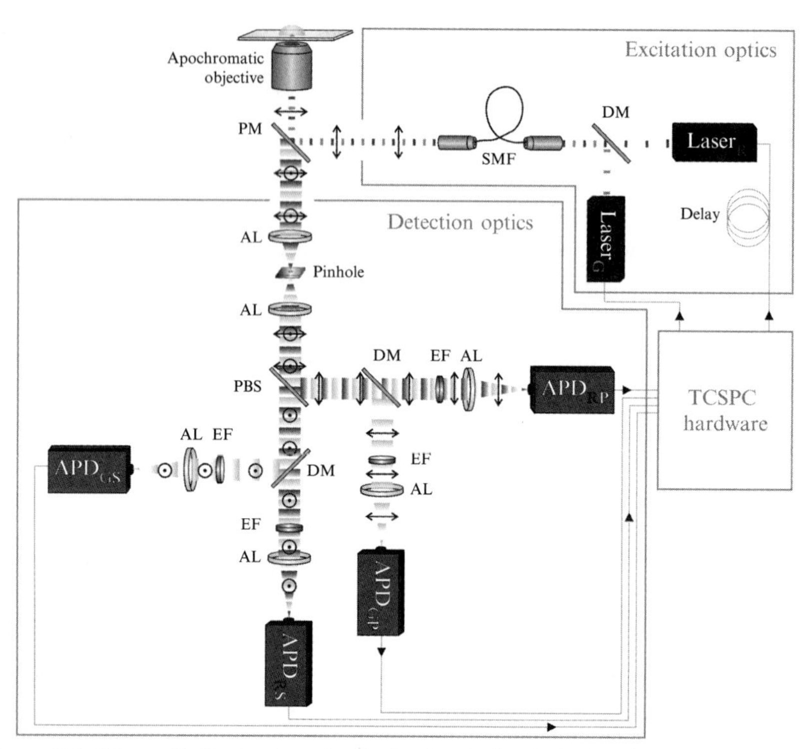

Figure 9.7 Schematic illustration of a dual-color dual-polarization MFD–PIE setup. The main differences with respect to the PIE setup in Fig. 9.2 are that the lasers must be linearly polarized and are usually left unexpanded before entering the objective. This under filling of the objective back aperture creates a larger probe volume, while the detection path uses the full resolving power of the objective. A polarizing beam splitter (PBS) placed in the detection path also splits the emission light into P- and S-polarized components, to allow sensitive measurements of anisotropy, in addition to intensity and lifetime. For further explanation of the setup and most of the abbreviations, see Fig. 9.2. (See Color Insert.)

In analysis of spFRET with MFD, the FRET efficiency can be defined pragmatically as the number of photons emitted in all acceptor detection channels, relative to the number of photons detected in all detection channels, after excitation of the donor. This parameter is called the proximity ratio E_{PR} and is a semiquantitative uncorrected equivalent of the FRET efficiency:

$$E_{PR} = \frac{F_{GR}}{F_{GR} + F_{GG}} \qquad [9.26]$$

where F_{GG} and F_{GR} are the total fluorescence intensity in all green and all red detection channels, respectively, after green excitation. Different calibration and correction factors, such as the direct excitation parameter α, the crosstalk parameter β, the relative detection efficiency parameter γ, and the anisotropy correction factor G, which improves the estimation of the orientational factor κ^2, have to be taken into account to convert E_{PR} into the absolute FRET efficiency (Eq. (9.18)). Typically, several control samples are needed to perform these calibrations. Using MFD–PIE, all of these calibrations can be performed from data collected during the actual measurement.

With MFD–PIE, an extra parameter can be calculated on a single-molecule basis as the total number of photons detected after green excitation relative to the total number of detected photons. This parameter is called the uncorrected stoichiometry ratio S_{PR} and reflects the stoichiometry of the fluorophores on a double-labeled molecule:

$$S_{PR} = \frac{F_{GR} + F_{GG}}{F_{GR} + F_{GG} + F_{RR}} \qquad [9.27]$$

where F_{RR} is the total fluorescence intensity in all red detection channels after red excitation. By carefully adjusting the relative excitation intensities, $S_{PR} \approx 0.5$ for double-labeled single molecules, while $S_{PR} \sim 1$ for donor-only and $S_{PR} \sim 0$ for acceptor-only molecules. Thus, while incomplete labeling might compromise quantitative analyses in ensemble measurements, in spFRET with MFD–PIE, molecules can be sorted according to the label content given by the stoichiometry value. This was originally shown by Kapanidis and coworkers with ALEX (Kapanidis et al., 2004). Because of the sorting capability of MFD–PIE, the important calibration factors for spFRET-MFD analysis cited above (α, β, γ, G, κ^2), as well as other important parameters can be calculated directly from different subpopulations present in the sample (Kudryavtsev et al., 2012).

To illustrate MFD–PIE, we shortly discuss the study of the conformational states of fluorescently labeled DnaK, the major bacterial Hsp70 chaperone, in the presence of substrate. DnaK labeled stochastically at two locations with donor and acceptor fluorophores to probe the distance between the nucleotide-binding domain and the substrate-binding domain. After calculating the semiquantitative S_{PR} and E_{PR} parameters for DnaK in the presence of substrate, different subpopulations such as the D-only, A-only, and double-labeled species can be observed (Fig. 9.8A). By filtering the data according to stoichiometry and subsequently plotting E_{PR} versus the fluorescence lifetime of the A-fluorophore, it becomes clear that the acceptor is slightly quenched in the low-FRET species (Fig. 9.8B). Acceptor quenching is due to binding of the substrate to DnaK. Because of acceptor

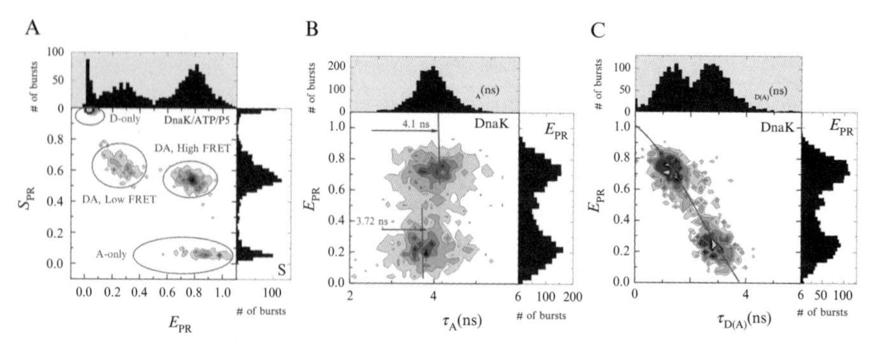

Figure 9.8 2D histograms of a spFRET–MFD–PIE measurement of fluorescently labeled DnaK (318,425). (A) Identification of D-only, A-only, and DA species. A 2D histogram of the uncorrected stoichiometry values, S_{PR}, versus the uncorrected FRET efficiency, E_{PR}, also known as the proximity ratio for spFRET experiments. Donor-only (D-only) labeled molecules, acceptor-only (A-only) labeled molecules, and donor–acceptor labeled (DA) molecules can be observed in different regions of the 2D histogram. The D-only species has high stoichiometry ($S_{PR} \approx 1$) and low proximity ratio ($E_{PR} < 0.2$); the A-only species has a low stoichiometry ($S_{PR} < 0.1$) and an intermediate to high proximity ratio ($E_{PR} > 0.2$); DA species have middle stoichiometry values. (B) A 2D histogram of uncorrected FRET efficiency, E_{PR}, versus directly excited acceptor lifetime, τ_A, of the DA species ($0.2 < S_{PR} < 0.8$). (C) A 2D histogram of uncorrected FRET efficiency, E_{PR}, versus donor lifetime, $\tau_{D(A)}$, of the DA species ($0.2 < S_{PR} < 0.8$). The static FRET line (shown in red) is given by a third-order polynomial with a donor-only fluorescence lifetime of $\tau_{D(0)} = 3.8$ ns. *Adapted from Kudryavtsev et al. (2012).* (See Color Insert.)

quenching, a different γ factor is necessary for the high-FRET and low-FRET species (Fig. 9.8C). Taking this information into account and extracting all calibration parameters necessary for analysis from the single experiment, an accurate assessment of the distance distributions of DnaK could be performed using the photon distribution analysis (Antonik, Felekyan, Gaiduk, & Seidel, 2006). Measurements were performed as a function of nucleotide (ADP or ATP) and in the absence or presence of substrate (Kudryavtsev et al., 2012; Mapa et al., 2010). The detection of acceptor quenching would not have been possible without PIE, underpinning its crucial importance for accurate MFD experimentation.

Photobleaching of the acceptor fluorophore will result in artifacts during MFD analysis and has to be corrected for. Typically, bursts with acceptor photobleaching are removed by thresholding the average burst arrival time in the F_{GG} and F_{GR} channels (Eggeling et al., 2006). Monitoring the donor and acceptor fluorescence independently with MFD–PIE, however, allows thresholding based on the average burst arrival time in the F_{GG+GR} and F_{RR} channels. This allows removal of bursts where the acceptor has photobleached while keeping molecules that exhibit FRET dynamics during the diffusion through the probe volume, which would be removed when thresholding based on the average burst arrival in the F_{GG} and F_{GR} channels (Kudryavtsev et al., 2012).

In summary, spFRET with MFD–PIE really combines the best of two worlds: (i) the specific species resolving power of ALEX and (ii) the robust analysis power of spFRET with MFD. With PIE, information of the intensity, lifetime, and anisotropy is available for both the donor and the acceptor fluorophores. Restriction of burst analysis to dilute concentrations can also be reduced to some extent by performing measurements in nanoapertures (Fore, Yuen, Hesselink, & Huser, 2007).

7. SCANNING FCCS WITH PIE

Measuring concentrations, dynamics, and interactions of biomolecules accurately in confined, microscopic environments, such as a single cell, presents a great challenge for fluorescence fluctuation spectroscopy. One of the reasons for this is the limited photostability of the fluorophore. First, diffusion is often much slower in such an environment, exposing the fluorophores to prolonged illumination when in the probe volume. Second, eukaryotic cells, for example, have a roughly 5 order of magnitude smaller volume than a typical *in vitro* sample, providing a smaller total number of molecules for ensemble averaging. Thirdly, it is often necessary to use FPs for labeling in live-cell experiments. Finally, besides being less photostable than the dyes typically used for *in vitro* FFS experiments, the FPs are known to exhibit reversible or irreversible light-enhanced dark states (Haupts, Maiti, Schwille, & Webb, 1998; Hendrix, Flors, Dedecker, Hofkens, & Engelborghs, 2008; Schwille, Kummer, Heikal, Moerner, & Webb, 2000). Both their overall lower photostability and complex photophysics can considerably hinder measurements of concentrations, dynamics and interactions with point FFS techniques in cells. One way to overcome these difficulties is to employ scanning FFS techniques. Scanning FCS, originally developed in the 1980s (Petersen, 1986), has been used to study dynamics in membranes (Ries & Schwille, 2006) and to measure absolute diffusion coefficients (Petrášek & Schwille, 2008). Raster image correlation spectroscopy (RICS) has been developed to measure the dynamics of proteins in cells (Digman et al., 2005). In these scanning techniques, the probe volume is continuously moved in a fast and defined manner through the sample, which decreases the illumination time of the fluorophore in the probe volume drastically. Thus, the effective "light stress" that the probe experiences is significantly reduced while retaining the statistical accuracy of the measurement by averaging over a larger area in the sample. Besides accurate determination of slow diffusion coefficients (Jaffiol, Blancquaert, Delon, & Derouard, 2006; Petrášek & Schwille, 2008; Ries & Schwille, 2006), it has also been shown that excitation intensity-related artifacts decrease considerably when using scanning based FFS techniques. This is particularly true when the dynamics of the measured fluorophores are very slow, as is often the case in cells. This makes scanning methods favorable over point-based techniques for measuring protein dynamics and interactions in cells.

The combination of PIE with scanning FCS allows multicolor scanning FCCS experiments to be performed with all the advantages of standard PIE–FCCS, that is, the straightforward analysis of data without any difficulties introduced by crosstalk, FRET or direct excitation. Beam scanning can be introduced on a point-FCS microscope by the introduction of a fast galvo-mirror X/Y-scanner in the excitation path (Fig. 9.9).

As a proof of principle, circle scanning PIE–FCCS has been applied *in vitro* to a DNA molecule double-labeled with two dyes that exhibit considerable crosstalk, Alexa 488 and Alexa 568 (Höller, 2011). A high cross-correlation amplitude was observed, indicating a high degree of double labeling (Fig. 9.10A). As a negative control, equimolar amounts of similar dyes, ATTO488 and Alexa 555, freely diffusion in solution were measured with circle scanning FCCS. Logically, these dyes should not exhibit any codiffusion. Without the application of PIE, however, a high false-positive cross-correlation signal was observed (Fig. 9.10B). When PIE is applied, the cross-correlation is reduced to zero (Fig. 9.10C). This simple application proves the power of PIE for removing artifacts also in scanning FCCS experiments.

The benefits of PIE are particularly useful for cellular measurements with FPs, where the spectral overlap may be more prevalent than for synthetic dyes. The FP pair, eGFP and mCherry, is widely used for dual-color microscopy and spectroscopy in cells. Examples of such studies include point-based FCCS (Baudendistel, Müller, Waldeck, Angel, & Langowski, 2005; Hendrix et al., 2011; Saito, Wada, Tamura, & Kinjo, 2004; Shi et al., 2009), dual-color PCH (Chen et al., 2005), image-based ccRICS (Digman, Wiseman, Horwitz, & Gratton, 2009), and fluorescence lifetime imaging (Padilla-Parra et al., 2009), to name just a few. Despite the relatively large spectral separation of these proteins, eGFP bleedthrough into the red detection channel is still a problem that can

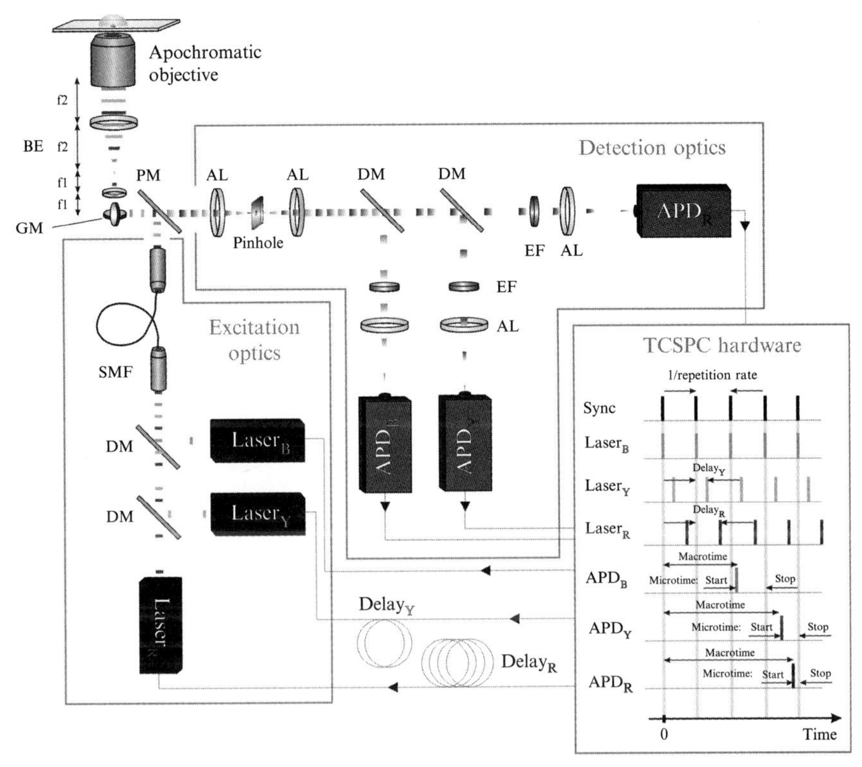

Figure 9.9 Schematic illustration of a triple-color excitation triple-color detection scanning PIE–FCCS microscope. Image scanning and descanning is possible by placing a two-dimensional fast galvanometric mirror (GM) scanner in the beam path. Furthermore, by placing the two focal planes of a Keplerian telescope at the GM and the back aperture of the objective, it is possible to illuminate the objective in the center of the back aperture at any position of the scanning mirrors. For further explanation of the setup and most of the abbreviations, see Fig. 9.2. (See Color Insert.)

render quantitative dual-color analysis difficult to impossible. Here, we applied PIE to image-based dual-color RICS (Digman et al., 2005, 2009). A mixture of eGFP and mCherry was expressed in HeLa cells, the cells were imaged and auto- and cross-correlation RICS analyses were performed. When calculating the confocal fluorescence images from all photons detected on the green or red detector and calculating the two-dimensional RICS CCF, a false-positive cross-correlation amplitude was observed due to spectral crosstalk (Fig. 9.11A, "CCF, no PIE"). However, calculation of the autocorrelation and cross-correlation functions for RICS from the crosstalk-free $F_{GG}(t)$ and $F_{RR}(t)$ PIE channels, the false cross-correlation disappeared completely (Fig. 9.11A,

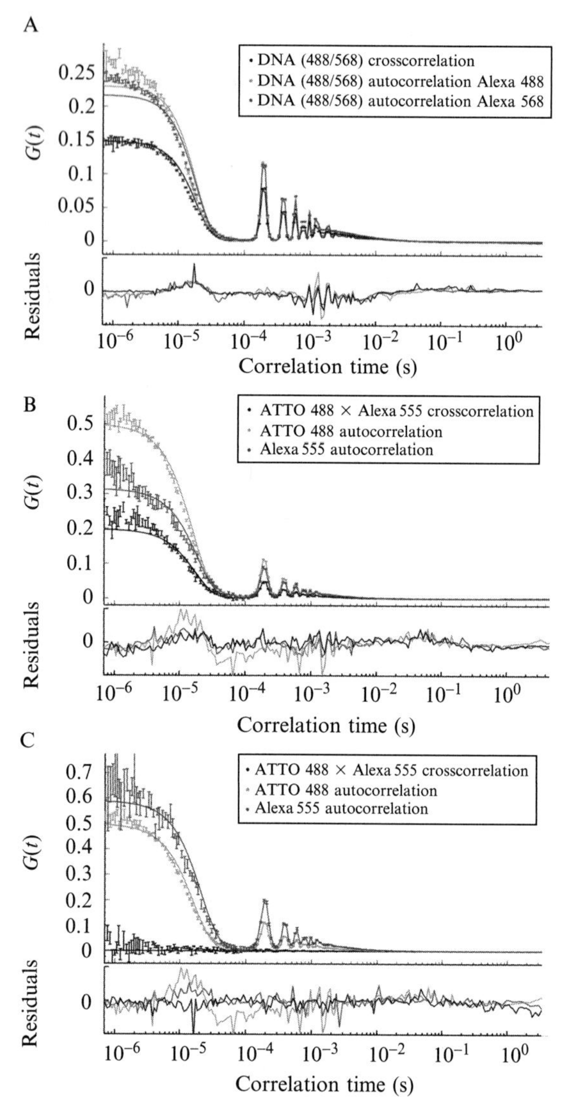

Figure 9.10 (A) Scanning FCCS curves and fits for a double-stranded DNA labeled with Alexa 488 and Alexa 568 in aqueous solution. Light grey curve: Alexa 488 autocorrelation. Dark gray curve: Alexa 568 autocorrelation. Black curve: Alexa 488/Alexa 568 cross-correlation function. The two fluorophores always diffuse as a pair due to their colocalization on the DNA strand, causing a non-zero cross-correlation. (B) Scanning FCS curves and fits for a mixture of ATTO 488 and Alexa 555 in aqueous solution analyzed without PIE. Light grey curve: ATTO 488 ACF. Dark gray curve: Alexa 555 ACF. Black curve: ATTO488/Alexa 555 CCF. The cross-correlation curve has a non-zero amplitude due to spectral cross-talk. (C) Scanning FCS curves and fits for a mixture of ATTO 488 and Alexa 555 in aqueous solution analyzed with PIE. Light gray curve: ATTO 488

"CCF, PIE"). Thus, for noninteracting molecules such as eGFP and mCherry, there is no cross-correlation. When applying PIE–RICS to cells expressing an eGFP–mCherry tandem fusion construct, a clear cross-correlation was observed, even when the calculation is based on the crosstalk-free PIE channels (Fig. 9.11B). Thus, also for quantitative image-based spectroscopy techniques, PIE holds great promise.

Finally, *Z*-scanning FCS is a very good technique to study diffusion in two-dimensional environments such as the membrane and recently, a dual-color *Z*-scan FCS with PIE has been used to study the complete independent diffusion patterns of membrane-bound phospholipids and proteins (Štefl, Kułakowska, & Hof, 2009).

8. GLANCING AT THE FUTURE FOR MULTICOLOR SPECTROSCOPY

In the previous parts of this chapter, the principles and applications of PIE were, for the most part, explained for dual-color spectroscopy and microscopy. In this final section, the possibilities for multicolor applications are reviewed. In principle, any multicolor fluorescence spectroscopy technique can benefit from the advantages of alternating excitation. The most straightforward, but at the same time, the most prevalent example, is everyday multicolor imaging of fluorescently-labeled samples. To avoid spectral bleedthrough, such imaging is normally performed by exciting and detecting each fluorophore separately. However, real-time multicolor imaging of molecules exhibiting dynamics on a timescale faster than the imaging time requires the different fluorophores to be imaged simultaneously. With fast laser alternation techniques, multicolor bleedthroughless imaging is possible without sacrificing imaging time or dynamical information.

With pulsed lasers becoming all the more common and with the advantages of the alternating excitation techniques clearly proven in literature, it makes sense to incorporate such technology also in conventional (commercial) confocal laser scanning microscopes employed by large number of scientists from many fields. Additionally, as explained before, next to temporal separation of emission, the alternating laser scheme also allows separation of

ACF. Dark gray curve: Alexa 555 ACF. Black curve: ATTO 488/Alexa 555 CCF. The cross-correlation curve has zero amplitude. *Adapted from Höller (2011).* (For interpretation of the references to color in this figure legend, the reader is referred to the online version of this chapter.)

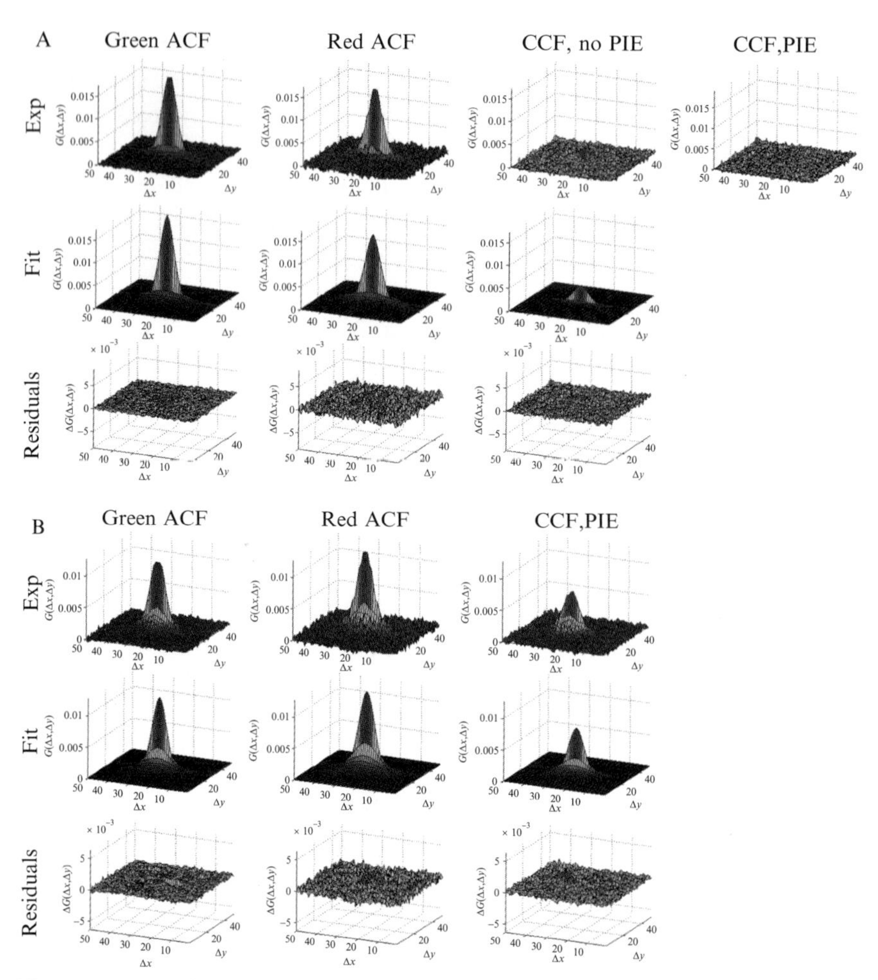

Figure 9.11 Proof of principle for measuring protein–protein interactions in cells with dual-color RICS. (A) Coexpression of eGFP and mCherry in HeLa cells and (B) expression of a genetic fusion construct, eGFP–mCherry. Measurements were performed on a home-built scanning PIE microscope similar to Fig. 9.9. More specifically, an Er-doped fiber laser with a fixed repetition rate of 27.4 MHz (Toptica Photonics, Munich, Germany) was used to excite mCherry at 561 nm and was delayed ~16 ns with respect to a pulsed 475 nm diode laser (Picoquant, Berlin, Germany) that was used to excite eGFP. Laser powers of 2 μW as measured between the excitation dichroic and objective (~1 μW in solution) were used. The confocal pinhole was 80 μm, the image size was 12.5 μm with a 300 × 300-pixel resolution, and image time was 1 s. The displayed RICS images (*upper panels*) are averages of RICS images from 100 consecutive image frames, corrected with a 10-frame moving average to remove the immobile fraction and with the (0,0) point omitted. Analysis was performed with a standard model for ccRICS analysis (Digman et al., 2009) in the 50 × 50-pixel central region of the RICS image. For the "CCF, no PIE" image, the macrotimes for photons from $F_{GG+RG}(t)$ and $F_{GR+RR}(t)$ PIE

fluorophores via the excitation source. This makes it possible to detect different fluorophores on the same detector, allowing for multiplexing of different fluorophores. The recent report of multiparameter wide-field FLIM with PIE also holds great promise for imaging multiple fluorophores in a time-resolved manner (Vitali et al., 2011). Recent technological advances in site-specific protein labeling with unnatural amino acids and the continuous development of new fluorophores and FPs will further smooth the way for multicolor applications.

With respect to quantitative FFS, we have given a detailed description of quantitative FCCS analysis and have shown that PIE increases the robustness of experimentation considerably. The large majority of FCCS (and similar dual-color FFS) experiments in literature have been performed without the use of PIE. Indeed, we have shown that quantitative analysis is still possible by measuring the necessary control parameters and by limiting experimental nonidealities with ingenious microscope design and appropriate experimental conditions and controls. The situation is quite different when going from dual- to triple-color FCCS. Besides the need for a more complicated instrumentation, the inherent complexity caused by the same nonidealities as for dual-color FCCS render quantitative analysis without the use of PIE impractical. The application of triple-color FCCS has thus mostly been limited to a qualitative description of interactions (Heinze, Jahnz, & Schwille, 2004; Hwang, Gosch, Lasser, & Wohland, 2006; Ridgeway, Millar, & Williamson, 2012). By choosing fluorophores appropriately, triple-color PIE–FCCS can be applied in the same straightforward manner as its dual-color variant. As a proof of principle, we shortly describe the concomitant interaction of the DNA methyltransferase 1 (Dnmt1) with different DNA substrates. To investigate whether Dnmt1 can interact with more than one DNA molecule at the same time, a purified GFP-labeled Dnmt1 was incubated with ATTO 565- and ATTO 700-labeled DNA and triple-color PIE–FCCS experiments were performed. Clearly, the positive cross-correlation observed between the three different PIE detection channels allows one to directly conclude that Dnmt1 can bind more than one DNA substrate at the same time (Fig. 9.12).

With fluorophores spanning the entire visual spectrum and even expanding into the near-IR, four- or five-color PIE–FCCS is also imaginable.

channels were used to calculate the "green" and "red" images needed for ccRICS. For the "CCF, PIE" image, the macrotimes for photons from $F_{GG}(t)$ and $F_{RR}(t)$ PIE channels were used to calculate the "green" and "red" images for ccRICS. The fits (*middle panels*) and residuals (*lower panels*) are shown. (See Color Insert.)

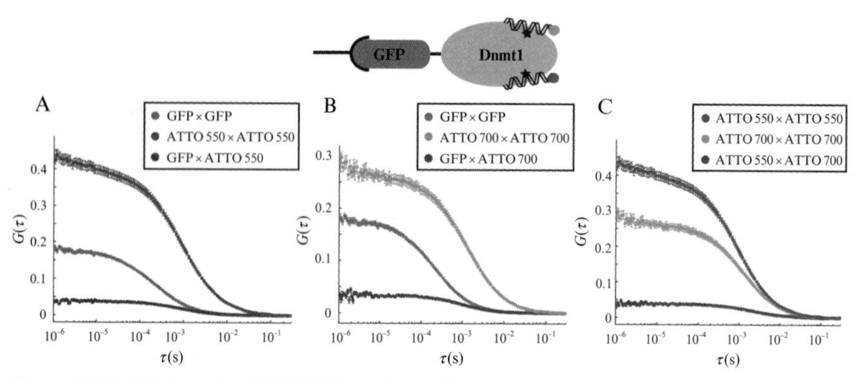

Figure 9.12 Triple-color PIE–FCCS analysis of GFP–Dnmt1 incubated with DNA-ATTO 550 and DNA-ATTO 700. The nonzero CCF amplitude between: (A) GFP and ATTO 550, (B) GFP and ATTO 700, and (C) ATTO 550 and ATTO 700 unambiguously prove the presence of protein–DNA complexes with at least two different DNA substrates bound. Autocorrelation curves are shown for amplitude comparison. *Adapted from Höller (2011).* (For color version of this figure, the reader is referred to the online version of this chapter.)

A clever choice of excitation wavelengths, pulse sequence, clock rate, and interleaving time depending on the spectrum and lifetime of the fluorophores makes such applications well possible, without sacrificing the unique strength of PIE for avoiding emission crosstalk.

In recent years, the spFRET technique with μsALEX has also expanded to three- and even four-color analysis, which provides the possibility for sorting molecules according to the fluorophore content (Lee et al., 2007, 2010). Since in three- and four-color FRET, four and six different FRET pairs are formed, respectively, interfluorophore distances might be hard to quantify in all-fluorophore-labeled molecules. Here, the ability of alternating excitation techniques to sort molecules according to the presence of certain labels, interfluorophore distances can be measured on a fluorophore labeling specific basis. Also here, an important place for multicolor PIE exists, since, on the one hand, MFD with PIE at present is one of the most robust spFRET methodologies and on the other hand, since it is amenable to multicolor molecule sorting without adding unnecessary complexity to the analysis.

ACKNOWLEDGMENTS

We thank Dr. Carina Frauer and Prof. Heinrich Leonhardt for providing the Dnmt1 protein and Dr. Matthias Höller for performing the three-color PIE–FCCS experiments. We gratefully acknowledge the financial support of the Deutsche Forschungsgemeinschaft through the Excellence Cluster Nanosystems Initiative Munich (NIM), and the Collaborate Research Center (SFB646), and the Ludwig-Maximilians-University, Munich (LMU innovativ BioImaging Network). J. H. wishes to thank the Research Foundation—Flanders (FWO) for a postdoctoral fellowship and travel grant.

REFERENCES

Antonik, M., Felekyan, S., Gaiduk, A., & Seidel, C. A. (2006). Separating structural heterogeneities from stochastic variations in fluorescence resonance energy transfer distributions via photon distribution analysis. *The Journal of Physical Chemistry B, 110*, 6970–6978.

Aragon, S. R., & Pecora, R. (1975). Fluorescence correlation spectroscopy and Brownian rotational diffusion. *Biopolymers, 14*, 119–137.

Baudendistel, N., Müller, G., Waldeck, W., Angel, P., & Langowski, J. (2005). Two-hybrid fluorescence cross-correlation spectroscopy detects protein-protein interactions, in vivo. *ChemPhysChem, 6*, 984–990.

Becker, W., Hickl, H., Zander, C., Drexhage, K. H., Sauer, M., Siebert, S., et al. (1999). Time-resolved detection and identification of single analyte molecules in microcapillaries by time-correlated single-photon counting (TCSPC). *The Review of Scientific Instruments, 70*, 1835.

Böhmer, M., Wahl, M., Rahn, H. J., Erdmann, R., & Enderlein, J. (2002). Time-resolved fluorescence correlation spectroscopy. *Chemical Physics Letters, 353*, 439–445.

Chakraborty, K., Chatila, M., Sinha, J., Shi, Q., Poschner, B. C., Sikor, M., et al. (2010). Chaperonin-catalyzed rescue of kinetically trapped states in protein folding. *Cell, 142*, 112–122.

Chen, Y., Müller, J. D., So, P. T., & Gratton, E. (1999). The photon counting histogram in fluorescence fluctuation spectroscopy. *Biophysical Journal, 77*, 553–567.

Chen, Y., Tekmen, M., Hillesheim, L., Skinner, J., Wu, B., & Müller, J. D. (2005). Dual-color photon-counting histogram. *Biophysical Journal, 88*, 2177–2192.

Cheung, H. C. (1991). Resonance energy transfer. In J. R. Lakowicz (Ed.), *Topics in fluorescence spectroscopy*, Vol. 2, (pp. 127–176). New York: Plenum Press.

Clegg, R. M. (1992). Fluorescence resonance energy transfer and nucleic acids. *Methods in Enzmology, 211*, 353–388.

Deniz, A. A., Dahan, M., Grunwell, J. R., Ha, T. J., Faulhaber, A. E., Chemla, D. S., et al. (1999). Single-pair fluorescence resonance energy transfer on freely diffusing molecules: Observation of Forster distance dependence and subpopulations. *Proceedings of the National Academy of Sciences of the United States of America, 96*, 3670–3675.

Dertinger, T., Pacheco, V., von der Hocht, I., Hartmann, R., Gregor, I., & Enderlein, J. (2007). Two-focus fluorescence correlation spectroscopy: A new tool for accurate and absolute diffusion measurements. *ChemPhysChem, 8*, 433–443.

Digman, M. A., Brown, C. M., Sengupta, P., Wiseman, P. W., Horwitz, A. R., & Gratton, E. (2005). Measuring fast dynamics in solutions and cells with a laser scanning microscope. *Biophysical Journal, 89*, 1317–1327.

Digman, M. A., Wiseman, P. W., Horwitz, A. R., & Gratton, E. (2009). Detecting protein complexes in living cells from laser scanning confocal image sequences by the cross correlation raster image spectroscopy method. *Biophysical Journal, 96*, 707–716.

Eggeling, C., Berger, S., Brand, L., Fries, J. R., Schaffer, J., Volkmer, A., et al. (2001). Data registration and selective single-molecule analysis using multi-parameter fluorescence detection. *Journal of Biotechnology, 86*, 163–180.

Eggeling, C., Widengren, J., Brand, L., Schaffer, J., Felekyan, S., & Seidel, C. A. M. (2006). Analysis of photobleaching in single-molecule multicolor excitation and Forster resonance energy transfer measurement. *The Journal of Physical Chemistry A, 110*, 2979–2995.

Elson, E. L., & Magde, D. (1974). Fluorescence correlation spectroscopy. 1. Conceptual basis and theory. *Biopolymers*, *13*, 1–27.

Felekyan, S., Kalinin, S., Sanabria, H., Valeri, A., & Seidel, C. A. (2012). Filtered FCS: Species auto- and cross-correlation functions highlight binding and dynamics in biomolecules. *ChemPhysChem*, *13*, 1036–1053.

Ferreon, A. C. M., Gambin, Y., Lemke, E. A., & Deniz, A. A. (2009). Interplay of -synuclein binding and conformational switching probed by single-molecule fluorescence. *Proceedings of the National Academy of Sciences*, *106*, 5645–5650.

Foldes-Papp, Z. (2005). How the molecule number is correctly quantified in two-color fluorescence cross-correlation spectroscopy: Corrections for cross-talk and quenching in experiments. *Current Pharmaceutical Biotechnology*, *6*, 437–444.

Foo, Y. H., Naredi-Rainer, N., Lamb, D. C., Ahmed, S., & Wohland, T. (2012). Factors affecting the quantification of biomolecular interactions by fluorescence cross-correlation spectroscopy. *Biophysical Journal*, *102*, 1174–1183.

Fore, S., Yuen, Y., Hesselink, L., & Huser, T. (2007). Pulsed-interleaved excitation FRET measurements on single duplex DNA molecules inside C-shaped nanoapertures. *Nano Letters*, *7*, 1749–1756.

Förster, T. (1948). Zwischenmolekulare Energiewanderung und Fluoreszenz. *Annalen der Physik*, *437*, 55–75.

Garcia-Saez, A. J., Ries, J., Orzaez, M., Perez-Paya, E., & Schwille, P. (2009). Membrane promotes tBID interaction with BCL(XL). *Nature Structural and Molecular Biology*, *16*, U1178 U9.

Haupts, U., Maiti, S., Schwille, P., & Webb, W. W. (1998). Dynamics of fluorescence fluctuations in green fluorescent protein observed by fluorescence correlation spectroscopy. *Proceedings of the National Academy of Sciences of the United States of America*, *95*, 13573–13578.

Heinze, K. G., Jahnz, M., & Schwille, P. (2004). Triple-color coincidence analysis: One step further in following higher order molecular complex formation. *Biophysical Journal*, *86*, 506–516.

Hendrix, J., Flors, C., Dedecker, P., Hofkens, J., & Engelborghs, Y. (2008). Dark states in monomeric red fluorescent proteins studied by fluorescence correlation and single molecule spectroscopy. *Biophysical Journal*, *94*, 4103–4113.

Hendrix, J., Gijsbers, R., De Rijck, J., Voet, A., Hotta, J., McNeely, M., et al. (2011). The transcriptional co-activator LEDGF/p75 displays a dynamic scan-and-lock mechanism for chromatin tethering. *Nucleic Acids Research*, *39*, 1310–1325.

Hillger, F., Hänni, D., Nettels, D., Geister, S., Grandin, M., Textor, M., et al. (2008). Probing protein-chaperone interactions with single-molecule fluorescence spectroscopy. *Angewandte Chemie International Edition*, *47*, 6184–6188.

Hofmann, H., Hillger, F., Pfeil, S. H., Hoffmann, A., Streich, D., Haenni, D., et al. (2010). Single-molecule spectroscopy of protein folding in a chaperonin cage. *Proceedings of the National Academy of Sciences of the United States of America*, *107*, 11793–11798.

Höller, M. (2011). Advanced fluorescence fluctuation spectroscopy with pulsed interleaved excitation. Development and applications. Fakultät für Chemie und Pharmazie, Ludwig-Maximilians-Universität München, pp. 135. http://www.cup.uni-muenchen.de/pc/lamb/pdf/Hoeller_Matthias.pdf.

Hwang, L. C., Gosch, M., Lasser, T., & Wohland, T. (2006). Simultaneous multicolor fluorescence cross-correlation spectroscopy to detect higher order molecular interactions using single wavelength laser excitation. *Biophysical Journal*, *91*, 715–727.

Hwang, L. C., & Wohland, T. (2005). Single wavelength excitation fluorescence cross-correlation spectroscopy with spectrally similar fluorophores: Resolution for binding studies. *The Journal of Chemical Physics*, *122*, 114708.

Ivanchenko, S., & Lamb, D. C. (2011). Fluorescence correlation spectroscopy: Principles and developments. In J. Brnjas-Kraljevic & G. Pifat-Mrzljak (Eds.), *Supramolecular structure and function*, Vol. 10, (pp. 1–30). Berlin, Germany: Springer-Verlag.

Jaffiol, R., Blancquaert, Y., Delon, A., & Derouard, J. (2006). Spatial fluorescence cross-correlation spectroscopy. *Applied Optics*, *45*, 1225–1235.

Kapanidis, A. N., Laurence, T. A., Lee, N. K., Margeat, E., Kong, X., & Weiss, S. (2005). Alternating-laser excitation of single molecules. *Accounts of Chemical Research*, *38*, 523–533.

Kapanidis, A. N., Lee, N. K., Laurence, T. A., Doose, S., Margeat, E., & Weiss, S. (2004). Fluorescence-aided molecule sorting: Analysis of structure and interactions by alternating-laser excitation of single molecules. *Proceedings of the National Academy of Sciences of the United States of America*, *101*, 8936–8941.

Kapusta, P., Wahl, M., Benda, A., Hof, M., & Enderlein, J. (2007). Fluorescence lifetime correlation spectroscopy. *Journal of Fluorescence*, *17*, 43–48.

Kask, P., Palo, K., Ullmann, D., & Gall, K. (1999). Fluorescence-intensity distribution analysis and its application in biomolecular detection technology. *Proceedings of the National Academy of Sciences of the United States of America*, *96*, 13756–13761.

Kobitski, A. Y., Hengesbach, M., Helm, M., & Nienhaus, G. U. (2008). Sculpting an RNA conformational energy landscape by a methyl group modification—A single-molecule FRET study. *Angewandte Chemie International Edition*, *47*, 4326–4330.

Kudryavtsev, V., Sikor, M., Kalinin, S., Mokranjac, D., Seidel, C. A., & Lamb, D. C. (2012). Combining MFD and PIE for accurate single-pair Forster resonance energy transfer measurements. *ChemPhysChem*, *13*, 1060–1078.

Kukolka, F., Müller, B. K., Paternoster, S., Arndt, A., Niemeyer, C. M., Bräuchle, C., et al. (2006). A single-molecule Förster resonance energy transfer analysis of fluorescent DNA–protein conjugates for nanobiotechnology. *Small*, *2*, 1083–1089.

Lamb, D. C., Schenk, A., Rocker, C., Scalfi-Happ, C., & Nienhaus, G. U. (2000). Sensitivity enhancement in fluorescence correlation spectroscopy of multiple species using time-gated detection. *Biophysical Journal*, *79*, 1129–1138.

Laurence, T. A., Kong, X. X., Jager, M., & Weiss, S. (2005). Probing structural heterogeneities and fluctuations of nucleic acids and denatured proteins. *Proceedings of the National Academy of Sciences of the United States of America*, *102*, 17348–17353.

Lee, N., Kapanidis, A., Koh, H., Korlann, Y., Ho, S., Kim, Y., et al. (2007). Three-color alternating-laser excitation of single molecules: Monitoring multiple interactions and distances. *Biophysical Journal*, *92*, 303–312.

Lee, J., Lee, S., Ragunathan, K., Joo, C., Ha, T., & Hohng, S. (2010). Single-molecule four-color FRET. *Angewandte Chemie International Edition*, *49*, 9922–9925.

Magde, D., Webb, W. W., & Elson, E. (1972). Thermodynamic fluctuations in a reacting system—Measurement by fluorescence correlation spectroscopy. *Physical Review Letters*, *29*, 705–708.

Mapa, K., Sikor, M., Kudryavtsev, V., Waegemann, K., Kalinin, S., Seidel, C. A. M., et al. (2010). The conformational dynamics of the mitochondrial Hsp70 chaperone. *Molecular Cell*, *38*, 89–100.

Marcinowski, M., Höller, M., Feige, M. J., Baerend, D., Lamb, D. C., & Buchner, J. (2011). Substrate discrimination of the chaperone BiP by autonomous and cochaperone-regulated conformational transitions. *Nature Structural and Molecular Biology*, *18*, 150–158.

Margittai, M., Widengren, J., Schweinberger, E., Schroder, G. F., Felekyan, S., Haustein, E., et al. (2003). Single-molecule fluorescence resonance energy transfer reveals a dynamic equilibrium between closed and open conformations of syntaxin 1. *Proceedings of the National Academy of Sciences of the United States of America*, *100*, 15516–15521.

Müller, J. D. (2004). Cumulant analysis in fluorescence fluctuation spectroscopy. *Biophysical Journal*, *86*, 3981–3992.

Müller, B. K. (2007). Die gepulste alternierende Anregung in der konfokalen Fluoreszenzspektroskopie. Entwicklung und Anwendungen. Fakultät für Chemie und Pharmazie, Ludwig-Maximilians-Universität München, pp. 164. http://www.cup.uni-muenchen.de/pc/lamb/pdf/Mueller_Barbara_PhD_Thesis.pdf.

Müller, C. B., Loman, A., Pacheco, V., Koberling, F., Willbold, D., Richtering, W., et al. (2008). Precise measurement of diffusion by multi-color dual-focus fluorescence correlation spectroscopy. *Optics Express*, *16*, 14353–14368.

Müller, B. K., Reuter, A., Simmel, F. C., & Lamb, D. C. (2006). Single-pair FRET characterization of DNA tweezers. *Nano Letters*, *6*, 2814–2820.

Müller, B. K., Zaychikov, E., Bräuchle, C., & Lamb, D. C. (2005). Pulsed interleaved excitation. *Biophysical Journal*, *89*, 3508–3522.

Padilla-Parra, S., Auduge, N., Lalucque, H., Mevel, J. C., Coppey-Moisan, M., & Tramier, M. (2009). Quantitative comparison of different fluorescent protein couples for fast FRET-FLIM acquisition. *Biophysical Journal*, *97*, 2368–2376.

Petersen, N. O. (1986). Scanning fluorescence correlation spectroscopy. I. Theory and simulation of aggregation measurements. *Biophysical Journal*, *49*, 809–815.

Petrášek, Z., & Schwille, P. (2008). Precise measurement of diffusion coefficients using scanning fluorescence correlation spectroscopy. *Biophysical Journal*, *94*, 1437–1448.

Price, E. S., DeVore, M. S., & Johnson, C. K. (2010). Detecting intramolecular dynamics and multiple Forster resonance energy transfer states by fluorescence correlation spectroscopy. *The Journal of Physical Chemistry B*, *114*, 5895–5902.

Ridgeway, W. K., Millar, D. P., & Williamson, J. R. (2012). The spectroscopic basis of fluorescence triple correlation spectroscopy. *The Journal of Physical Chemistry B*, *116*, 1908–1919.

Ries, J., & Schwille, P. (2006). Studying slow membrane dynamics with continuous wave scanning fluorescence correlation spectroscopy. *Biophysical Journal*, *91*, 1915–1924.

Rigler, R., Foldes-Papp, Z., Meyer-Almes, F. J., Sammet, C., Volcker, M., & Schnetz, A. (1998). Fluorescence cross-correlation: A new concept for polymerase chain reaction. *Journal of Biotechnology*, *63*, 97–109.

Rosenkranz, T., Schlesinger, R., Gabba, M., & Fitter, J. (2011). Native and unfolded states of phosphoglycerate kinase studied by single-molecule FRET. *ChemPhysChem*, *12*, 704–710.

Rothwell, P. J., Berger, S., Kensch, O., Felekyan, S., Antonik, M., Wohrl, B. M., et al. (2003). Multiparameter single-molecule fluorescence spectroscopy reveals heterogeneity of HIV-1 reverse transcriptase:primer/template complexes. *Proceedings of the National Academy of Sciences of the United States of America*, *100*, 1655–1660.

Saito, K., Wada, I., Tamura, M., & Kinjo, M. (2004). Direct detection of caspase-3 activation in single live cells by cross-correlation analysis. *Biochemical and Biophysical Research Communications*, *324*, 849–854.

Schwille, P. (2001). Cross-correlation analysis in FCS. In E. S., Elson, and R. Rigler (Eds.), In Fluorescence correlation spectroscopy. Theory and applications., (pp. 360–378). Berlin. Springer-Verlag.

Schwille, P., Kummer, S., Heikal, A. A., Moerner, W. E., & Webb, W. W. (2000). Fluorescence correlation spectroscopy reveals fast optical excitation-driven intramolecular dynamics of yellow fluorescent proteins. *Proceedings of the National Academy of Sciences of the United States of America*, *97*, 151–156.

Schwille, P., Meyer-Almes, F. J., & Rigler, R. (1997). Dual-color fluorescence cross-correlation spectroscopy for multicomponent diffusional analysis in solution. *Biophysical Journal*, *72*, 1878–1886.

Sharma, S., Chakraborty, K., Müller, B. K., Astola, N., Tang, Y.-C., Lamb, D. C., et al. (2008). Monitoring protein conformation along the pathway of chaperonin-assisted folding. *Cell*, *133*, 142–153.

Shi, X., Foo, Y. H., Sudhaharan, T., Chong, S.-W., Korzh, V., Ahmed, S., et al. (2009). Determination of dissociation constants in living zebrafish embryos with single wavelength fluorescence cross-correlation spectroscopy. *Biophysical Journal*, *97*, 678–686.

Sikor, M. (2011). Single-molecule fluorescence studies of protein folding and molecular chaperones. Fakultät für Chemie und Pharmazie, Ludwig-Maximilians-Universität München, pp. 142. http://www.cup.uni-muenchen.de/pc/lamb/pdf/Sikor_Martin_Dissertation.pdf.

Štefl, M., Kułakowska, A., & Hof, M. (2009). Simultaneous characterization of lateral lipid and prothrombin diffusion coefficients by Z-scan fluorescence correlation spectroscopy. *Biophysical Journal, 97*, L1–L3.

Takahashi, Y., Nishimura, J., Suzuki, A., Ishibashi, K., Kinjo, M., & Miyawaki, A. (2008). Cross-talk-free fluorescence cross-correlation spectroscopy by the switching method. *Cell Structure and Function, 33*, 143–150.

Thews, E., Gerken, M., Eckert, R., Zäpfel, J., Tietz, C., & Wrachtrup, J. (2005). Cross talk free fluorescence cross correlation spectroscopy in live cells. *Biophysical Journal, 89*, 2069–2076.

Thompson, N. L. (1991). Fluorescence correlation spectroscopy. In J. R. Lakowicz (Ed.), *Topics in fluorescence spectroscopy*, Vol. 1. New York: Plenum Press.

Torres, T., & Levitus, M. (2007). Measuring conformational dynamics: A new FCS-FRET approach. *The Journal of Physical Chemistry B, 111*, 7392–7400.

Vitali, M., Picazo, F., Prokazov, Y., Duci, A., Turbin, E., Gotze, C., et al. (2011). Wide-field multi-parameter FLIM: Long-term minimal invasive observation of proteins in living cells. *PLoS One, 6*, e15820.

Widengren, J., Kudryavtsev, V., Antonik, M., Berger, S., Gerken, M., & Seidel, C. A. (2006). Single-molecule detection and identification of multiple species by multiparameter fluorescence detection. *Analytical Chemistry, 78*, 2039–2050.

Widengren, J., Schweinberger, E., Berger, S., & Seidel, C. A. M. (2001). Two new concepts to measure fluorescence resonance energy transfer via fluorescence correlation spectroscopy: Theory and experimental realizations. *The Journal of Physical Chemistry A, 105*, 6851–6866.

Zander, C., Sauer, M., Drexhage, K. H., Ko, D. S., Schulz, A., Wolfrum, J., et al. (1996). Detection and characterization of single molecules in aqueous solution. *Applied Physics B-Lasers and Optics, 63*, 517–523.

Zarrabi, N., Ernst, S., Duser, M. G., Golovina-Leiker, A., Becker, W., Erdmann, R., et al. (2009). Simultaneous monitoring of the two coupled motors of a single F(o)F(1)-ATP synthase by three-color FRET using duty cycle-optimized triple-ALEX. *Single Molecule Spectroscopy and Imaging II, 7185.*

CHAPTER TEN

Image Correlation Spectroscopy: Mapping Correlations in Space, Time, and Reciprocal Space

Paul W. Wiseman[1]
Department of Physics, McGill University, Montreal, Quebec, Canada
Department of Chemistry, McGill University, Montreal, Quebec, Canada
[1]Corresponding author: e-mail address: paul.wiseman@mcgill.ca

Contents

Abstract

This chapter presents an overview of two recent implementations of image correlation spectroscopy (ICS). The background theory is presented for spatiotemporal image correlation spectroscopy and image cross-correlation spectroscopy (STICS and STICCS, respectively) as well as *k*-(reciprocal) space image correlation spectroscopy (kICS). An introduction to the background theory is followed by sections outlining procedural aspects for properly implementing STICS, STICCS, and kICS. These include microscopy image collection, sampling in space and time, sample and fluorescent probe requirements, signal to noise, and background considerations that are all required to properly implement the ICS methods. Finally, procedural steps for immobile population removal and actual implementation of the ICS analysis programs to fluorescence microscopy image time stacks are described.

Methods in Enzymology, Volume 518
ISSN 0076-6879
http://dx.doi.org/10.1016/B978-0-12-388422-0.00010-8

1. INTRODUCTION

Image correlation spectroscopy (ICS) refers to a group of fluorescence fluctuation spectroscopy techniques that are based on analysis of fluorescence microscopy image data (Kolin & Wiseman, 2007) (i.e., from the spatial and temporal domains) in contrast to fluorescence correlation spectroscopy (FCS) which analyzes time domain fluorescence fluctuations excited from a single laser spot focused in the sample (Thompson, 1991). The fluorescence intensity fluctuations from the microscopy image series are analyzed using correlation functions in the different domains of space and time. An advantage of ICS is that fluorescence fluctuation data are sampled from multiple points in space and hence takes full advantage of the image in terms of mapping transport phenomena in cells. An inherent disadvantage is that time resolution is limited by the image acquisition frequency (usually 0.1–10 Hz range) so most ICS methods are restricted to sampling slower transport processes in cells, whereas FCS can measure faster transport since the fluorescence fluctuations are collected in time bins on the order of microseconds. An exception to this limitation is the raster scan ICS method (RICS) which does resolve the faster timescale dynamics by taking advantage of inherent rapid pixel-to-pixel sampling in a laser scanning microscope (Digman, Brown, et al., 2005; Digman, Sengupta, et al., 2005). However, the spatial sampling of fluorescence fluctuations in the image enables ICS to extract information from immobile populations and chemically fixed cells which is not possible with the exclusively time domain FCS.

The ICS methods are subclassified according to which domains of fluorescence fluctuation information are analyzed from the image series data. The original variant, now called spatial ICS, employed spatial autocorrelation functions to analyze the surface density and aggregation state distributions of fluorescently tagged cell surface receptors imaged by confocal laser scanning microscopy (CLSM) on chemically fixed cells (Petersen, Hoddelius, Wiseman, Seger, & Magnusson, 1993; Wiseman & Petersen, 1999). Extension to correlation analysis of the fluorescence fluctuations in time recorded in pixels of the image time series to measure slow transport of membrane receptors was next done for the temporal ICS method (Srivastava & Petersen, 1998; Wiseman, Squier, Ellisman, & Wilson, 2000). Image cross-correlation spectroscopy (ICCS) (Comeau, Costantino, & Wiseman, 2006) is the imaging extension of dual color fluorescence cross-correlation spectroscopy

(Schwille, Meyer-Almes, & Rigler, 1997) which can be used to measure colocalization fractions for two different fluorescently tagged species by cross-correlation analysis in space or time between the two different imaging detection channels.

Recently, the ICS methods have been extended for combined spatiotemporal image correlation spectroscopy (STICS) which correlates fluorescence intensity fluctuations in space and time to add sensitivity to flow directions of fluorescently tagged macromolecules in cells enabling the measurement of vector maps in cells (Brown et al., 2006; Hebert, Costantino, & Wiseman, 2005). As well, the extension to reciprocal (*k*-) space–time correlation was introduced as kICS which has advantages for measurement of transport dynamics in cases where the nonstationary photophysics of the fluorophore complicates the fluctuation measurements (Kolin, Ronis, & Wiseman, 2006). This review of the ICS methods will provide an introduction to the background theory of ICS and focus on the implementation of the more recent STICS and kICS approaches.

2. THEORY OF ICS

All variants of ICS are based on correlation analysis of fluorescence fluctuations of intensity recorded in the pixels of images acquired by some variant of fluorescence microscopy. The analysis involves calculation of auto and cross-correlation functions in space and/or time or reciprocal space–time correlation functions from the fluorescence intensity record in the image series. As such, there are certain basic concepts that are essential to understanding and applying all forms of ICS. The following sections will provide the required background.

2.1. General theory of ICS and ICCS

The seminal measurement for all forms of ICS is an image or image time series measured using some form of fluorescence microscopy including confocal or two-photon laser scanning microscopy and total internal reflection fluorescence (TIRF) microscopy (Fig. 10.1A). The image is made up of pixels where the pixel intensity is the integrated digital conversion of the analog output from a photomultiplier tube (PMT) or bin counts from a CCD camera (in some more recent commercial CLSM instruments, actual photon counting detectors are used, and in this case, the pixel represents a true count of detected photons as opposed to an intensity value). It is very important to remember that integrated fluorescence intensity from a point

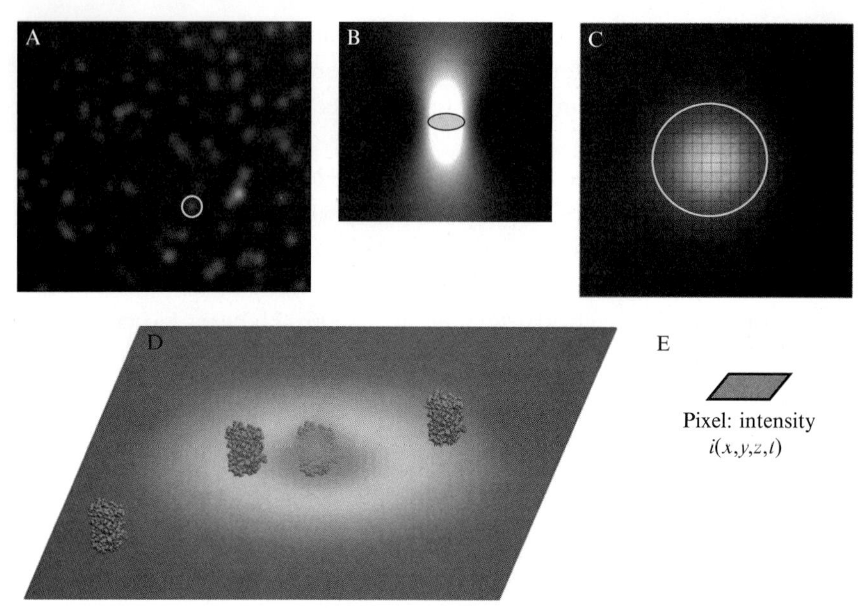

Figure 10.1 (A) Typical fluorescence microscopy image used as input for ICS analysis with the yellow circle highlighting fluorescence signal within a focal spot region which would contribute a positive intensity fluctuation. (B) The 3D point spread function (PSF) for a focused Gaussian beam. The gray disk illustrates the 2D cross section at focus. (C) A representation of a fluorescence signal fluctuation that is oversampled onto pixels in the image. The yellow circle shows the 2D PSF size illustrating that there will be spatial correlation between adjacent pixels over a scale defined by the PSF. (D) A cartoon representation of the molecular contributions of fluorescent molecules within the PSF-defined region to an integrated fluorescence signal as per (C). Note that the PSF would be about 100 × larger for standard fluorescence microscopy. (E) A cartoon representation of an image pixel showing the integrated intensity value derived from photons emitted and detected from across the PSF-defined region. (See Color Insert.)

source fluorescent molecule will be spread out in 3D space upon detection due to the inherent diffraction of light. The mapping of the spatial extent of this diffraction pattern is referred to as the optical point spread function (PSF) of the microscope and is shown schematically in Fig. 10.1B for a plot of a simulated 3D Gaussian laser focus. In the focal plane, the PSF cross section will be Gaussian for a TEM_{00} laser beam at focus, as well we typically approximate the optical PSF in TIRF microscopy or for measurements on a quasi-2D membrane system as a Gaussian function (Fig. 10.2C). Thus, any detected fluorescent molecular point source will always appear to be the size of the PSF diffraction spot for an optimally aligned fluorescence

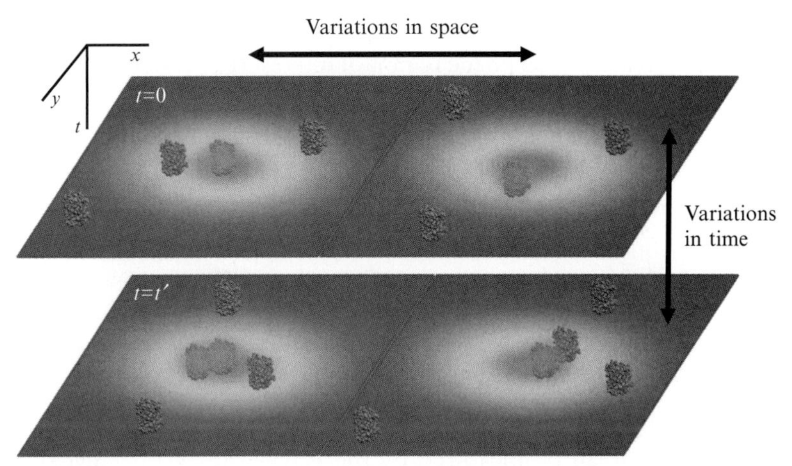

Figure 10.2 A molecular cartoon representing variations of molecules in space and time within the PSF. The variations in space at a given time would lead to fluorescence intensity fluctuations as the PSF area moves from pixel to pixel in a single image and the photons emitted by fluorophores in the PSF are integrated and converted to a pixel value. The variations in time would represent changes in the number of emitting fluorophores in the same PSF area due to molecular transport or photophysical changes in emission during the period between pixel integration for a given image pixel at different times. Note that the PSF would be about 100 × larger for standard fluorescence microscopy. (See Color Insert.)

microscope (Fig. 10.1A, yellow circle enclosing a single spot). If there are multiple fluorescent species in a region defined by the optical PSF (Fig. 10.2D), then the integrated intensity at an image pixel $i(x, y, z, t)$ (Fig. 10.2E) will represent the PSF-weighted emission detected during the pixel dwell time (LSM) or integration time (CCD) (illustrated in 2D for simplicity). To encode spatial correlations in ICS measurements, it is essential that the image pixel size be smaller than the diameter of the PSF in order for spatial intensity correlations to exist between adjacent pixels in an image (Fig. 10.2, pixel grid vs. yellow e^{-2} Gaussian PSF radius).

In any ICS measurement, correlations in fluctuations of the fluorescence intensity are analyzed by some form of correlation function. An intensity fluctuation, δi, is defined simply as the difference between the pixel intensity at some point in space and time and the mean intensity:

$$\delta i_{\mathrm{a}}(x, y, t) = i_{\mathrm{a}}(x, y, t) - \langle i_{\mathrm{a}} \rangle \qquad [10.1]$$

where a represents a single detection channel and the angular brackets represent an average. In practice, the mean intensity can be defined in a number of different ways depending on the form of ICS that is employed including spatial mean intensity of an image region of interest (ROI) or temporal mean intensity from a pixel stack in a time series.

Fluctuations can be classified as spatial intensity fluctuations across a given image or temporal intensity fluctuations between images taken at different times in a time series stack (Fig. 10.2). Spatial fluctuations across an image reflect the variations in the numbers of static or slowly moving fluorophores distributed in space, while temporal fluctuations arise from dynamic changes in the number of fluorophores at a given spot over time as the fluorescent molecules undergo transport (diffusion, flow, anomalous diffusion, etc.) or undergo nonstationary emission photophysics (photobleaching, probe blinking, etc.). In principle, a coupling of spatial and temporal fluctuations can be measured by the RICS variant on LSM systems which build up the image sequentially pixel by pixel by raster scanning the laser beam repeatedly across the sample (Digman, Brown, et al., 2005). If there is rapid molecular transport or dynamics that occurs on the timescales of the fast beam raster scan (horizontal pixel-to-pixel dwell time; typically microseconds) and slower line scan (vertical line to line; typically ms), then RICS can detect and measure it. Slower transport dynamics can be captured and measured on the imaging timescale which is typical of other variants of ICS. Thus, the image acquisition time will impose a temporal sampling limit, but faster imaging often has to be balanced against other factors such as photobleaching.

The single-channel ICS methods can be extended to two-color ICCS where different macromolecules labeled with fluorophores of different emission wavelengths are imaged simultaneously or iteratively in two detection channels using a fluorescence microscopy imaging system (Fig. 10.3). In the two-color cross-correlation measurements, we define a fluorescence intensity fluctuation for a pixel sampled in space and time in each of the two detection channels a and b:

$$\delta i_{\mathrm{a/b}}(x,y,t) = i_{\mathrm{a/b}}(x,y,t) - \langle i_{\mathrm{a/b}} \rangle \qquad [10.2]$$

Cross-correlations will be detected in space and/or time between the two detection channels for cases where there are different labeled species interacting as a part of a common complex that are imaged in the two detection channels (Fig. 10.3A and B). Detection of the interacting species against a background of noninteracting species depends on the relative densities of the interacting species, the densities of the noninteracting species, and the sampling of fluctuations in space and/or time in the ICCS analysis.

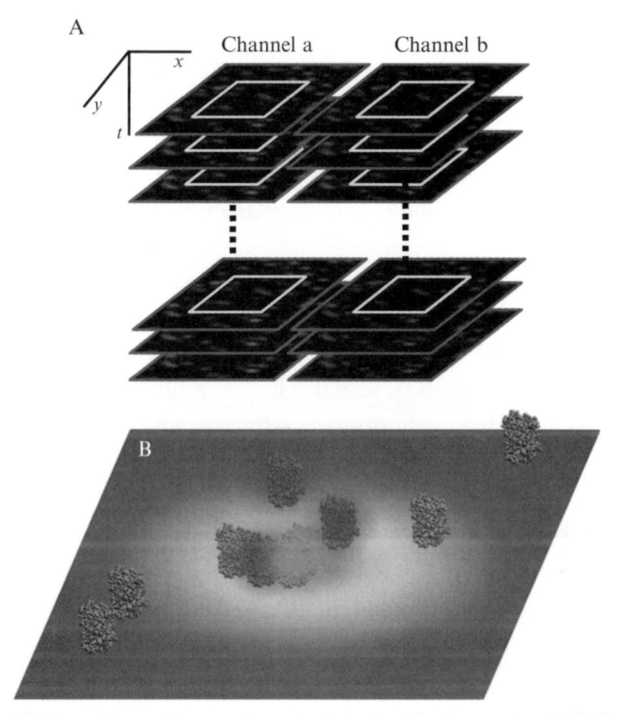

Figure 10.3 (A) Two-channel image time stack typical for two-color STICCS with yellow box showing region of interest (ROI) for correlation analysis. (B) A molecular cartoon illustrating the emission of two species labeled with green and red fluorophores from within the PSF-defined focus that forms the basis of cross-correlation measurements. Interacting species can be measured by cross-correlating the signals between the two channels integrated over all pixels in the ROI. Note that the PSF would be about 100 × larger for standard fluorescence microscopy. (See Color Insert.)

Given input image time series collected in the two detection channels a and b on a fluorescence microscope, we can calculate a generalized space–time intensity fluctuation correlation function from the data sets:

$$r_{\mathrm{ab}}(\xi,\eta,\tau) = \frac{\left\langle \langle \delta i_{\mathrm{a}}(x,y,t)\delta i_{\mathrm{b}}(x+\xi,y+\eta,t+\tau)\rangle_{xy} \right\rangle_t}{\langle i_{\mathrm{a}}\rangle_t \langle i_{\mathrm{b}}\rangle_{t+\tau}} \qquad [10.3]$$

If a = b, then Eq. (10.3) defines an autocorrelation function for the single detection channel, otherwise we calculate a cross-correlation function between the two channels. The correlation function depends on spatial lags (ξ and η) and temporal lag (τ) variables which simply represent shifts in space and time for the calculation of correlations between different pixels.

The angular brackets in the numerator represent the calculation of correlation functions in space (xy) and time (t), while in the denominator they represent the averaging of image pixel intensities for a and b from the image ROI in the time series frame where the given pixel is recorded. In practice, there are several ways to calculate the correlation functions and typically the analysis is applied to a substack of interest from the image series for a given ROI in space and time (TOI).

2.2. Theory of spatiotemporal ICS and ICCS

The spatiotemporal variants of ICS and ICCS (STICS and STICCS, respectively) analyze input from a fluorescence microscopy time series collected in one channel (STICS) from a cellular sample containing a single macromolecular species labeled with one type of fluorophore or a two-color paired image series (STICCS) collected in two detection channels from a sample with two different macromolecules tagged with different fluorophores (Fig. 10.3A). An ROI/TOI substack from the image series selected for analysis is used to calculate a set of space–time correlation functions using Eq. (10.3) (Fig. 10.3A, yellow boxes). STICS analysis of a single image series yields one space–time autocorrelation function (r_{aa}), while the two-color STICCS analysis of a dual detection channel image series yields four space–time correlation functions: autocorrelation functions for each channel (r_{aa} and r_{bb}) as well as two cross-correlation functions between the detection channels with the order of the spatial correlation reversed in the calculation (r_{ab} and r_{ba}).

The output of the STICS or STICCS analysis will be a space–time correlation function stack which will reveal the time evolution of the spatial correlations of the fluorophores in the ROI/TOI substack analyzed which depend on the underlying molecular transport dynamics as well as probe photophysics. For time lag zero ($\tau = 0$), each image will be spatially correlated with itself to yield a correlation function ($r_{ab}(\zeta,\eta,\tau = 0)$) that is a maximum at zero spatial lags (Fig. 10.4A and B) and decays as a Gaussian due to the spatial profile of the PSF (Fig. 10.2C). The e^{-2} radius of $r_{ab}(\zeta,\eta,\tau = 0)$ will be a factor of $2^{1/2}$ larger than that of the PSF due to the properties of correlation of Gaussian functions. As the underlying macromolecular dynamics unfold and are sampled in the image time series, the correlation function at longer time lags will evolve due to a change in the number of macromolecules that are correlating at the same location in space at later times (Fig. 10.2). For 2D diffusion, the motion is isotropic as the macromolecules follow random walks so we would observe the $r_{ab}(\zeta,\eta,\tau)$ decay in amplitude and spread in a radially

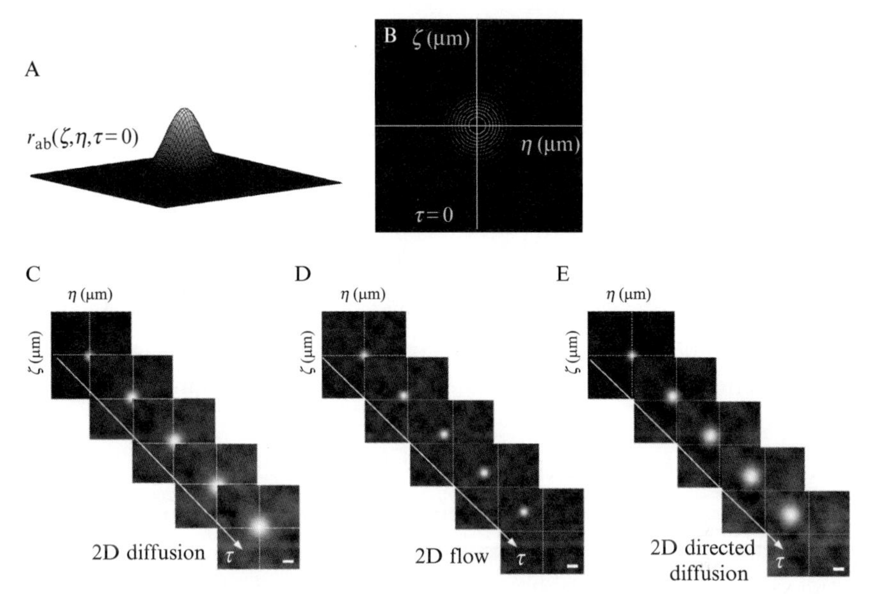

Figure 10.4 Spatiotemporal image correlation spectroscopy (STICS). (A) A spatiotemporal autocorrelation function at time lag $\tau=0$ calculated from an ROI of a simulated image with Gaussian PSF. (B) Contour plot of the same correlation function as shown in (A) illustrating that the Gaussian correlation peak amplitude is centered at zero spatial lags at $\tau=0$. (C–E) All show density contour plots of the spatiotemporal autocorrelation function as a function of increasing time lag as calculated from computer-simulated image series with particles undergoing 2D diffusion, 2D directed flow, and 2D directed or biased diffusion, respectively. The spread and translation of the correlation function are evident in the time lag series for each of the transport cases. (See Color Insert.)

symmetric fashion as a function of time lag (Fig. 10.4C). If the macromolecules are undergoing directed flow in the cells, then there will be a directional correlation of the fluctuations due to the population flow. In this case, the $r_{ab}(\zeta,\eta,\tau)$ peak would translate from the zero spatial lags origin as a function of lag time at the same rate and in a direction defined by the flow (Fig. 10.4D). If the macromolecules are undergoing a flow biased diffusion, then the correlation function peak will translate in space and simultaneously spread as a function of lag time (Fig. 10.4E). If there is a population of immobile macromolecules, their correlation function peak will remain centered at the zero spatial lags origin over all time lags with amplitude determined by the density of this static population. Such a peak can obscure the short time lag dynamics; however, it is possible to filter the immobile population for the analysis (see Section 3.2.1).

The space–time evolution of the correlation function is captured by locating the maximum of the $r_{ab}(\zeta,\eta,\tau)$ peak and fitting this with a translationally and radially free Gaussian function:

$$r'_{ab}(\Delta x,\Delta y,\Delta t) = g_{ab}(\xi-\Delta x,\eta-\Delta y,\Delta t) \exp\left\{-\frac{(\xi-\Delta x)^2+(\eta-\Delta y)^2}{\omega(\Delta t)^2}\right\}+r_\infty \quad [10.4]$$

where the best fit parameters to the correlation maximum peak are the time-dependent amplitude g_{ab}, the position of the peak (Δx, Δy) in lag space, the time-dependent peak e^{-2} radius ω, and a spatial offset parameter r_∞.

2.3. Theory of reciprocal space (*k*-space) ICS

The reciprocal or *k*-space variant of image correlation spectroscopy (kICS) (Kolin et al., 2006) takes the same fluorescence microscopy image series input as other ICS variants; however, each image in the time series is transformed to a spatial frequency representation by calculating a discrete 2D fast Fourier transform. This operation yields an image stack that is a function of spatial frequencies (k_x, k_y) and time (Fig. 10.5). This frequency space representation of the time series is then correlated in time to generate a *k*-space–time correlation function $r(\mathbf{k},\tau)$:

$$r(\mathbf{k},\tau) = \left\langle \tilde{i}(\mathbf{k},t)\tilde{i}^*(\mathbf{k},t+\tau)\right\rangle_t \quad [10.5]$$

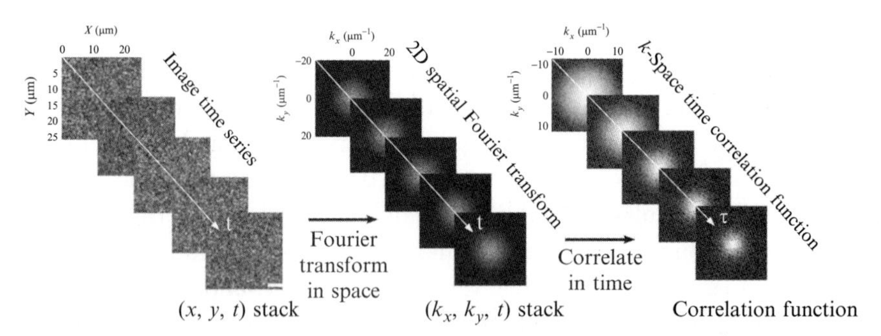

Figure 10.5 Schematic figure showing the steps involved in generating a *k*-space–time correlation function. The basic steps involve a 2D fast Fourier transform (2D-FFT) of each image frame in the fluorescence microscopy time series to transform to an image time stack in reciprocal or *k*-spatial frequencies. The *k*-space–time stack is then time correlated to generate the final *k*-space–time correlation function output. (See Color Insert.)

where the ~ indicates Fourier-transformed intensities from the image series and * is the complex conjugate. The next step involves calculating a radial average of the correlation function $r(\mathbf{k},\tau)$ for each time lag τ and then calculation of the natural logarithm of this function (Fig. 10.6A and B). If we assume that fluctuations in the fluorescence intensity can arise from 2D transport (diffusion and flow of fluorescent molecules moving in and out of a region) and from photophysics (intermittent emission blinking of a fluorophore or photobleaching of the probe) and further assume that these two processes are statistically independent and integrated with a Gaussian PSF with the microscopy imaging system, then the k-space–time correlation function is:

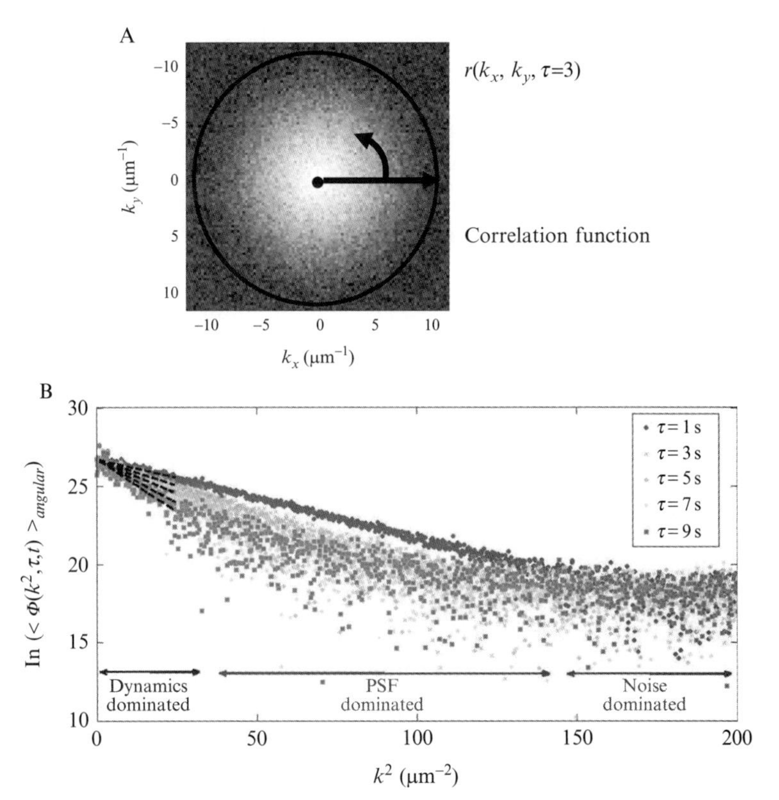

Figure 10.6 (A) Single frame for one time lag ($\tau = 3$) of the k-space–time correlation function that is radially averaged and natural logarithm transformed to yield (B) the output of Eq. (10.7). This plot shows **k**-vectors over which there are fluorescence correlations (dynamics and PSF dominate regimes) which decay to a noise floor at a certain threshold **k**-vector for the given time lag. The correlation function data are calculated from computer-simulated images of a diffusing population in 2D. (See Color Insert.)

$$r(\mathbf{k},\tau) = N\frac{q^2 I_o^2 \omega^4 \pi^2}{4}\langle\theta(t)\theta(t+\tau)\rangle \exp\left[i\mathbf{k}\cdot v\tau - |\mathbf{k}|^2\left(\frac{\omega^2}{4}+D\tau\right)\right] \quad [10.6]$$

where N is the number of particles in the image ROI, q is the quantum yield, I_o is the peak incident laser excitation at focus, v is the velocity of the fluorescent particles, D is the diffusion coefficient, ω is the e^{-2} beam radius, and $\theta(t)$ is the photophysics emission function which is 1 for emitting fluorophores and 0 for nonemitting (Kolin et al., 2006). The angular brackets indicate correlation over all fluorescent particles in the image ROI. If the system only contains diffusing fluorescent particles, then we can radially average and take the natural logarithm of the k-space–time correlation function to obtain the following:

$$\ln\left[\gamma\left(|\mathbf{k}|^2,\tau\right)\right] = \ln\left[N\frac{q^2 I_o^2 \omega^4 \pi^2}{4}\langle\theta(t)\theta(t+\tau)\rangle\right] - \left(\frac{\omega^2}{4}+D\tau\right)|\mathbf{k}|^2 \quad [10.7]$$

If the photophysics fluorescence fluctuations are independent of the fluctuations due to material transport in and out of the PSF, then the natural logarithm operation serves to separate the correlation function into a linear function of independent terms for each. For each time lag, t, a plot of $\ln[r(|\mathbf{k}|^2,\tau)]$ versus $|\mathbf{k}|^2$ will decay linearly with a slope related to the diffusion coefficient and a y-intercept that depends on the number of particles, optical constants, and the time correlation of their photophysical emission properties (Fig. 10.6B). In practice, the decay for $|\mathbf{k}|^2 > 0$ is linear until reaching a noise floor established by a $\mathbf{k}$-vector threshold on a length scale where they are no longer correlated due to loss of particle–particle correlations (see Fig. 10.6B). This single lag correlation function is fit by a first-order spline with three knots, in which the first segment fits the linear decay due to diffusion and the second segment fits the noise floor (this is usually done after PSF/immobile population removal). We can use kICS to build photophysical time correlation functions from the intercept data as a function of time lag and thus independently study the photophysics of the fluorescent probe (Kolin et al., 2006).

Additionally, we can chose to plot the slope of each $\ln[r(|\mathbf{k}|^2,\tau)]$ versus $|\mathbf{k}|^2$ curve for each time lag τ (so-called slope-of-slopes plot), then we can obtain the diffusion coefficient as the slope of this plot without needing to know the beam radius and independent of the fluorophore photophysics (assuming there is sufficient sampling of transport fluctuations in the image ROI series). For the slope-of-slopes plot, the y-intercept is $-\omega^2/4$ and the slope is simply $-D$ so the diffusion coefficient can be obtained with having to measure the beam profile radius (Fig. 10.7).

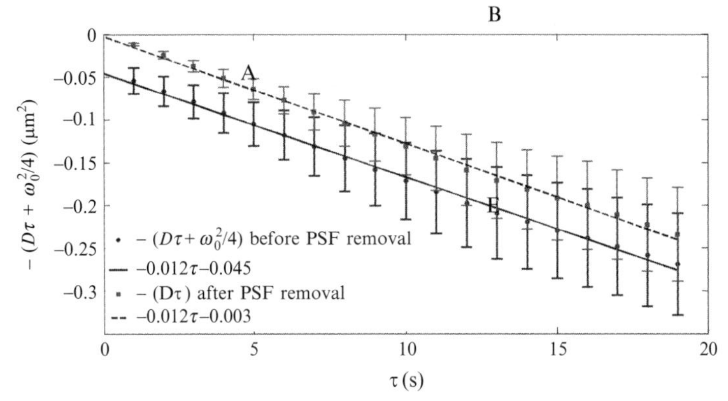

Figure 10.7 Log-of-log plots for all time lags from the same correlation function and image simulation data as was used to generate Fig. 10.6. The plot shows the decay before and after PSF removal (which is equivalent to immobile filtering). The slope of the plot yields $-D$. (For color version of this figure, the reader is referred to the online version of this chapter.)

The separation of photophysical fluorescence fluctuations from transport fluctuations is a key advantage of kICS and it can be used to measure transport properties of cell membrane receptors labeled with fluorescent probes with complex photophysics. This has been demonstrated for kICS measurements of quantum-dot-tagged GPI receptors on fibroblasts where diffusion coefficients unbiased by the power law emission blinking of the quantum dot labels could be measured by kICS (Durisic et al., 2007). As is the case for STICS, flow of the fluorescent population can be measured by kICS via numerical calculation of the gradient of the phase of the complex correlation function which is obtained from the real and imaginary components (Kolin et al., 2006); however, STICS is more intuitive in this case as it serves to visually show the flow direction and magnitude directly in real space and requires less spatial sampling (see Section 3).

3. PROCEDURES FOR STICS AND KICS

3.1. Microscopy image series collection

The starting point for ICS analysis is the collection of an image series on a fluorescence microscope. ICS analysis has been applied to image series collected on confocal and two-photon laser scanning microscopes as well as TIRF microscopy image series. There are a number of important considerations in terms of image collection that should be adhered to for ICS analysis, no matter which modality of imaging is used.

3.1.1 Sampling in space and time

For most physical measurements, it is important to adhere to the Nyquist–Shannon sampling theorem and sample at two times the highest expected frequency component in a band-limited signal. For microscopy imaging, an image can be considered to be composed of spatial frequencies with the highest spatial frequency being set by the inverse of the PSF size. Standard Nyquist sampling would require an image pixel spatial dimension of half the PSF size. However, for STICS, we would typically chose a smaller pixel size (e.g., through image zoom on a CLSM) in order to oversample the PSF in space (see Fig. 10.1C). The reason for oversampling is to generate sufficient spatial sampling for calculating, fitting, and tracking of the spatial correlation function (see Fig. 10.4A and B). The PSF and the size of the image ROI set the spatial sampling for STICS since the PSF area in the focal plane establishes the size of a spatial fluctuation. If fluorescence intensity per fluorescent particle is high, then the signal-to-noise ratio of the spatial correlation function will roughly scale as the square root of the number of independent fluorescence fluctuations sampled in the image ROI. The number of spatial fluctuations sampled will be given by the ratio of the image ROI area to the PSF focal area (focal area $=\pi\omega^2$) so the spatial sampling can be increased by increasing the ROI size. However, the drawback to this is that this reduces the mapping resolution of the STICS generated transport maps across the cell, and additionally averages transport over a larger region in the cell. Since cells are heterogeneous, we usually wish to generate maps at the highest mapping grid resolution possible. For applications of STICS to adhesion protein transport in focal adhesions using microscopy collection with high NA immersion objectives, a minimum ROI size of 16×16 pixels yielded a detectable correlation function, and reasonably high mapping resolution. In practice, adjacent ROIs shifted by 4 pixels were typically analyzed to yield correlated vector maps for comparison between adjacent vectors (Brown et al., 2006).

Since STICS couples spatial and temporal correlation, the sampling of temporal fluctuations will factor in the overall signal to noise for the space–time correlation function. It is important to note that the sampling of temporal fluctuations will depend on the imaging rate and the rate of transport. For diffusion measurements, the important timescale is set by the characteristic diffusion time which is the average time it takes the fluorescently labeled species to diffuse out of a PSF-defined focal region. The imaging rate has to be at least two times greater than the inverse of the characteristic diffusion time of the fluorescent species; if STICS is to capture diffusion, otherwise correlation will be lost as molecules diffuse out of the PSF in the time between sequential images (see Fig. 10.2). However, in most applications, STICS has been used for flow measurements and the temporal sampling restrictions are less severe in this case. For detection and measurement of flow, the imaging rate has to be high enough so that the flowing species does not traverse and exit an image ROI (as opposed to the PSF) in the time between sequential images. The rate of flow, the image frame acquisition rate, and the size of the ROI will establish a limit for flow velocities that can be detected.

In contrast to STICS, spatial sampling in kICS will dictate which **k**-vectors are present in the *k*-space–time correlation function, and their decorrelation or decay will depend on the ROI size and the underlying dynamics of the fluorescent species measured. The dynamic range will, in some sense, be opposite to that of STICS since this is a frequency (*k*-) space technique. The largest **k**-vectors in the analysis are set by the inverse of the PSF dimension, while the smallest **k**-vectors will be the inverse of the ROI dimension. As fluorescent particles diffuse out of PSF-defined regions, the largest **k**-vectors will decorrelate, whereas the smallest **k**-vectors will only decorrelate when the particles have exited the image ROI (see Fig. 10.6). The kICS method typically requires a coarser ROI spatial sampling than STICS in order to obtain a suitable dynamic range of **k**-vectors. In practice, kICS has been applied to ROIs of a minimum diameter of 32 pixels for standard high NA fluorescence imaging.

3.1.2 Density limits

The intensity normalization defined in Eq. (10.3) for the generalized spatio-temporal correlation function is the same as is used in FCS as the zero lags value becomes the square relative intensity fluctuation (essentially, the intensity variance divided by the mean intensity squared). The significance of this is that the zero lags amplitude is equal to the inverse of the mean number of independent fluorescent entities within the PSF volume (assuming noise contributions have been corrected and fluorophores are independent). This amplitude dependence means that the fluorescent particle concentration or density will establish an upper detection limit. If the concentration is too high, the relative fluorescence fluctuations will be too small to be detected against random background fluctuations, and STICS analysis will not yield a correlation function that can be fit and tracked. Our previous characterization of limits in ICS established that a density of 100 particles μm^{-2} in a 2D system could be measured in computer-simulated images and this would also hold for STICS measurements (Costantino, Comeau, Kolin, & Wiseman, 2005).

At the other extreme of low density, there has to be a high enough concentration of fluorescent molecules that some are present on average within the image ROI. However, there is a threshold density where single molecules become resolvable in the images and more information can be acquired using single-molecule tracking methods since STICS averages over trajectories, while single-molecule methods give complete trajectory information within the limits set by the imaging time resolution.

3.1.3 Signal-to-noise ratio and background

In any FCS experiment, the signal to noise depends on the number of photons collected per fluorophore, per unit time. However, since ICS methods are applied to analyze CLSM or TIRF microscopy images, the pixels record intensities that are measured from analog photomultiplier tubes or binned photoelectrons in the case of CCD cameras (the exception would be for microscopes equipped with digital photon counting detectors). In the case of analog detection, it is assumed that the recorded pixel intensity is proportional to the number of photons and the signal response is linear. For quantitative experiments, it is important to test the linearity of response of the detector. This can be done by measuring the back reflected signal of a glass slide as a function of PMT gain voltage. It is also important that detector saturation is avoided so that there are no saturated pixels in the image. In terms of intensity signal-to-noise ratio, the sample fluorescence needs to be measureable above background which entails a minimum signal-to-noise ratio of 3 for the microscopy images.

Following image collection on the microscope, it is necessary to correct for white noise background that includes detector shot noise. This is done by selecting an ROI that encompasses only off-cell regions as there should be no real signal fluorescence in this area. The intensity values within this background ROI are averaged, and the average value is subtracted from every pixel in the image to remove background contributions. In principle, this can be done for every image in the time series if the background is constant (the background region can be monitored in time using ImageJ software to see if it remains constant).

Fluorescence photobleaching can be a problem for any quantitative ICS measurement that involves the analysis of correlation function amplitudes to obtain densities. However, it is less problematic for STICS measurements of flow velocities if the timescale of the photobleaching is longer than the characteristic transport timescale. This is usually the case when the imaging is optimized to minimize bleaching. As well, transport measurements using kICS are insensitive to the contribution of bleaching (it affects the photophysical correlations that map to the intercept) unless the bleaching is incredibly severe and signal is lost within a few images (Kolin et al., 2006). Such extremes are generally avoided under routine microscopy imaging collection.

3.2. Image data analysis procedures

The STICS and kICS analyses can be done using MATLAB-based graphical user interface (GUI) programs available from the Wiseman Research Group at McGill University. The program interface is straightforward as it allows the user to upload one or two image time series collected via one or two detection channels and output from the microscope as TIFF image stacks. The STICS and kICS methods ultimately differ in terms of fitting functions and their output (see Section 2), but there are procedures common to both that are required before correlation function analysis is initiated. These procedures are described below.

3.2.1 Immobile background filtering

In STICS and kICS, we are usually interested in measuring the transport properties of a mobile population. Cells are naturally heterogeneous and will contain immobile structures and tethered macromolecules. The presence of an immobile population can cause problems for the basic correlation function fitting so it is important to filter and remove the static population. This can be achieved by applying the "immobile filter" feature before carrying out the correlation analysis step. Once the image data set is loaded, the immobile filter can be applied by clicking the immobile filter button (see Fig. 10.8) after choosing the type of filtering that is desired. The program gives the option for Fourier filtering of the minimum frequency component in time from every pixel in the image frame (Fourier option in the filter options box (see Fig. 10.8)) or moving average filtering (moving average (images) option in the filter options box). We usually employ the Fourier option. It is important to realize that the Fourier filter can be applied to the whole image series or to a time subset of images (TOI) selected via the program options. It is important to apply the filter to the complete data set when using the Fourier option since the filter is based on removing the minimum time frequency component which depends on the number of temporal frequency bins that are present. A time subset of the complete image stack will result in changing the binning of the time frequencies and may result in filtering of slow dynamics. So, in practice, we always apply the Fourier filter before doing image region and subtime stack selection for correlation function analysis.

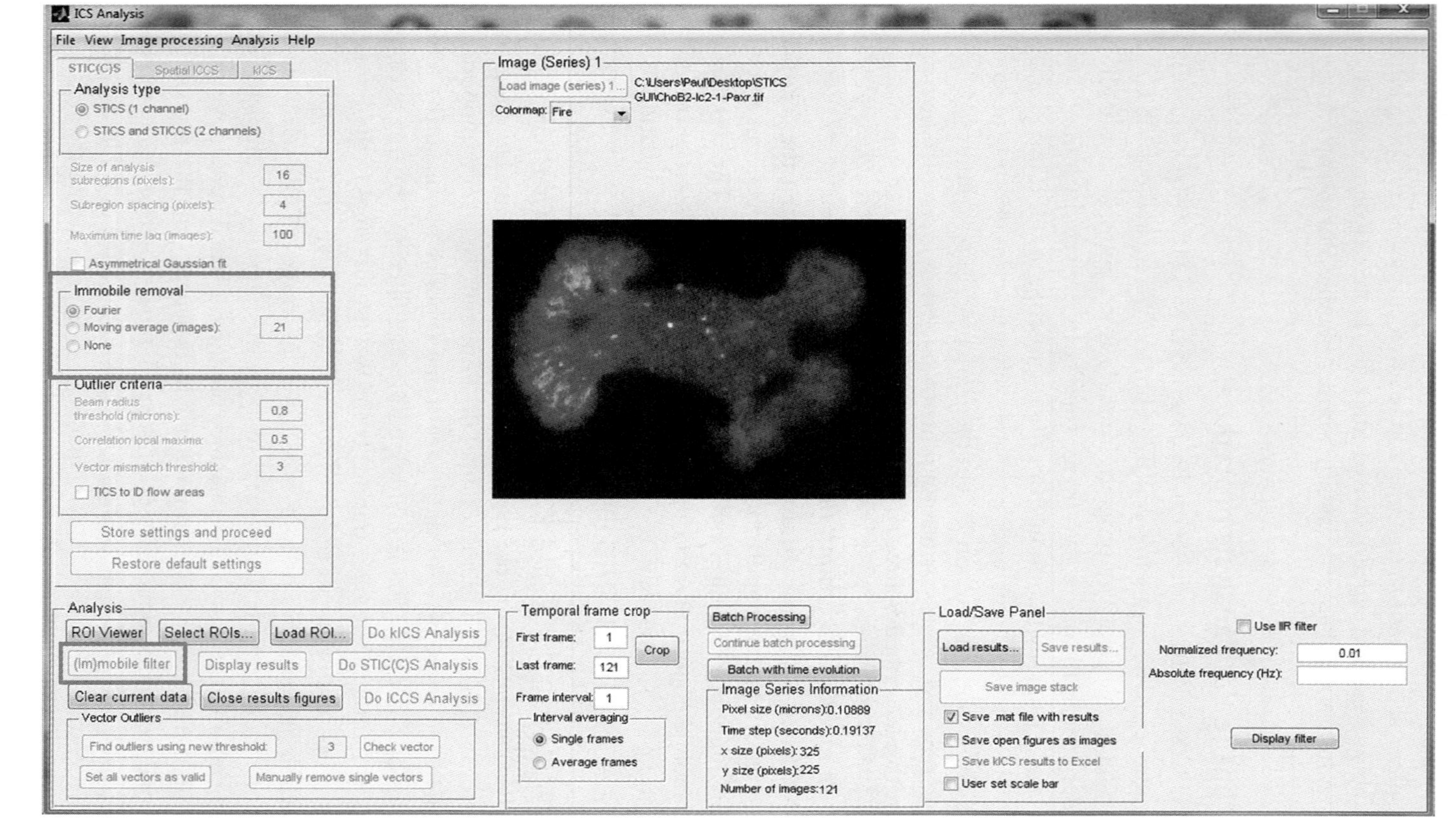

Figure 10.8 Screen capture image of the ICS GUI showing highlighted control interface for the immobile population filtering. (See Color Insert.)

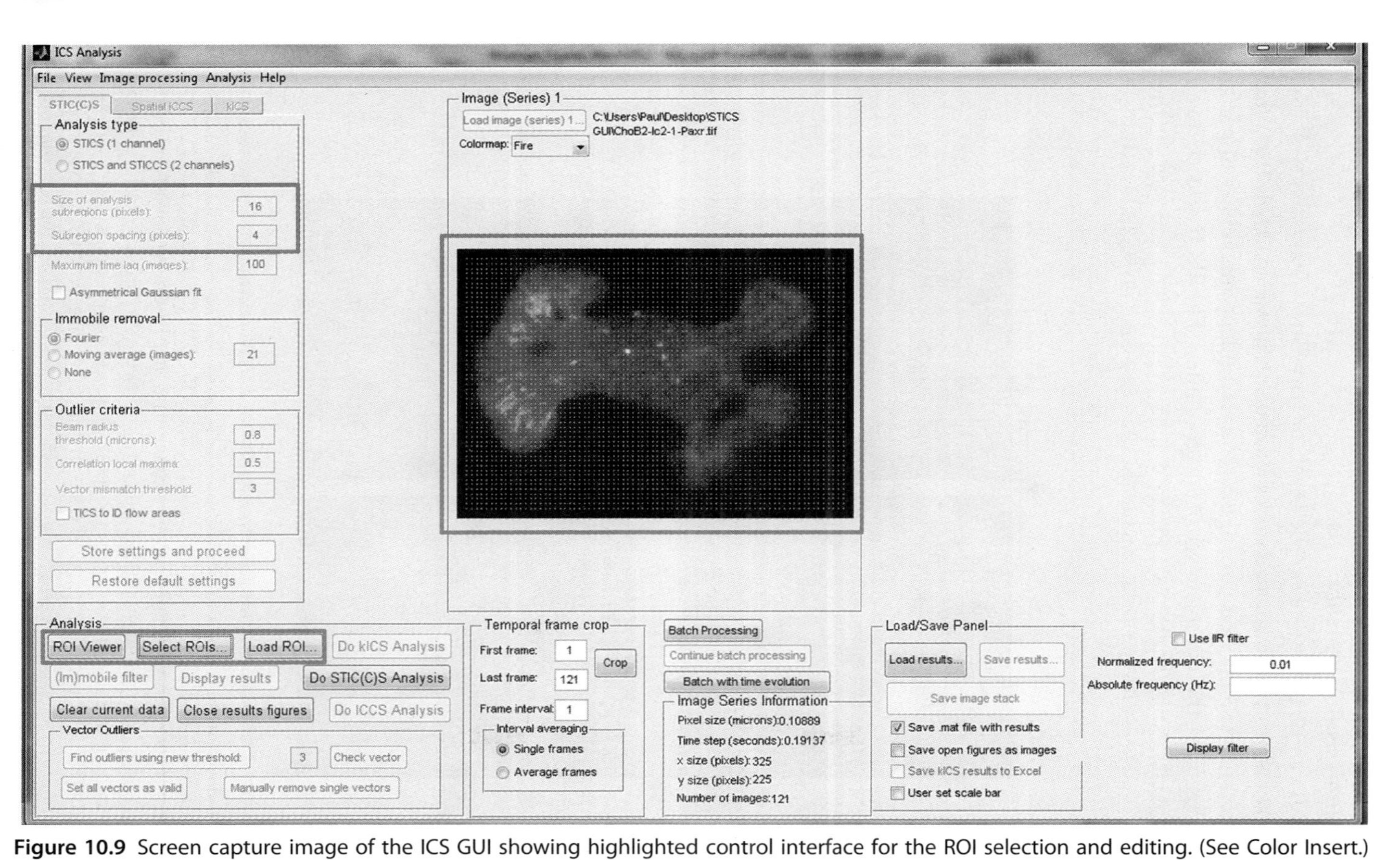

Figure 10.9 Screen capture image of the ICS GUI showing highlighted control interface for the ROI selection and editing. (See Color Insert.)

3.2.2 Analysis ROI selection

The GUI can be used to select the ROI for correlation analysis as well as a temporal frame crop to select a subset of image frames (the TOI) from the series for analysis. The temporal frame crop is evident and applies to select subframes from the complete set of image frames. The ROI selection requires some explanation. As indicated in Section 3.1.1, the size of the ROI is important for determining the number of spatial fluctuations sampled

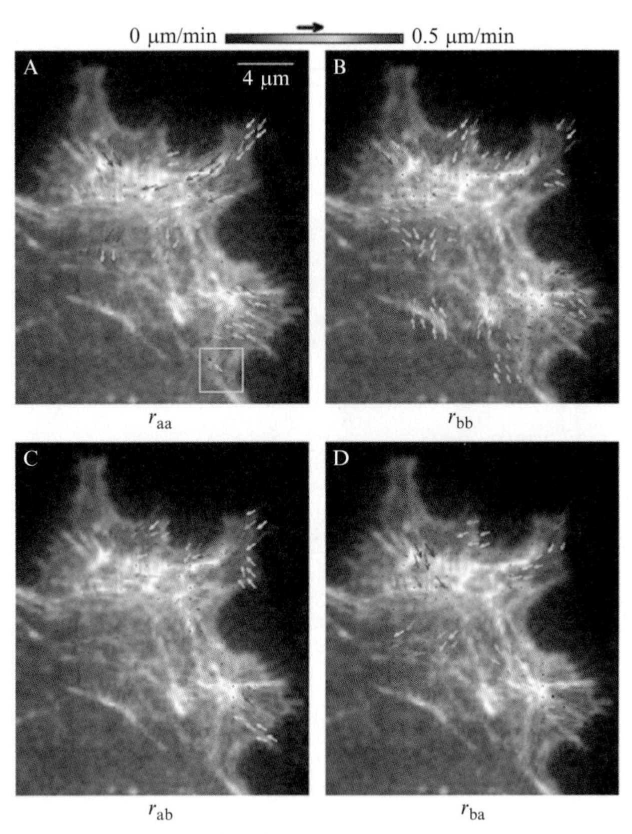

Figure 10.10 STICCS analysis results of frames 60–90 of a TIRF microscopy image series of actin-GFP and α-actinin-mCherry expressed in CHO cells that were platted on 5 μg/ml fibronectin. (A) Autocorrelation vector map for the actin-GFP channel. (B) Autocorrelation vector map for the α-actinin-mCherry channel. (C and D) Cross-correlation vector maps r_{ab} and r_{ba} calculated via STICCS between the two detection channels. Velocity vectors for correlation function flow peaks that were above noise threshold are displayed. The image time step was 6 s and the pixel size was 0.21 μm and the yellow square depicts the size of the 16 × 16 pixels ROI used in the calculations. (See Color Insert.)

per ROI and the resolution of the transport maps across the cell. The user has input control on the size and spacing of the ROIs via the analysis type input of "size of analysis subregions (pixels)" and "subregion spacing (pixels)" (see Fig. 10.9). The former simply sets the size in pixels of the square ROIs, while the latter dictates how many pixels adjacent ROIs are shifted to allow for oversampling of the transport maps in space. These inputs are set before selecting the ROIs (see Fig. 10.9). When selecting the ROIs, the user can choose to apply the sampling to the entire image frame or use mouse-driven input to select a closed polygon region on the image to analyze (with the spacing and square ROI size preset as described above). The "ROI viewer" gives an expanded image window frame view of the images for the time subframes selected. The "load ROI" option allows input of common analysis templates for repetitive analysis of data sets.

3.3. Example ICS results

When measuring flow, the STICS and STICCS versions of the GUI program output resultant vectors superimposed on the center of each ROI projected on the first image in the series (for one or both detection channels). Correlation functions that are fit above noise threshold that are within the detection criteria are displayed for the ROIs selected for analysis. Figure 10.10 displays STICCS measured vector maps calculated from TIRF microscopy image series of actin-GFP and α-actinin-mCherry expressed in CHO cells that were platted on 5 μg/ml fibronectin. The cross-correlation channel (Fig. 10.10C) shows vectors in regions where there is cotransport of the two proteins. The kICS analysis can be initiated in the GUI program using a different tab, but only a single-channel image series can be analyzed.

4. CONCLUSIONS

The ubiquity of fluorescence imaging in the biomedical sciences is promising in that there is an established imaging infrastructure that can be used for quantitative biophysical analysis. Image correlation techniques have been expanding in terms of the modalities available to extract details of molecular and macromolecular transport properties in living cells. The STICS and STICCS methods allow relatively rapid and facile analysis of transport in cellular systems expressing standard fluorescent proteins. The kICS method allows similar measurements to be made but has the advantage that it can be applied to systems labeled with a fluorescent probe with complex photophysics (which can perturb standard correlation spectroscopy methods based on fluorescence fluctuation analysis). However, it is important to remember that STICS, STICCS, and kICS involve averaging over the ensemble of fluorescent molecules in the image ROI so behavior of outlier single molecules will be obscured by the averaging. These ensemble-based image analysis methods are useful though as they allow a biomedical researcher to obtain molecular scale transport information, averaged over the population of labeled molecules in the image analysis region, that may not be evident from direct visual examination of the image movies. The ICS methods continue to be extended and improvements made in the GUI to make them more widely available to interested biomedical researchers with access to standard fluorescence imaging systems.

ACKNOWLEDGMENTS

The author gratefully acknowledges Elvis Pandzic, Tim Toplak, and Laurent Potvin-Trottier for preparing some of the figures used in this chapter as well as microscopy image series collected in Prof. Rick Horwitz's lab at the University of Virginia. The author acknowledges funding support from the Natural Science and Engineering Research Council (NSERC) of Canada Discovery Grant program for research support in the development of the image correlation techniques.

REFERENCES

Brown, C. M., Hebert, B., Kolin, D. L., Zareno, J., Whitmore, L., Horwitz, A. R., et al. (2006). Probing the integrin-actin linkage using high-resolution protein velocity mapping. *Journal of Cell Science*, *119*, 5204–5214.

Comeau, J. W., Costantino, S., & Wiseman, P. W. (2006). A guide to accurate fluorescence microscopy colocalization measurements. *Biophysical Journal*, *91*, 4611–4622.

Costantino, S., Comeau, J. W., Kolin, D. L., & Wiseman, P. W. (2005). Accuracy and dynamic range of spatial image correlation and cross-correlation spectroscopy. *Biophysical Journal*, *89*, 1251–1260.

Digman, M. A., Brown, C. M., Sengupta, P., Wiseman, P. W., Horwitz, A. R., & Gratton, E. (2005). Measuring fast dynamics in solutions and cells with a laser scanning microscope. *Biophysical Journal*, *89*, 1317–1327.

Digman, M. A., Sengupta, P., Wiseman, P. W., Brown, C. M., Horwitz, A. R., & Gratton, E. (2005). Fluctuation correlation spectroscopy with a laser-scanning microscope: Exploiting the hidden time structure. *Biophysical Journal*, *88*, L33–L36.

Durisic, N., Bachir, A. I., Kolin, D. L., Hebert, B., Lagerholm, B. C., Grutter, P., et al. (2007). Detection and correction of blinking bias in image correlation transport measurements of quantum dot tagged macromolecules. *Biophysical Journal*, *93*, 1338–1346.

Hebert, B., Costantino, S., & Wiseman, P. W. (2005). Spatiotemporal image correlation spectroscopy (STICS) theory, verification, and application to protein velocity mapping in living CHO cells. *Biophysical Journal*, *88*, 3601–3614.

Kolin, D. L., Ronis, D., & Wiseman, P. W. (2006). k-Space image correlation spectroscopy: A method for accurate transport measurements independent of fluorophore photophysics. *Biophysical Journal*, *91*, 3061–3075.

Kolin, D. L., & Wiseman, P. W. (2007). Advances in image correlation spectroscopy: Measuring number densities, aggregation states, and dynamics of fluorescently labeled macromolecules in cells. *Cell Biochemistry and Biophysics*, *49*, 141–164.

Petersen, N. O., Hoddelius, P. L., Wiseman, P. W., Seger, O., & Magnusson, K. E. (1993). Quantitation of membrane receptor distributions by image correlation spectroscopy: Concept and application. *Biophysical Journal*, *65*, 1135–1146.

Schwille, P., Meyer-Almes, F. J., & Rigler, R. (1997). Dual-color fluorescence cross-correlation spectroscopy for multicomponent diffusional analysis in solution. *Biophysical Journal*, *72*, 1878–1886.

Srivastava, M., & Petersen, N. O. (1998). Diffusion of transferrin receptor clusters. *Biophysical Chemistry*, *75*, 201–211.

Thompson, N. L. (1991). Fluorescence correlation spectroscopy. In J. R. Lakowicz (Ed.), *Topics in fluorescence spectroscopy. Techniques*, Vol. 1, (pp. 337–378). New York: Plenum Press.

Wiseman, P. W., & Petersen, N. O. (1999). Image correlation spectroscopy. II. Optimization for ultrasensitive detection of preexisting platelet-derived growth factor-beta receptor oligomers on intact cells. *Biophysical Journal*, *76*, 963–977.

Wiseman, P. W., Squier, J. A., Ellisman, M. H., & Wilson, K. R. (2000). Two-photon image correlation spectroscopy and image cross-correlation spectroscopy. *Journal of Microscopy*, *200*, 14–25.

AUTHOR INDEX

Note: Page numbers followed by "*f*" indicate figures, "*t*" indicate tables, and "*b*" indicate boxes.

A

B

C

N

O

P

Q

R

S

SUBJECT INDEX

Note: Page numbers followed by "*f*" indicate figures, and "*t*" indicate tables.

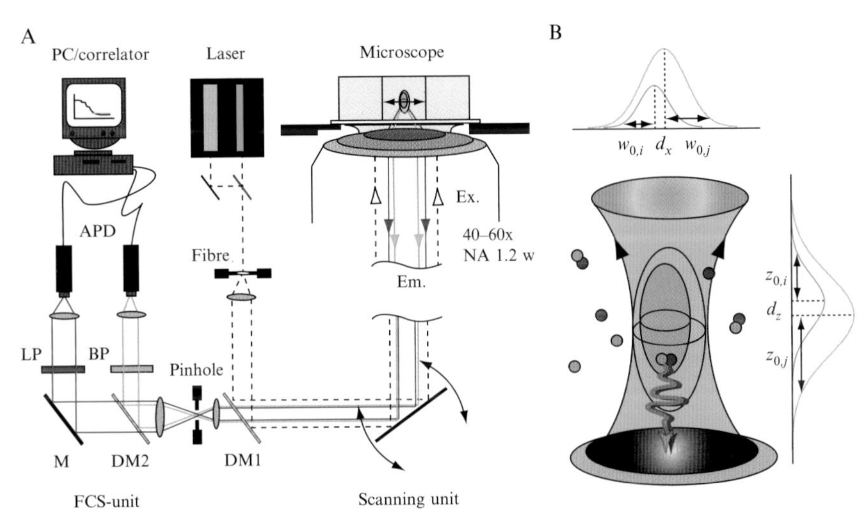

Thomas Weidemann and Petra Schwille, Figure 3.1 Optical setup as used in this chapter for the eGFP/Alexa647 dye pair. (A) Coaxial laser beams (488, 633 nm) are reflected by a dichroic mirror (DM1, 488/633 nm), guided to the back aperture of the objective and focused into the sample. From there the emitted fluorescence is collected, passes DM1 and the detection pinhole (70 μm) after which a second dichroic (DM2, transmission >635 nm) distributes the light into the two APD terminated detection channels. A band pass (BP, transmission 505–560 nm) and a long pass (LP, transmission >655 nm) filter were used to further narrow the spectral range of the fluorescence signal. (B) Gaussian-shaped intensity distributions of the superimposed foci. The centers exhibit a certain width and displacement for each color.

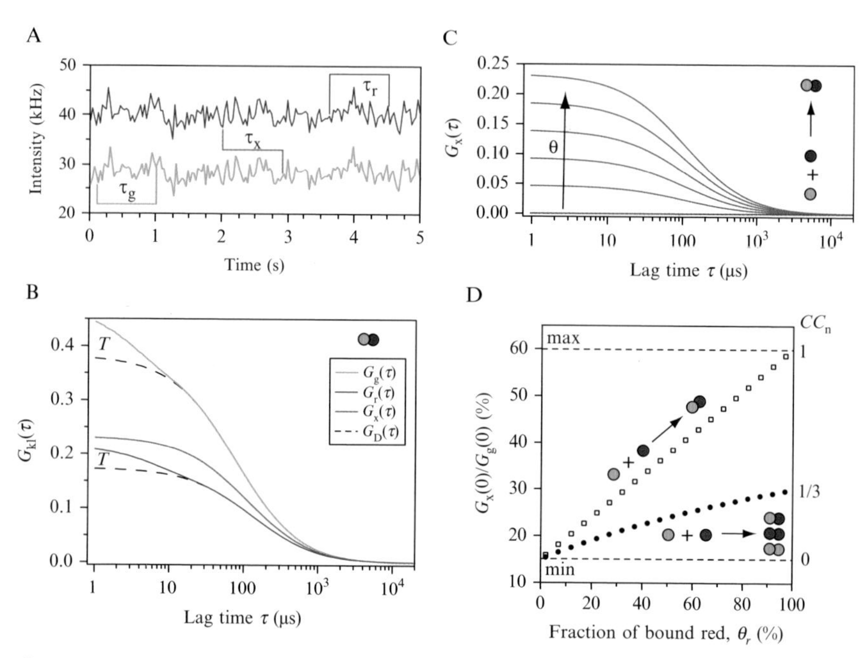

Thomas Weidemann and Petra Schwille, Figure 3.2 Example correlation curves. (A) Intensity traces as simultaneously measured in the green and the red channel. For correlation, pairs of values separated by a lag time τ are multiplied and normalized with the square of the mean intensity. (B) Numerical correlation functions for a hypothetical double-labeled standard ($D=200\ \mu m^2/s$, $c=10$ nM, $\tau_{trip,g}=3$ µs, and $\tau_{trip,r}=6$ µs, $f_{nf}=20\%$) with a typical focal geometry ($w_{0,g}=0.25$ µm, $w_{0,r}=0.325$ µm, $S=5$). The CCF contains displacement of the centers by 20% of the 488 nm $1/e^2$ radii in all dimensions. Dashed curves show the purely diffusion related decay. (C) Increase of the cross-correlation amplitude with increasing fractions of ligand bound θ (heterotypic binding, no cross talk). (D) Progression of the ratio between cross- and autocorrelations for a heterotypic (open squares) and homotypic (dots) dimerization. In the latter case the curve is nonlinear and depressed due to double-labeled, single-colored species.

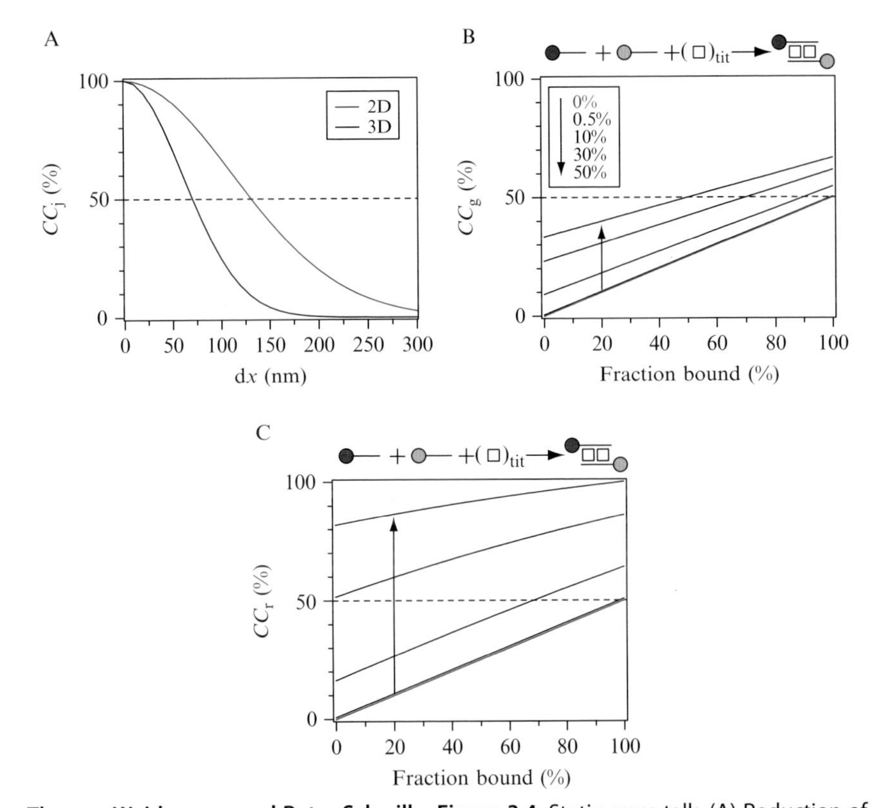

Thomas Weidemann and Petra Schwille, Figure 3.4 Static cross talk. (A) Reduction of CC_j for increasing displacement (dx) (Eq. 3.10) and a typical observation volume with $w_0 = 224$ nm for the two-dimensional (e.g., diffusion in a horizontal membrane, red) and the three-dimensional case (e.g., diffusion in the cytoplasm, black). Dimensions and displacement along the optical axis z were scaled with the structure parameter ($S = 5$). (B) Effect of successive cross talk from the green into the red channel (legend, arrows) for a static binding experiment in which the concentrations c_i and molecular brightness η_j of labeled ligands are constant (here $c_g = c_r$ and $\eta_g = \eta_r = 1$), while a non-fluorescent reaction partner (open square) is titrated. The examples assumes a chromatic displacement of 50%. (C) Same plot as in (B) but now the red, cross talk affected channel was used for normalization. The predicted progression of CC_j with 0% cross talk is shown for reference ((B) and (C), red).

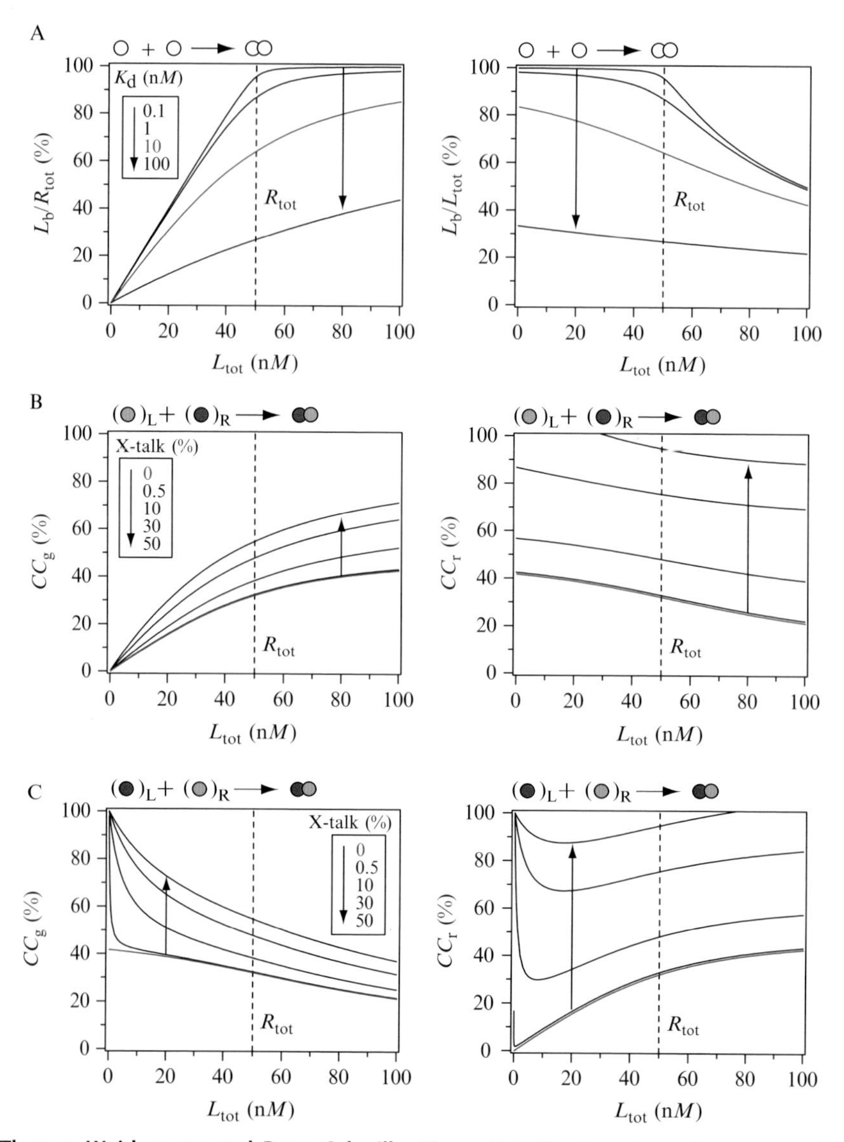

Thomas Weidemann and Petra Schwille, Figure 3.5 Titration-dependent cross talk. (A) Binding isotherms calculated for a heterotypic reaction. The "fraction receptor occupied" (left panel) and the "fraction ligand bound" (right panel) are plotted for increasing K_d-values (legend, arrows) and a fixed receptor concentration ($R_0 = 50$ nM, dashed vertical line). The 10 nM curve (red) was used for (B) and (C). (B) Titration with a green-labeled ligand. Effect of successive cross talk from the green into the red channel (legend, arrows) assuming a constant molecular brightness η_j of the labeled ligands ($\eta_g = \eta_r = 1$), a chromatic displacement of 50%, and a 1.5-fold enlarged detection volume in the red channel. The predicted progression of CC_j with 0% cross talk is shown for reference (red). (C) Same as in (B), but titration with a red labeled ligand.

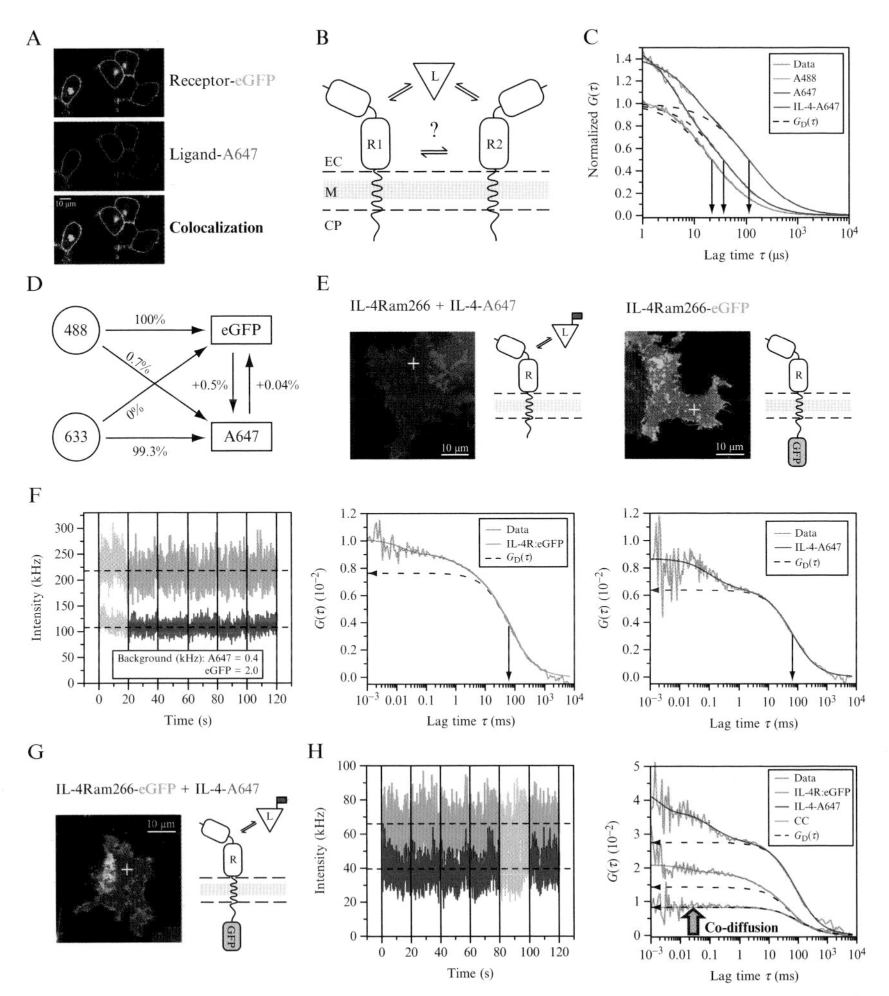

Thomas Weidemann and Petra Schwille, Figure 3.6 Experimental outline to quantify ligand–receptor interactions. (A) Colocalization of a Alexa647 mono-labeled Interleukin-4 (kindly donated by Walter Sebald) and a GFP-tagged IL-4 receptor construct (Weidemann et al., 2011). (B) Problems which can be addressed: (i) which of the receptors is bound by the ligand, (ii) which of the receptors is recruited into a ternary complex, and (iii) what are the associated affinity constants? (C) Characterization of the optical setup with diffusion standards A488 and A647 in free solution, as well as the free ligand. (D) Cross-excitation and cross talk scheme. (E) Confocal images of single-stained cells and (F) corresponding intensity traces measured at indicated positions (crosses). Runs not used for correlation are gray shaded. Arrows point out the slow, receptor related diffusion time and the associated amplitude (dashed line with arrow head). (G) Dual-color FCCS experiment. The positive CCF amplitude indicates codiffusion.

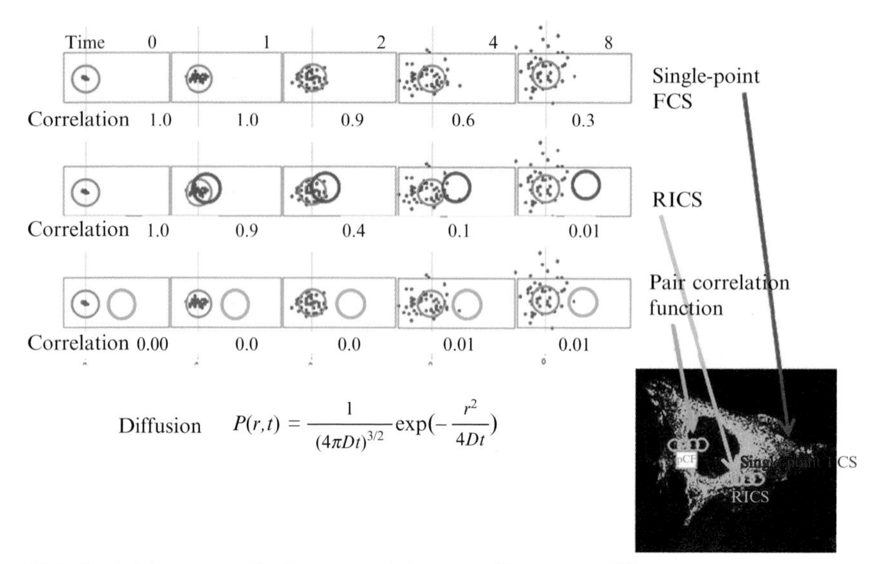

Michelle A. Digman *et al.*, Figure 6.1 Schematic illustration of fluctuation spectroscopy experiments. In single point FCS, the intensity fluctuations at a given observation volume are measured as a function of time due to diffusion. The correlation is the product of the intensity at time 0 (points at left in the figure) to the intensity at a later time. The correlation decreases as a function of time due to diffusion. This time depends on the diffusion coefficient of the molecule and on the size of the observation volume. In the RICS method, the position of the volume of observation is changed with time producing a correlation that is dependent on how fast molecules are moving as well as how fast the position of the volume of observation is changing with time. In the pair correlation approach, two (or more) volumes of observations are used at a given distance. Initially, a molecule is in one of the observation volume. Due to diffusion, a molecule could appear in the other volume at a later time. The time it takes to appear in the other volume depends on how fast the molecule moves, how far the two volumes are, and if there are obstacles to diffusion in between the two volumes.

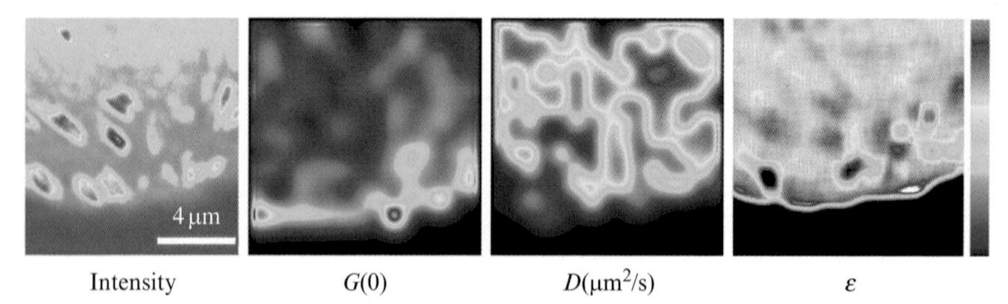

Michelle A. Digman *et al.*, Figure 6.3 Intensity image, (G0) map, Diffusion map and brightness map of paxillin–EGFP in focal adhesions. The same color scheme at right is used for all figures. The corresponding full scale are intensity 0–489 photon/s; $G(0) = 0.0064$, $D = 0$–$46\ \mu m^2/s$, 28,000–52,000 counts/dwell time/molecule. The maps from the fluctuation experiments were obtained using a region of analysis of 32 pixels and sliding this region in steps of 16 pixels across the entire image.

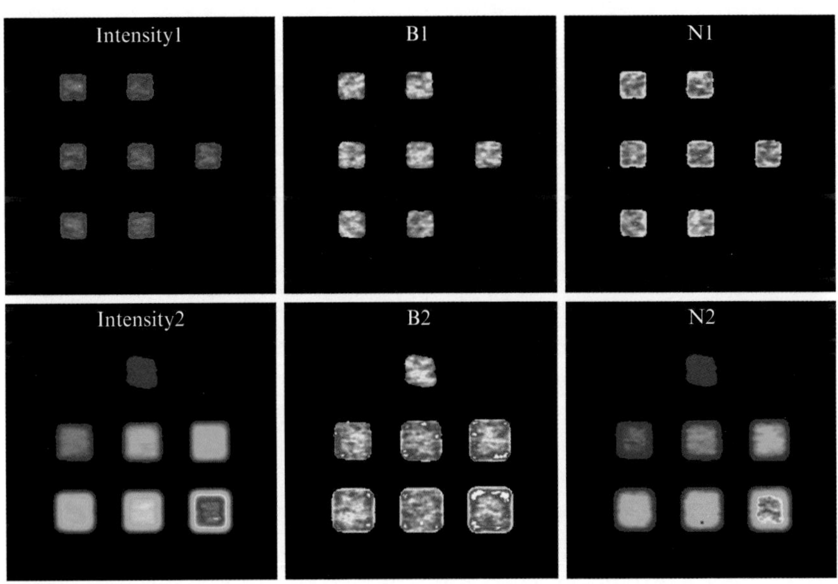

Michelle A. Digman *et al.*, Figure 6.5 According to expected values from Table 6.1, we recover a uniform intensity in all spots for intensity 1, a variable intensity for the spots as seen in channel 2, the same brightness in all spots for channel 1 and channel 2, the same number of molecules for channel 1 in all spots, and a variable number of molecules in the various spots for channel 2.

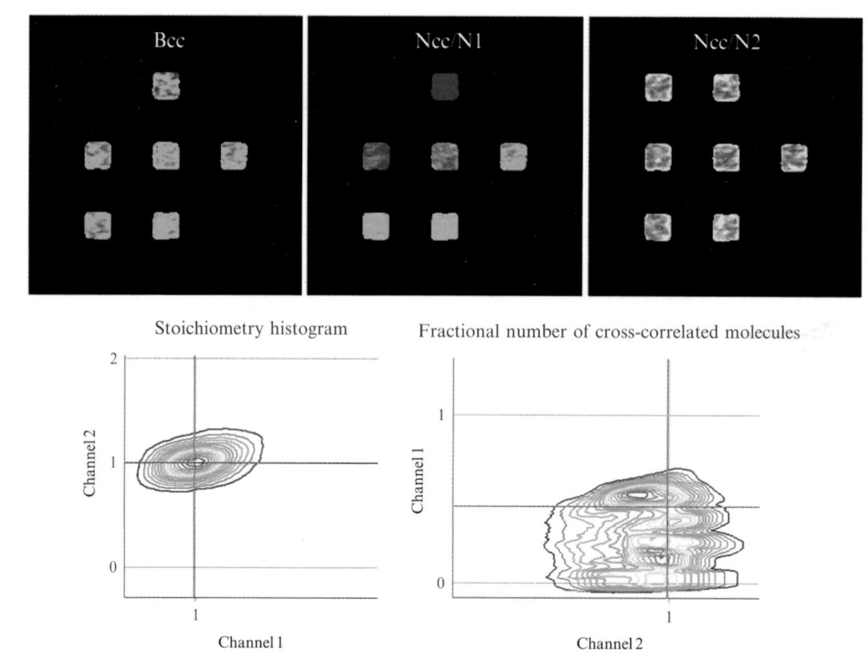

Michelle A. Digman *et al.*, Figure 6.6 The cross-brightness is the same for all molecules since the molecules that are in the complexes have all the same brightness. The ratio of molecules that are in the complex with respect to the total molecules in each channels varies in channel 1 in the different spots, but it is the same (all molecules are cross-correlated) in channel 2. The stoichiometry histogram reports the relative number of molecules of each kind in the complex. In this case, all the complexes have a stoichiometry 1:1. The histogram of the fractional number of cross-correlated molecules varies only along the channel 1 axis, since all the molecules in channel 2 are cross-correlated.

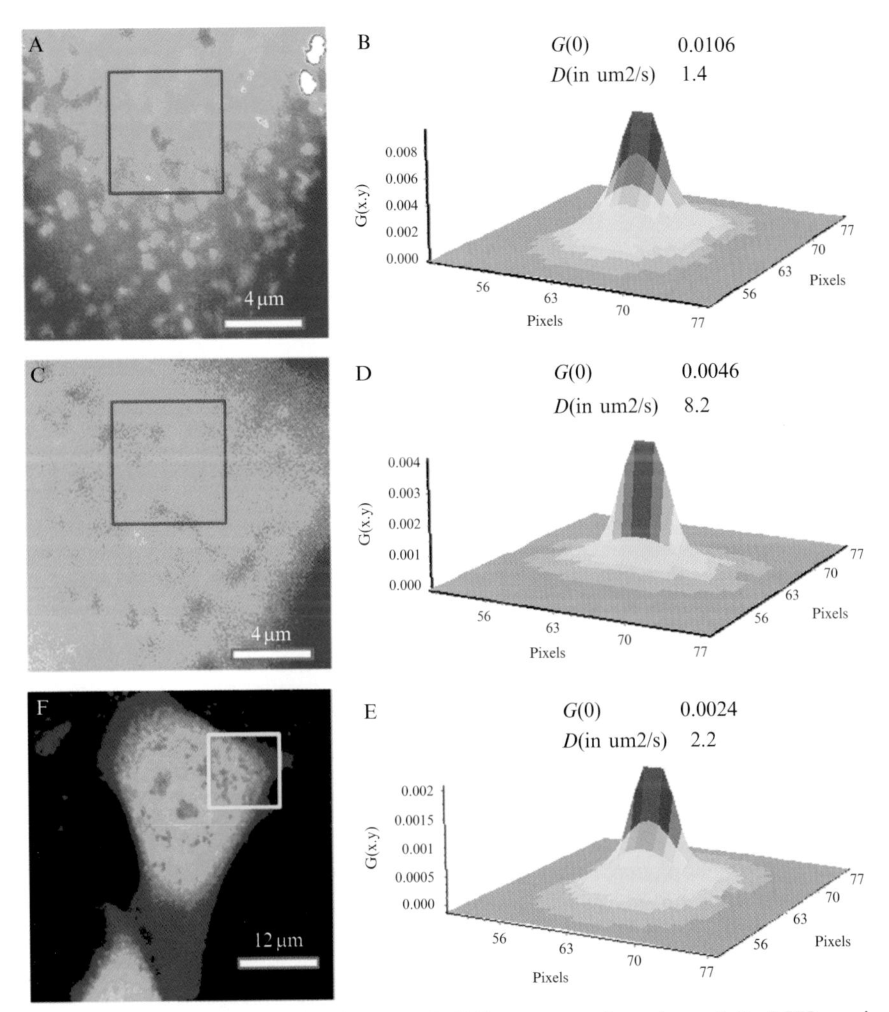

Michelle A. Digman *et al.*, Figure 6.7 Cell co-expressing dynamin2a-EGFP and endophilin-mCherry. Pixel size is 0.065 μm. Optical section in the cytoplasm. Only the part in the red square was used for the RICS analysis. (A) Channel 1 (EGFP) and (C) Channel red (mCherry). (B and C) RICS autocorrelation function and fit using one component diffusion for the green and red channels, respectively. (E) RICS cross-correlation. For the fit, the waist was measured to be 0.23 μm. The same waist was used for both wavelengths. (F) Unzoomed average intensity image of the red channel. The area analyzed in parts A–E is in the yellow square.

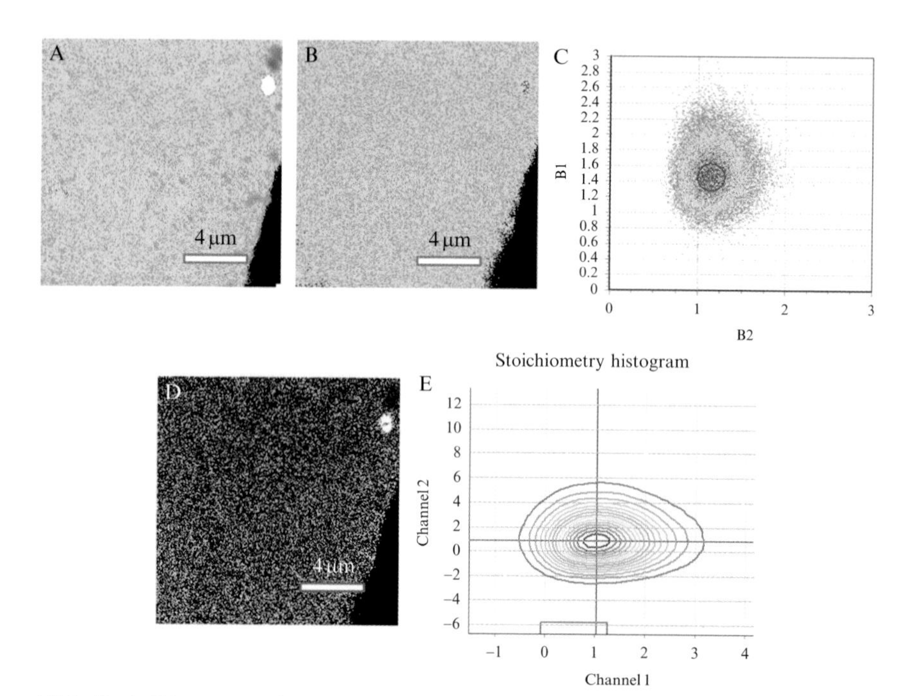

Michelle A. Digman *et al.*, Figure 6.8 Brightness analysis of the cell shown in Fig. 6.7. (A) Green channel. (B) Red channel. (C) 2-dimensional pixel histogram of the values of brightness for the two channels. The *x* and *y* numbers above are used for the calibration of the brightness of mCherry and EGFP, respectively. (D) Cross-correlation brightness map. (E) Contour map of the number of pixels in channel 1 of a given brightness which cross-correlate with channel 2 for all possible brightness. This is the so-called stoichiometry histogram. In this case, the most abundant value is at a ratio 1:1.

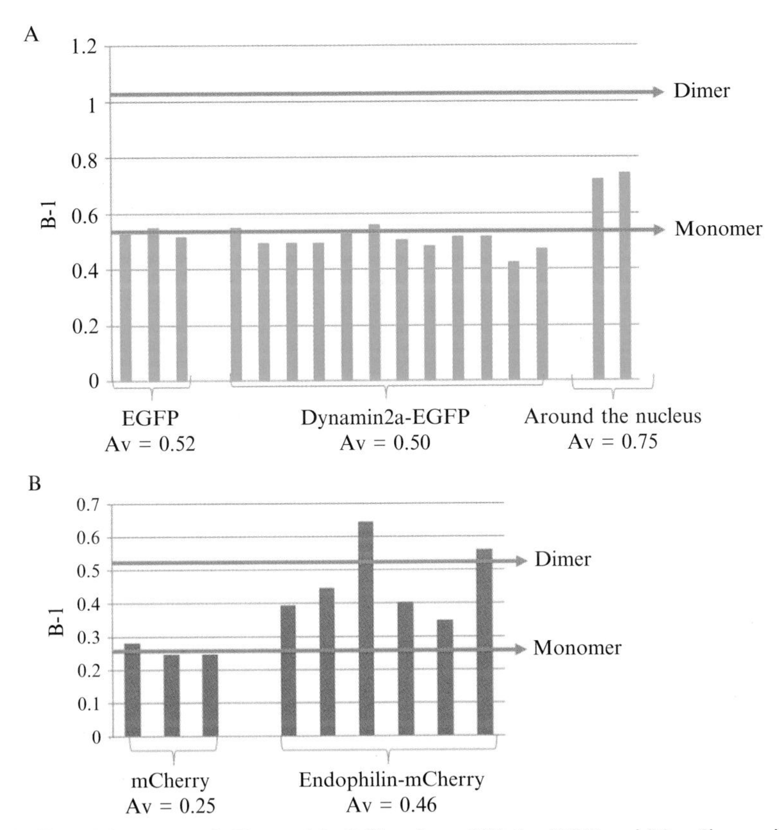

Michelle A. Digman *et al.*, Figure 6.9 Calibration of (A) the EGFP and (B) mCherry channel. Three cells expressing EGFP and mCherry were used to calibrate the brightness scale. The plane of focus was in the cytoplasm of the cells. For the dynamin2a-EGFP, four cells were measured with the plane of focus on the cytoplasm for a total of 12 planes. All cells and planes show the same brightness than for the EGFP only cell. For the same, four cells for a total of six focal planes in the cytoplasm of cells expressing the endophilin-mCherry show larger brightness than the mCherry alone, in the range expected for a dimer of the endophilin protein in accord with previous results. The brightness scale is in units of counts/dwell time/molecule where the dwell time was 12.6 μs.

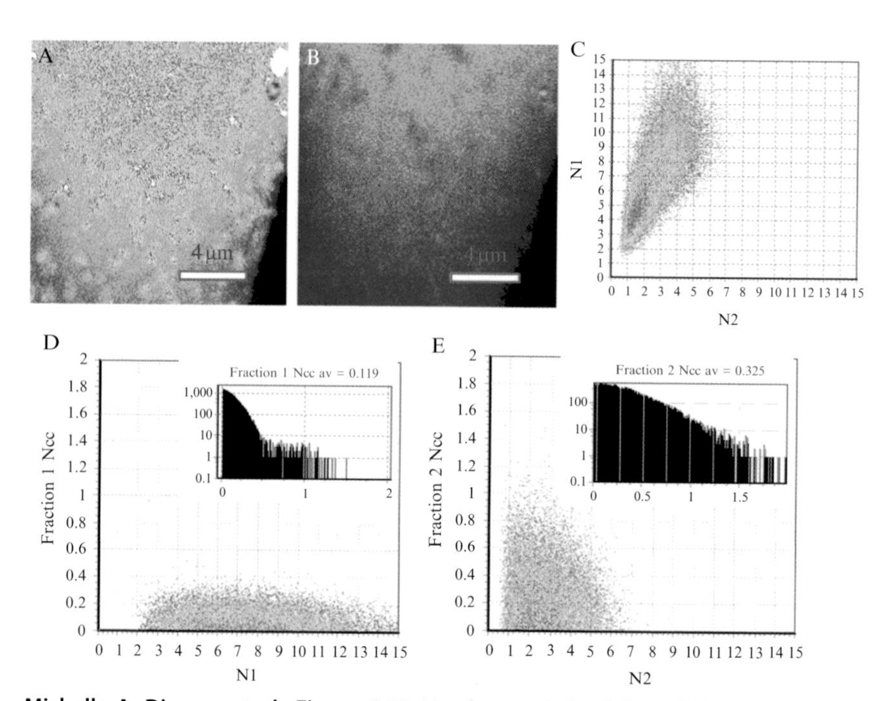

Michelle A. Digman *et al.*, Figure 6.10 Number analysis of the cell shown in Fig. 6.7. (A) Number of particles in the green channel. (B) Number of particles in the red channel. (C) Two-dimensional pixel histogram of the values of the number of particles for the two channels. The *x* and *y* numbers above are used for the calibration of the maximum number of particles of mCherry and EGFP, respectively. (D) Fraction of molecules cross-correlated in channel 1. The inset is the histogram. (E) Fraction of molecules cross-correlated in channel 2. The inset is the histogram.

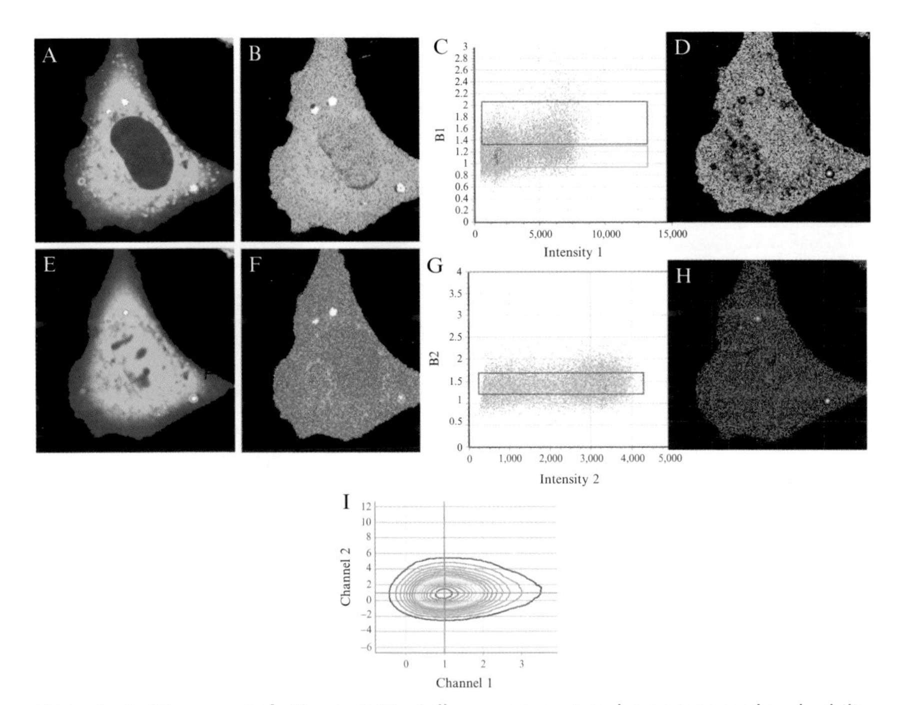

Michelle A. Digman *et al.*, Figure 6.11 Cell over expressing dynamin2a and endophilin. (A) Intensity image of green channel. (B) Brightness map of the green channel. (C) Selection map, green cursor corresponds to the brightness of monomers, and red cursor corresponds to higher aggregates. (D) Selection map coded with the color of the cursors in (C). (E) Intensity image of red channel. (F) Brightness map of the red channel. (G) Selection map, red cursor corresponds aggregates of 2 endophilin molecules. (H) Selection map coded with the color of the cursors in (G). (I) Stoichiometry map showing that the most populated species is the 1:1 molecular complex.

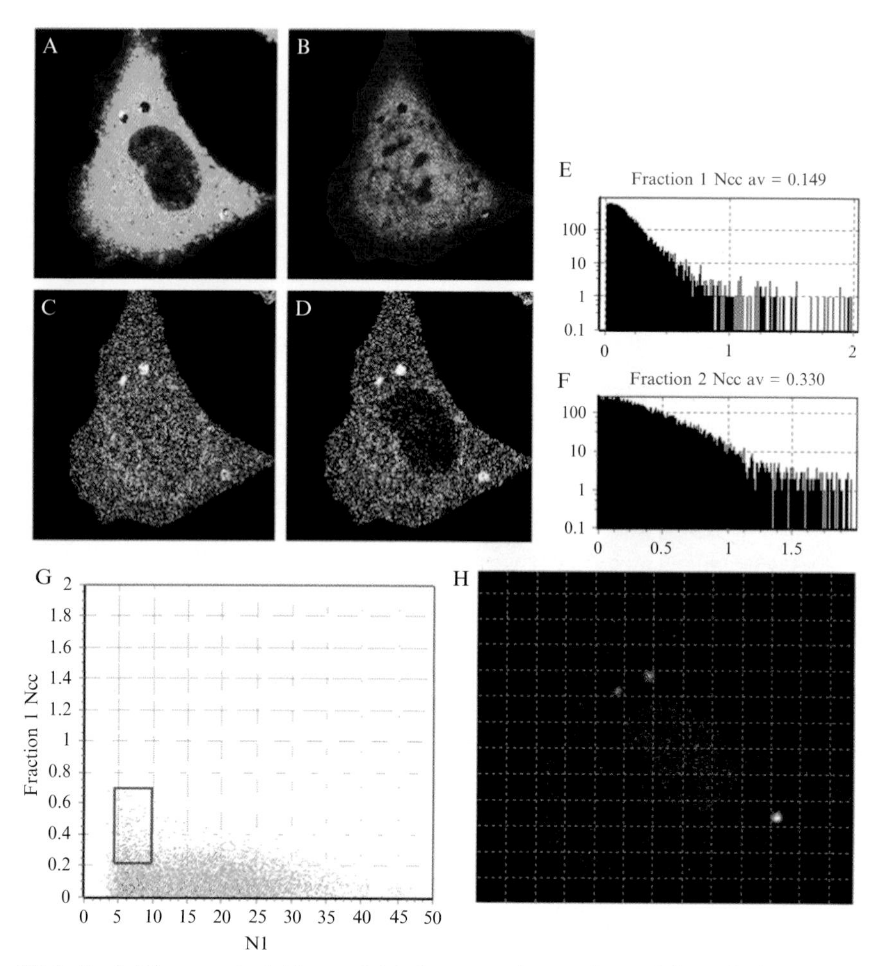

Michelle A. Digman *et al.*, Figure 6.12 Cross-number analysis of the cell showing in Fig. 6.11. (A) and (B) fluorescence intensity in the green and red channels, respectively. (C) and (D) Cross-number maps for the green and red channels, respectively. The scale in the figure is obtained from the histograms in (E) and (F), for the two channels. (G) Selection of the Ncc1 versus N1 histogram for low number N1 (in the nucleus) has the larger fraction of cross-correlated molecules. The region selected by the pixels in the red square is shown in panel H.

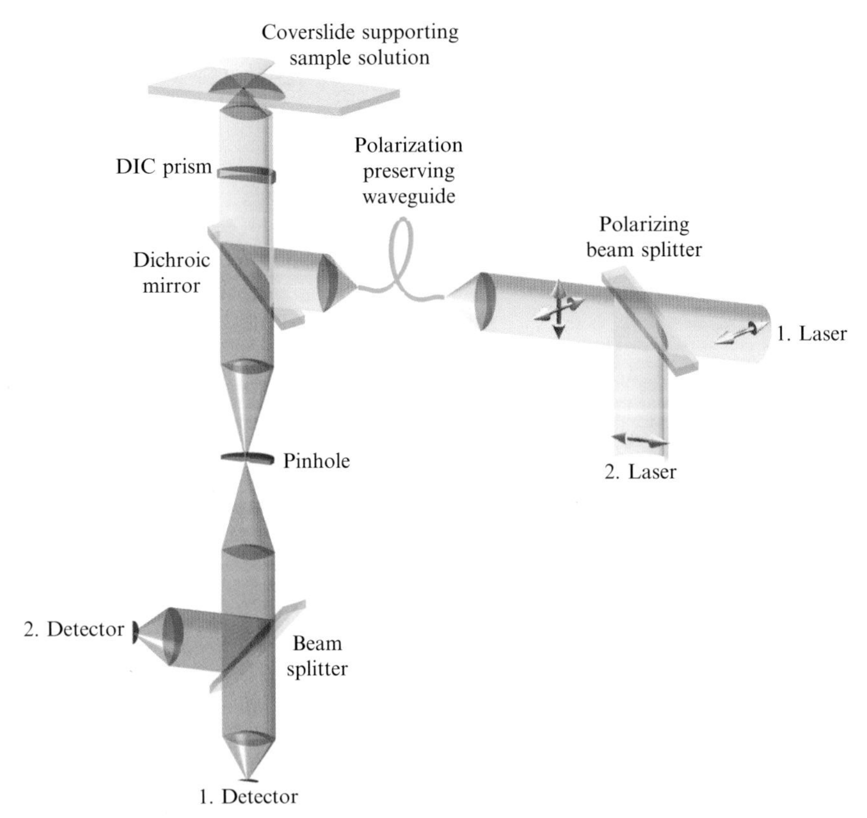

Christoph Pieper *et al.*, Figure 8.5 Schematic of the 2fFCS setup. Excitation is done by two interleaved pulsed lasers of the same wavelength. The polarizations of both lasers are linear and orthogonal to each other. Light is then combined by a polarizing beam splitter and coupled into a polarization-maintaining optical single-mode fiber. After exiting the fiber, the laser light is collimated by an appropriate lens and reflected by a dichroic beam splitter through a DIC prism. The DIC prism deflects the laser light into two beams according to the polarization of the incoming laser pulses. The microscope objective focuses the two beams into two laterally shifted foci. Fluorescence is collected by the same objective. The tube lens focuses the detected fluorescence from both excitation foci on a single pinhole. Subsequently, the fluorescence light is split by a 50/50 beam splitter and detected by two single-photon avalanche diodes.

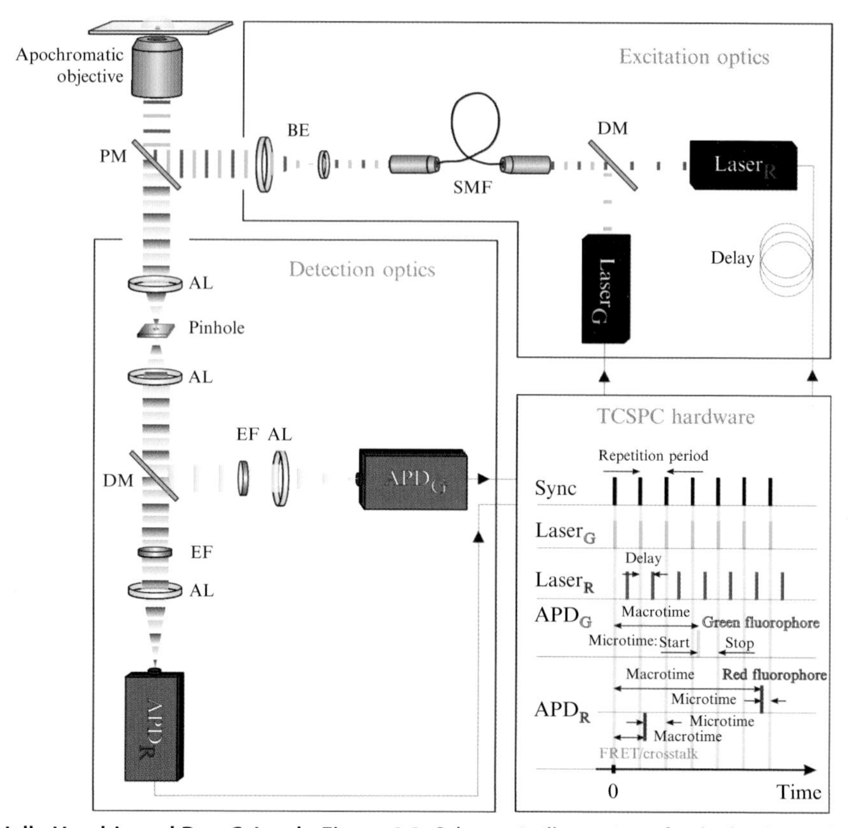

Jelle Hendrix and Don C. Lamb, Figure 9.2 Schematic illustration of a dual-color excitation dual-color detection PIE–FCCS microscope. (*Excitation optics*) The emission of the lasers is synchronized by the TCSPC hardware that sends electrical pulses with a fixed repetition rate. The red pulses interleave the green ones because of electronic delay (shown in the image) or optical delaying (not shown) of the red laser. The lasers are combined with a dichroic mirror (DM). The two beams are made concentric and their beam profile is cleaned up by coupling into a single-mode optical fiber (SMF) with an achromatic collimator. A Keplerian beam expander (BE) finally expands the beam to ensure overfilling of the objective back aperture, if desired. (*Detection optics*) A polychroic mirror (PM) spectrally and spatially separates excitation from emission light. The latter is focused through a confocal pinhole and collimated again by achromatic lenses (AL) to limit chromatic aberrations. The emission dichroic mirror splits up the emission light into two spectral ranges. An emission filter (EF) transmits the correct spectral band while blocking the lasers and a lens finally focuses the emission light on the active area of an avalanche photodiode (APD). (*TCSPC hardware*) The counting electronics register three parameters per photon: the detector, the microtime and the macrotime. The microtime is the arrival time of the photon with respect to the laser pulse and is measured with picosecond resolution in the reverse start–stop principle, irrespective of the laser that triggered its emission. The macrotime is the arrival time of the photon with respect to the start of the measurement and is measured as an integer multiple of clock ticks with respect to the previously detected photon.

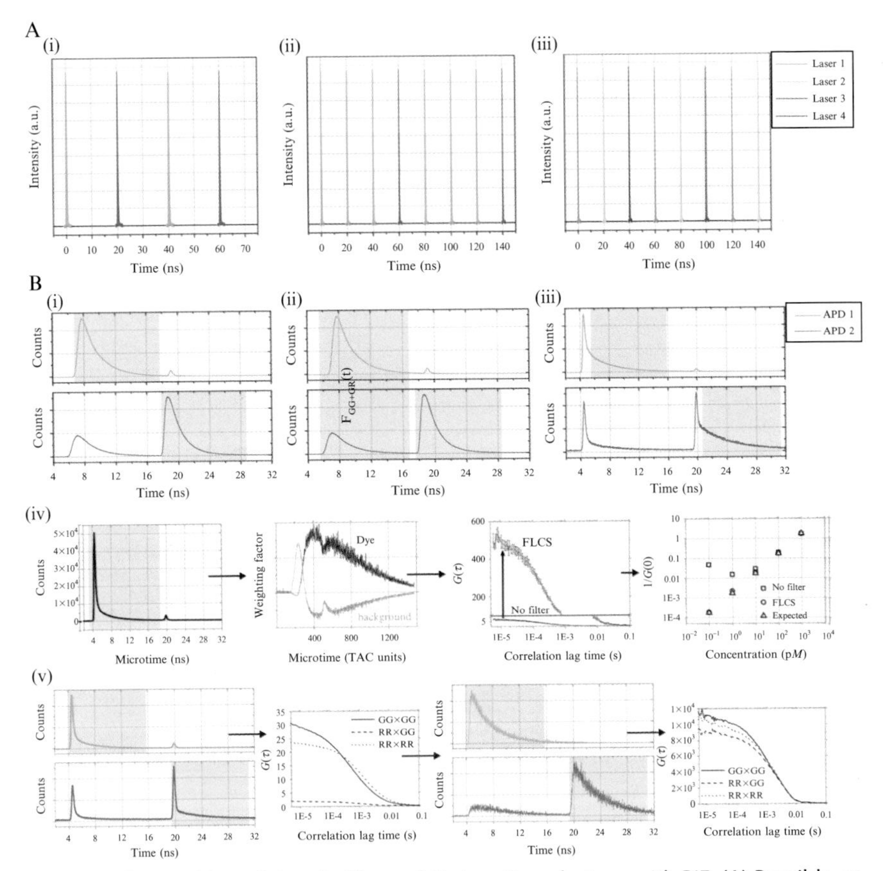

Jelle Hendrix and Don C. Lamb, Figure 9.3 Juggling photons with PIE. (A) Possible excitation sequences. (i) Standard dual-color PIE where green and red excitation alternate each other with a fixed repetition rate and fixed time delay between the lasers, (ii) FRET donor optimized excitation where the green laser is allowed to fire more often than the red laser, (iii) multicolor excitation, where the green laser has a fixed repetition rate that determines the master clock of the whole system, the blue laser precedes the green laser at the same frequency and the red and infrared lasers alternate and are pulsed at half the frequency of the blue and green laser. In this way, the photons needed to determine FRET are probed more often. (B) Possible macrotime cross-correlation strategies for dual-color PIE–FCCS. (i) Artifact-free and species-specific analysis. The macrotimes corresponding to $F_{GG}(t)$ and $F_{RR}(t)$ do not contain misguided photons and can be cross-correlated and analyzed with Eqs. (9.14) and (9.23) to verify if double-labeled species exist. (ii) Addition of microtime channels. The macrotimes corresponding to $F_{GG+GR}(t)$ and $F_{RR}(t)$ do not contain contributions from FRET and can be cross-correlated and analyzed with Eqs. (9.15), (9.21), and (9.16) to allow for absolute calculation of concentrations of single- and double-labeled species. (iii) Time gating. At low concentrations, time gating of the microtime histograms allows reduction of the background and scattering-related artifacts. (iv) FLCS. Lifetime filtering allows further reduction of background and scattering artifacts, making it possible to perform quantitative FCS/FCCS analyses at ultra-low concentrations. Illustrated is an actual measurement of ATTO 532 at different concentrations. The raw microtime data is shown in the left panel. The weighting functions used to separate scattered light form ATTO 532 fluorescence is shown in the second panel. The third panel shows the autocorrelation analysis of the same data once as normal FCS data and once using FLCS. With FLCS, the

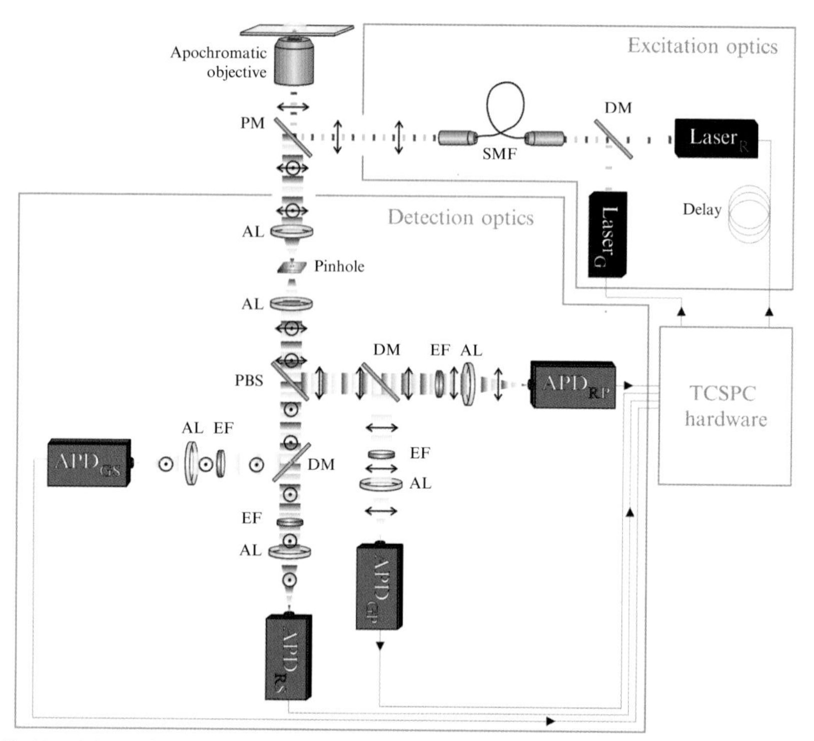

Jelle Hendrix and Don C. Lamb, Figure 9.7 Schematic illustration of a dual-color dual-polarization MFD–PIE setup. The main differences with respect to the PIE setup in Fig. 9.2 are that the lasers must be linearly polarized and are usually left unexpanded before entering the objective. This under filling of the objective back aperture creates a larger probe volume, while the detection path uses the full resolving power of the objective. A polarizing beam splitter (PBS) placed in the detection path also splits the emission light into P- and S-polarized components, to allow sensitive measurements of anisotropy, in addition to intensity and lifetime. For further explanation of the setup and most of the abbreviations, see Fig. 9.2.

---

measured concentration remains inversely proportional to the correlation function amplitude, even at a concentration of 100-fM (fourth panel). (v) Burstwise filtering. The left two panels illustrate unfiltered burst analysis data measured with MFD–PIE. The CCF has a very low amplitude due to the presence of an excess of single-labeled species. An *E–S* plot (Fig. 9.8A) allows sorting of fluorescent species according to the label content. This subsequently allows for a species-specific macrotime cross-correlation analysis. The right two panels illustrate the same data, now filtered according to stoichiometry. The near absence of scattering light in the microtime histograms and very high CCF amplitude is obvious.

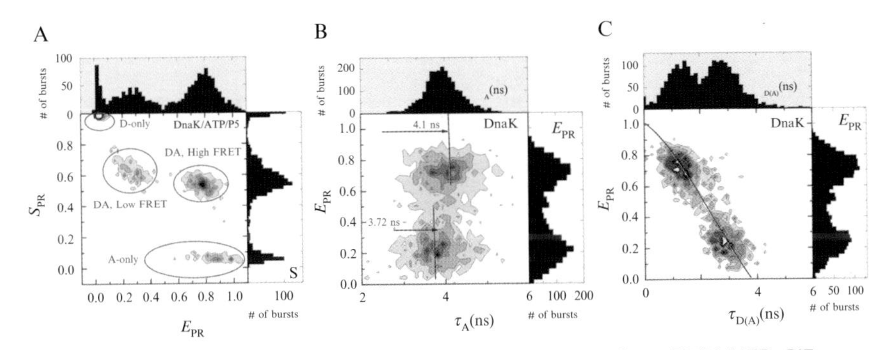

Jelle Hendrix and Don C. Lamb, Figure 9.8 2D histograms of a spFRET–MFD–PIE measurement of fluorescently labeled DnaK (318,425). (A) Identification of D-only, A-only, and DA species. A 2D histogram of the uncorrected stoichiometry values, S_{PR}, versus the uncorrected FRET efficiency, E_{PR}, also known as the proximity ratio for spFRET experiments. Donor-only (D-only) labeled molecules, acceptor-only (A-only) labeled molecules, and donor–acceptor labeled (DA) molecules can be observed in different regions of the 2D histogram. The D-only species has high stoichiometry ($S_{PR} \approx 1$) and low proximity ratio ($E_{PR} < 0.2$); the A-only species has a low stoichiometry ($S_{PR} < 0.1$) and an intermediate to high proximity ratio ($E_{PR} > 0.2$); DA species have middle stoichiometry values. (B) A 2D histogram of uncorrected FRET efficiency, E_{PR}, versus directly excited acceptor lifetime, τ_A, of the DA species ($0.2 < S_{PR} < 0.8$). (C) A 2D histogram of uncorrected FRET efficiency, E_{PR}, versus donor lifetime, $\tau_{D(A)}$, of the DA species ($0.2 < S_{PR} < 0.8$). The static FRET line (shown in red) is given by a third-order polynomial with a donor-only fluorescence lifetime of $\tau_{D(0)} = 3.8$ ns. *Adapted from Kudryavtsev et al. (2012).*

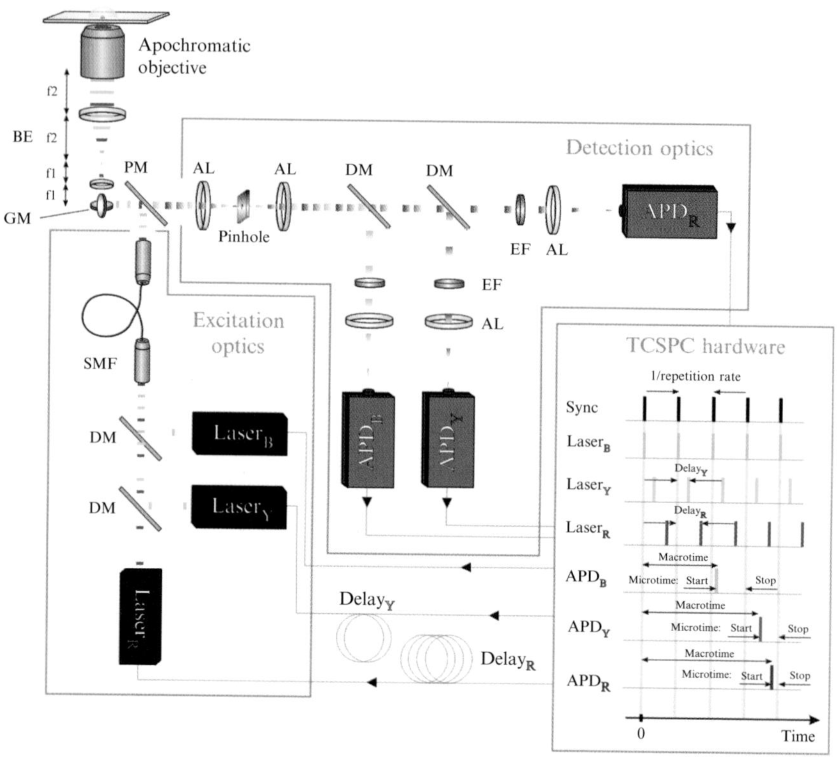

Jelle Hendrix and Don C. Lamb, Figure 9.9 Schematic illustration of a triple-color excitation triple-color detection scanning PIE–FCCS microscope. Image scanning and descanning is possible by placing a two-dimensional fast galvanometric mirror (GM) scanner in the beam path. Furthermore, by placing the two focal planes of a Keplerian telescope at the GM and the back aperture of the objective, it is possible to illuminate the objective in the center of the back aperture at any position of the scanning mirrors. For further explanation of the setup and most of the abbreviations, see Fig. 9.2.

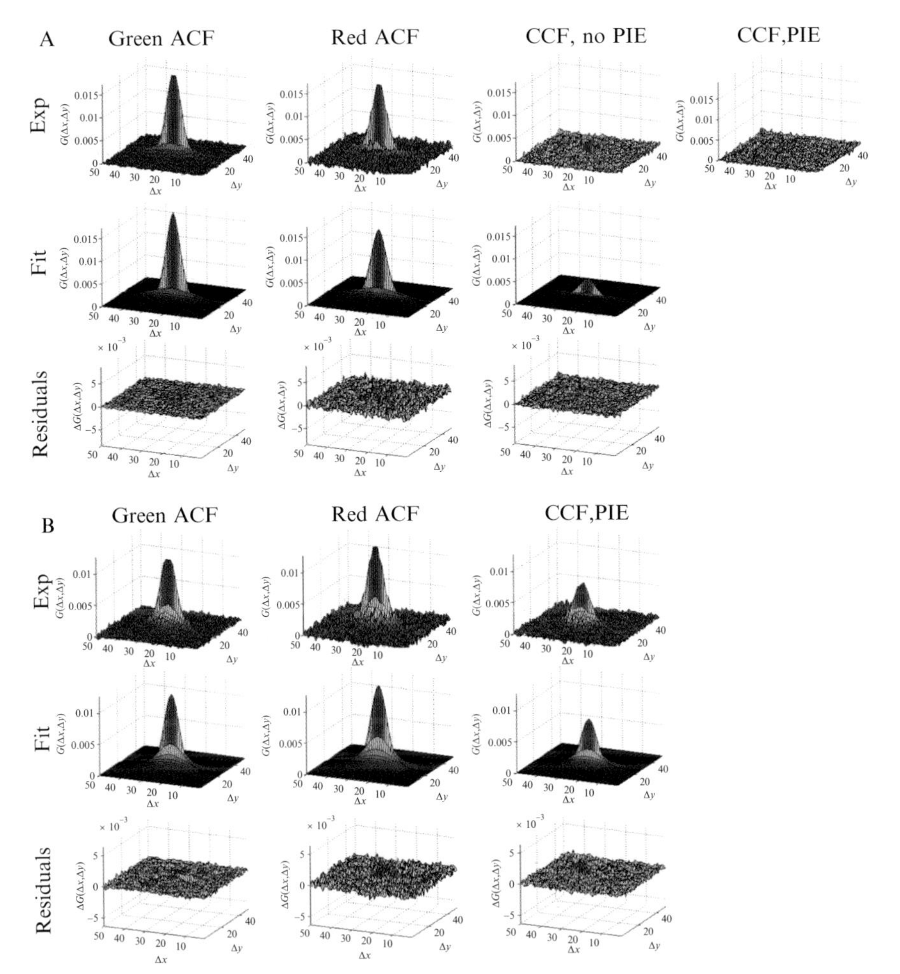

Jelle Hendrix and Don C. Lamb, Figure 9.11 Proof of principle for measuring protein–protein interactions in cells with dual-color RICS. (A) Coexpression of eGFP and mCherry in HeLa cells and (B) expression of a genetic fusion construct, eGFP–mCherry. Measurements were performed on a home-built scanning PIE microscope similar to Fig. 9.9. More specifically, an Er-doped fiber laser with a fixed repetition rate of 27.4 MHz (Toptica Photonics, Munich, Germany) was used to excite mCherry at 561 nm and was delayed $\sim$16 ns with respect to a pulsed 475 nm diode laser (Picoquant, Berlin, Germany) that was used to excite eGFP. Laser powers of 2 μW as measured between the excitation dichroic and objective ($\sim$1 μW in solution) were used. The confocal pinhole was 80 μm, the image size was 12.5 μm with a 300 $\times$ 300-pixel resolution, and image time was 1 s. The displayed RICS images (*upper panels*) are averages of RICS images from 100 consecutive image frames, corrected with a 10-frame moving average to remove the immobile fraction and with the (0,0) point omitted. Analysis was performed with a standard model for ccRICS analysis (Digman et al., 2009) in the 50 $\times$ 50-pixel central region of the RICS image. For the "CCF, no PIE" image, the macrotimes for photons from $F_{GG+RG}(t)$ and $F_{GR+RR}(t)$ PIE channels were used to calculate the "green" and "red" images needed for ccRICS. For the "CCF, PIE" image, the macrotimes for photons from $F_{GG}(t)$ and $F_{RR}(t)$ PIE channels were used to calculate the "green" and "red" images for ccRICS. The fits (*middle panels*) and residuals (*lower panels*) are shown.

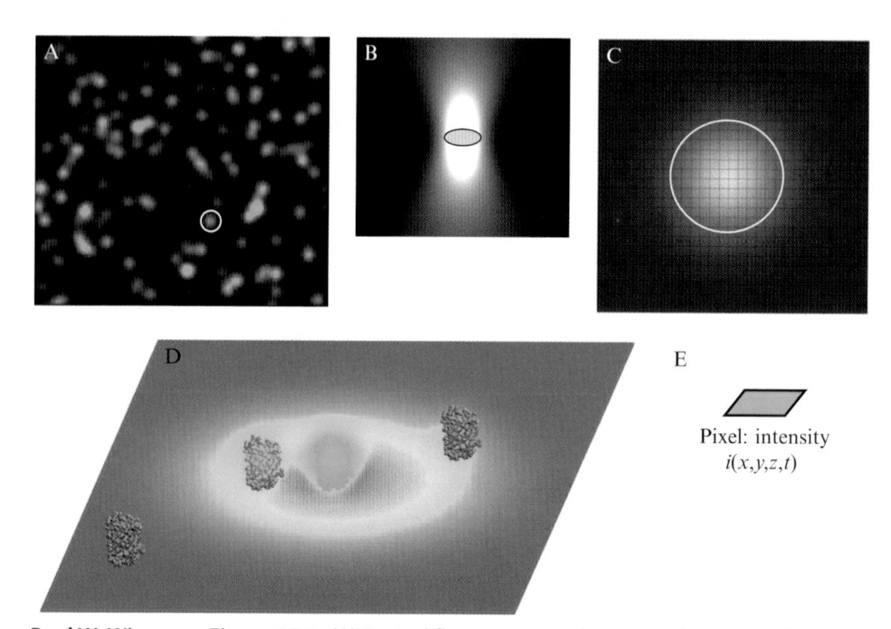

Paul W. Wiseman, Figure 10.1 (A) Typical fluorescence microscopy image used as input for ICS analysis with the yellow circle highlighting fluorescence signal within a focal spot region which would contribute a positive intensity fluctuation. (B) The 3D point spread function (PSF) for a focused Gaussian beam. The gray disk illustrates the 2D cross section at focus. (C) A representation of a fluorescence signal fluctuation that is oversampled onto pixels in the image. The yellow circle shows the 2D PSF size illustrating that there will be spatial correlation between adjacent pixels over a scale defined by the PSF. (D) A cartoon representation of the molecular contributions of fluorescent molecules within the PSF-defined region to an integrated fluorescence signal as per (C). Note that the PSF would be about 100 × larger for standard fluorescence microscopy. (E) A cartoon representation of an image pixel showing the integrated intensity value derived from photons emitted and detected from across the PSF-defined region.

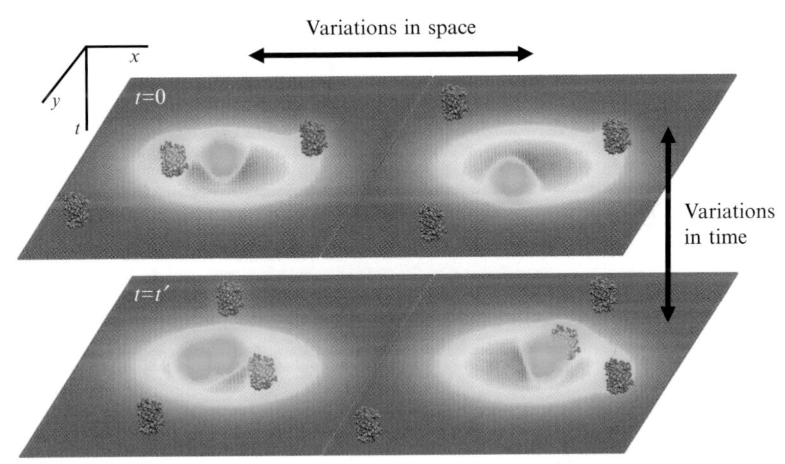

Paul W. Wiseman, Figure 10.2 A molecular cartoon representing variations of molecules in space and time within the PSF. The variations in space at a given time would lead to fluorescence intensity fluctuations as the PSF area moves from pixel to pixel in a single image and the photons emitted by fluorophores in the PSF are integrated and converted to a pixel value. The variations in time would represent changes in the number of emitting fluorophores in the same PSF area due to molecular transport or photophysical changes in emission during the period between pixel integration for a given image pixel at different times. Note that the PSF would be about 100 × larger for standard fluorescence microscopy.

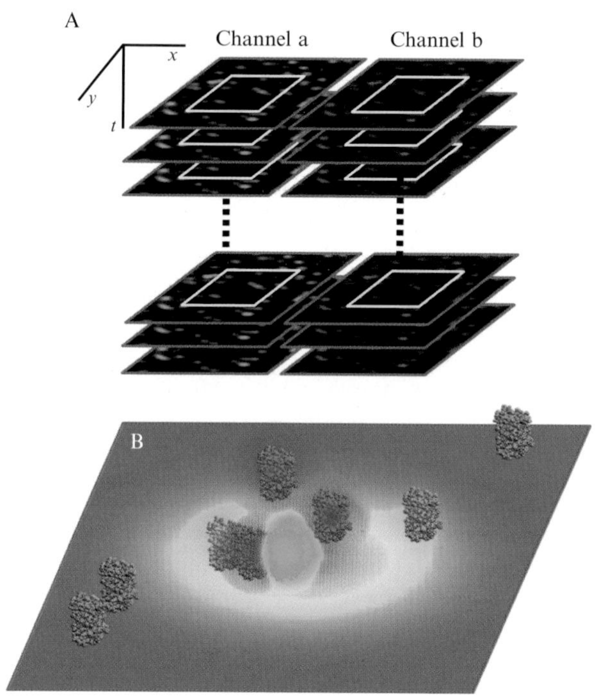

Paul W. Wiseman, Figure 10.3 (A) Two-channel image time stack typical for two-color STICCS with yellow box showing region of interest (ROI) for correlation analysis. (B) A molecular cartoon illustrating the emission of two species labeled with green and red fluorophores from within the PSF-defined focus that forms the basis of cross-correlation measurements. Interacting species can be measured by cross-correlating the signals between the two channels integrated over all pixels in the ROI. Note that the PSF would be about 100 $\times$ larger for standard fluorescence microscopy.

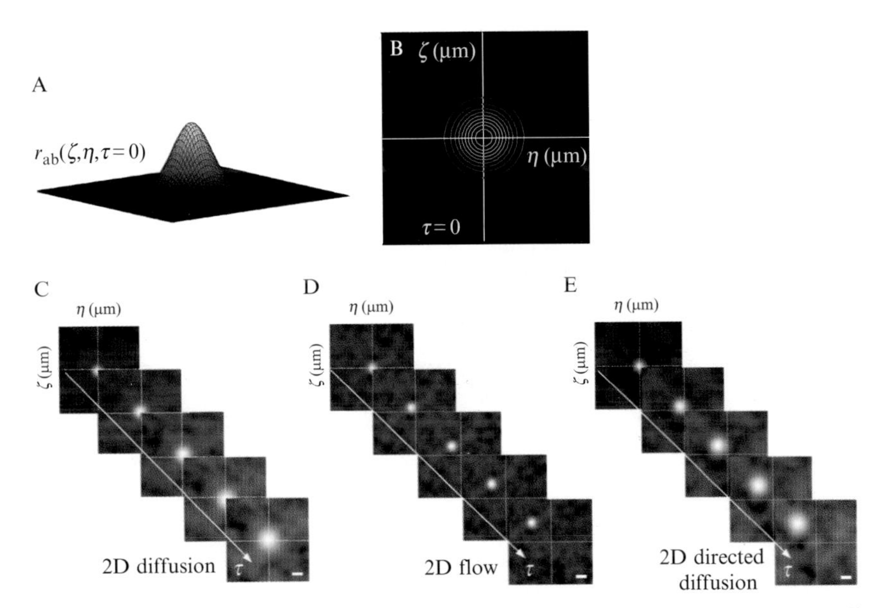

Paul W. Wiseman, Figure 10.4 Spatiotemporal image correlation spectroscopy (STICS). (A) A spatiotemporal autocorrelation function at time lag $\tau = 0$ calculated from an ROI of a simulated image with Gaussian PSF. (B) Contour plot of the same correlation function as shown in (A) illustrating that the Gaussian correlation peak amplitude is centered at zero spatial lags at $\tau = 0$. (C–E) All show density contour plots of the spatiotemporal autocorrelation function as a function of increasing time lag as calculated from computer-simulated image series with particles undergoing 2D diffusion, 2D directed flow, and 2D directed or biased diffusion, respectively. The spread and translation of the correlation function are evident in the time lag series for each of the transport cases.

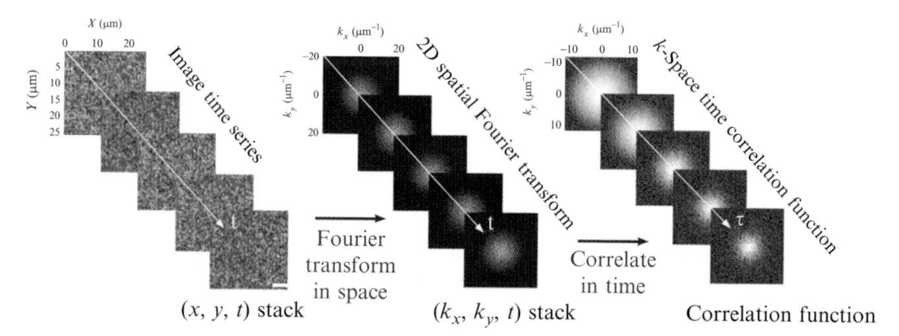

Paul W. Wiseman, Figure 10.5 Schematic figure showing the steps involved in generating a *k*-space–time correlation function. The basic steps involve a 2D fast Fourier transform (2D-FFT) of each image frame in the fluorescence microscopy time series to transform to an image time stack in reciprocal or *k*-spatial frequencies. The *k*-space–time stack is then time correlated to generate the final *k*-space–time correlation function output.

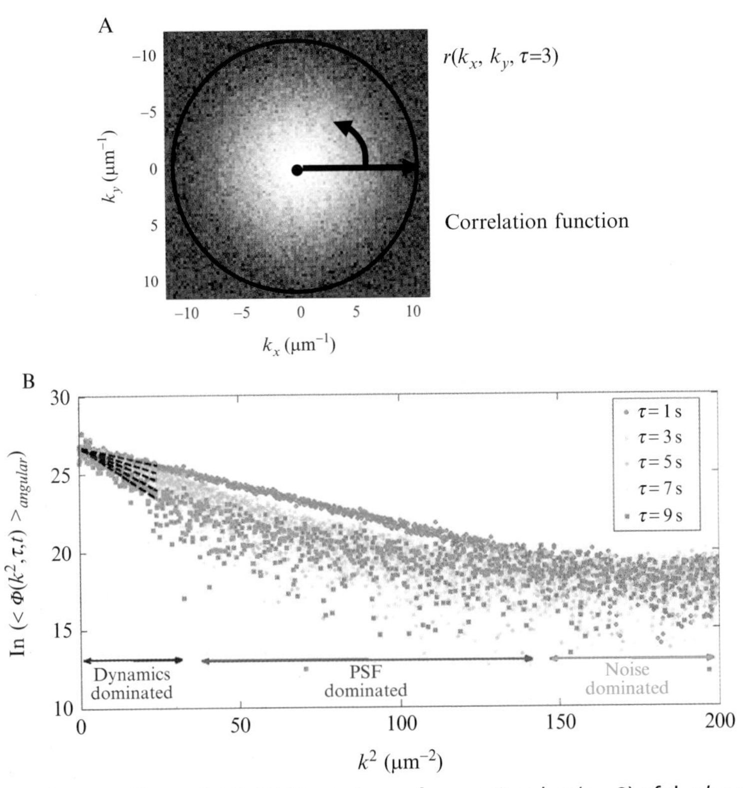

Paul W. Wiseman, Figure 10.6 (A) Single frame for one time lag ($\tau = 3$) of the k-space–time correlation function that is radially averaged and natural logarithm transformed to yield (B) the output of Eq. (10.7). This plot shows **k**-vectors over which there are fluorescence correlations (dynamics and PSF dominate regimes) which decay to a noise floor at a certain threshold **k**-vector for the given time lag. The correlation function data are calculated from computer-simulated images of a diffusing population in 2D.

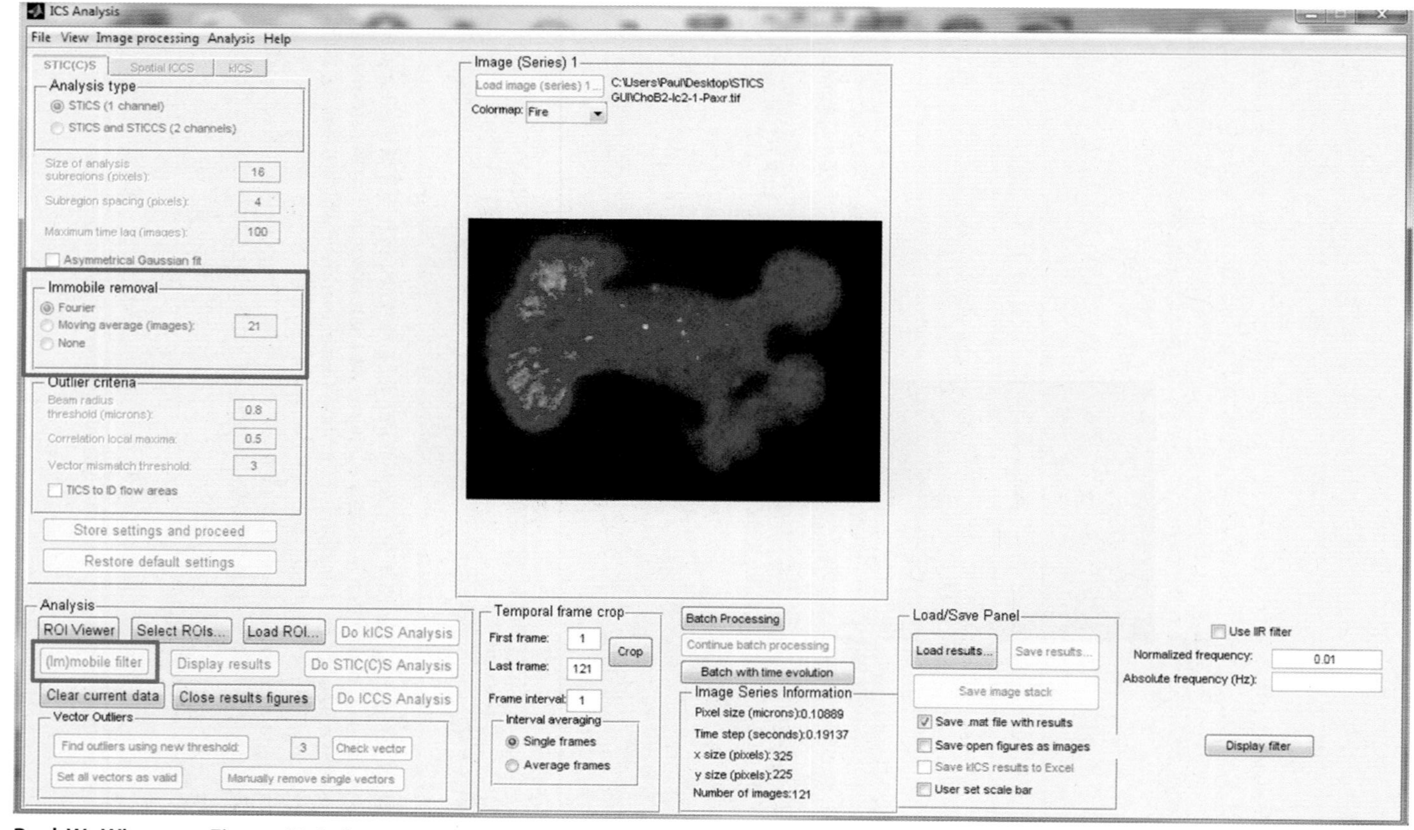

Paul W. Wiseman, Figure 10.8 Screen capture image of the ICS GUI showing highlighted control interface for the immobile population filtering.

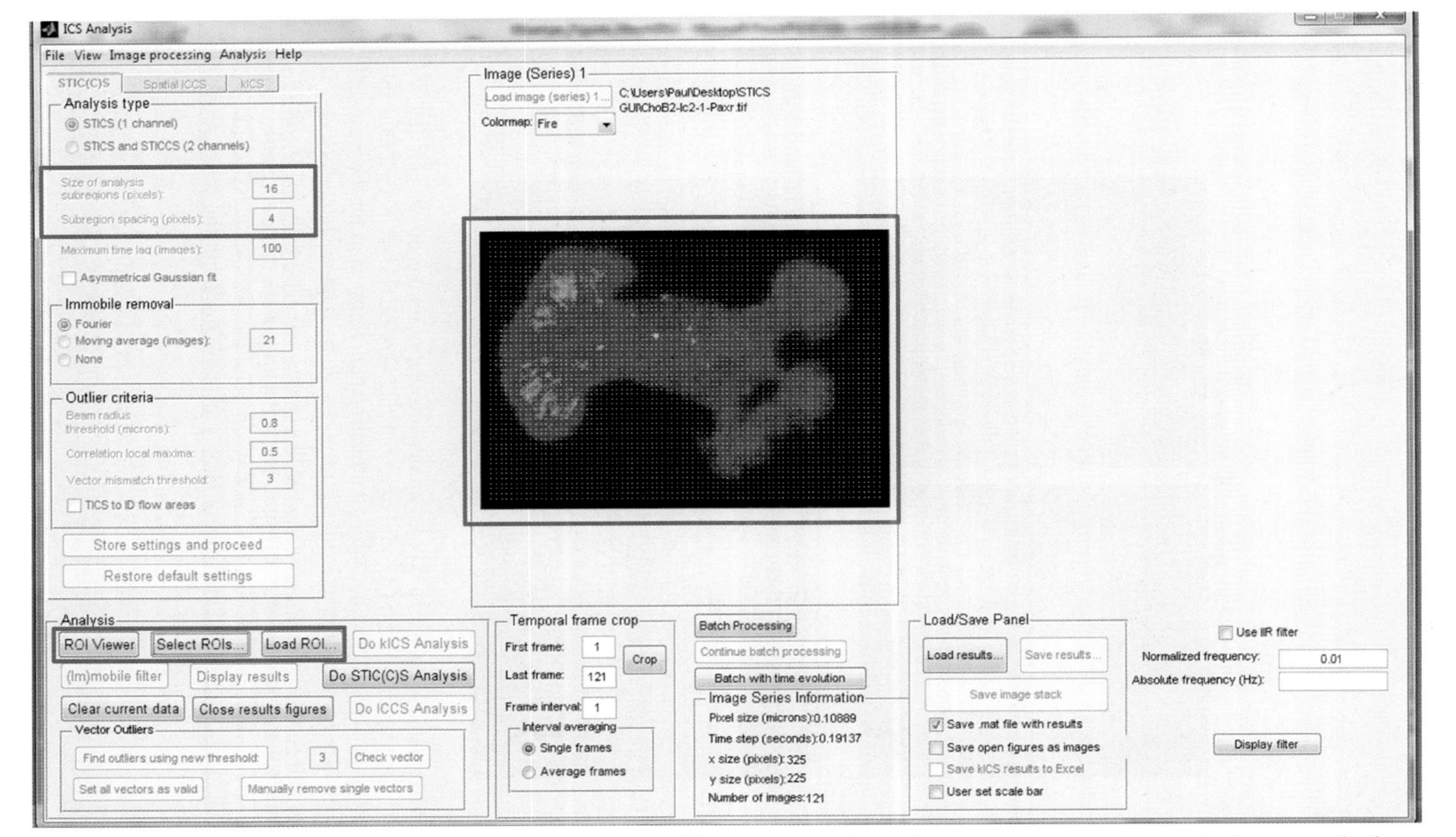

Paul W. Wiseman, Figure 10.9 Screen capture image of the ICS GUI showing highlighted control interface for the ROI selection and editing.

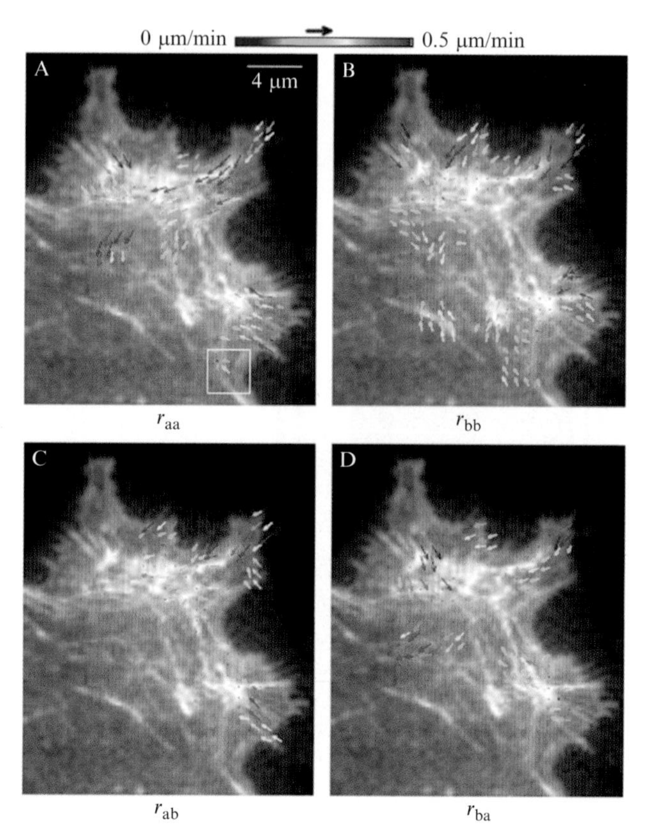

Paul W. Wiseman, Figure 10.10 STICCS analysis results of frames 60–90 of a TIRF microscopy image series of actin-GFP and α-actinin-mCherry expressed in CHO cells that were platted on 5 μg/ml fibronectin. (A) Autocorrelation vector map for the actin-GFP channel. (B) Autocorrelation vector map for the α-actinin-mCherry channel. (C and D) Cross-correlation vector maps r_{ab} and r_{ba} calculated via STICCS between the two detection channels. Velocity vectors for correlation function flow peaks that were above noise threshold are displayed. The image time step was 6 s and the pixel size was 0.21 μm and the yellow square depicts the size of the 16 × 16 pixels ROI used in the calculations.